T0361396

CONTRIBUTIONS TO THE THEORY OF ZETA-FUNCTIONS

The Modular Relation Supremacy

Series on Number Theory and Its Applications   Vol. 10

CONTRIBUTIONS TO THE THEORY OF ZETA-FUNCTIONS

The Modular Relation Supremacy

Shigeru Kanemitsu & Haruo Tsukada

Kinki University, Japan

 World Scientific

NEW JERSEY · LONDON · SINGAPORE · BEIJING · SHANGHAI · HONG KONG · TAIPEI · CHENNAI

Published by

World Scientific Publishing Co. Pte. Ltd.
5 Toh Tuck Link, Singapore 596224
USA office: 27 Warren Street, Suite 401-402, Hackensack, NJ 07601
UK office: 57 Shelton Street, Covent Garden, London WC2H 9HE

Library of Congress Cataloging-in-Publication Data
Kanemitsu, Shigeru.
    Contributions to the theory of zeta-functions : the modular relation supremacy / Shigeru Kanemitsu (Kinki University, Japan) and Haruo Tsukada (Kinki University, Japan).
        pages cm. -- (Series on number theory and its applications ; volume 10)
    Includes bibliographical references and index.
    ISBN 978-9814449618 (hardcover : alk. paper)
    1. Functions, Zeta. 2. Functional equations. 3. Mathematics. I. Tsukada, Haruo, 1961– II. Title.
    QA351.K33 2015
    515'.56--dc23

                                            2014041588

British Library Cataloguing-in-Publication Data
A catalogue record for this book is available from the British Library.

Printed in Singapore

To Professor Kok Khoo Phua

Preface

This is a book on zeta-functions as viewed from their symmetry—functional equations. There is an eternal conflict between the viewpoints on whether one assumes the Euler product or not. Without the Euler product, the class is wider and with it, the class gives rise to some more delicate information. Here we take the former viewpoint and do not assume the Euler product, whence no zero-free region is assumed. We try to exhaust a.a. identities that are equivalent to the functional equation, thus providing the reader or browser with a rather handy manual for a quick search. We try to make this as useful as the famous series of source books on special functions by Erdélyi [ErdH], [ErdT]. Our collection is in that spirit in thoroughness and unifiedness as well as in Hamburger [Ham2], Chandrasekharan and Narasimhan [ChN1], [ChN2], [ChN3] or the more recent [BK, Chapter 8] in diversity. That is, we shall try to locate the genesis of those identities which appeared hitherto in mathematical or scientific literature in the last 150 years. The most influential works which led us to this point are [Bo1], [ChN2], [ChN3], [HeckW], [MKn2], [KoshI]-[KoshIII] etc., which will be referred to as the story goes on.

Although the contents can be seen from the chapter headings, we give a brief description here. Chapter 1 contains pre-functional equations, i.e. those material which are related to the functional equation at least in implicit form. The reader is advised to read the Conventions carefully. Chapter 2 assembles material on special functions, which can be read together with [Vista], [Vi2]. This can be consulted when need occurs. Chapter 3 collects equivalent relations to the functional equation without the processing gamma factor. Chapter 4 is devoted to the presentation of the Fourier-Bessel expansion, i.e. the results on perturbed Dirichlet series. §4.3 contains the concrete version of Theorem 4.3, of the main theorem, Theorem 7.1.

Chapter 5 is devoted to the Ewald expansion, i.e. hypergeometric expansion and *a fortiori*, incomplete gamma expansion. Chapter 6 is devoted to the Riesz sums, some of which are also given in other chapters. Since the summability has been traced for a long time and therefore we cannot be exhaustive. Instead we incorporate integrals and integrated modular relations. Chapter 7 gives the main theorem, Theorem 7.1, on modular relations with proofs. This chapter needs to be consulted from time to time, although in most of the chapters we state the main formulas that are used in the beginning. Chapter 8 is a database of the Hecke type zeta-functions. A similar database for the Riemann type zeta-functions can be compiled and we have given several of them in §6.1. Chapter 9 is about the product of zeta-functions, where the results are still partial but we are able to elucidate the hitherto most far-reaching Wilton's Riesz sum. Chapter 10 collects miscellaneous future projects.

The first author would like to thank many of his colleagues who have given him explicit and implicit influence in the last forty years toward the completion of the present book. Thanks are above all due to S. Akiyama, K. Alladi, T. Aoki, R. Balasubramanian, B. C. Berndt, K. Chakraborty, H^2 Chan, V. N. Chubarikov, C. Deninger, S. Egami, K. Györy, M. Hata, Y. Hironaka, C.-H. Jia, J. Jimenes-Urroz, M. Kaneko, T. Kano, I. Kátai, A. Katsurada, K. Kawada, Y. Kitaoka, the late Professor J. Knopmacher, I. Kuchi, G. Lachaud, A. Lauričkas, C. Levesque, J.-Y. Liu, K. Miyake, C. Matsumoto, L. Murata, M. Nakahara, Y.-N. Nakai, R. Okazaki, F. Pappalardi, Y. Tanigawa, the late Professor K. Ramachandra, A. Schinzel, the late Professor K. Shiratani my supervisor, I. Wakabayashi, M. Waldschmidt, M. Yoshida and many others whose publications are cited, where H^2, A, C are shorthand for Heng Huat, Augustinus, Cozy, respectively. The first author also would like to express his special thanks to Professor K. K. Phua, the chairman of the World Scientific for his constant and generous encouragement which has been a great impetus to continue publishing. Last but not the least, the authors would like to thank Ms. Lai Fun Kwong for her patience and encouragement which helped them to complete this heavy-going task.

The Authors
October, 2014

Contents

Chapter 1

Prelude

1.1 Introduction

The year 1859 is a remarkable one in the history of science (or homo sapiens) because of two important works published in that year. One is Darwin's "The origin of species" which gave a tremendous impact on the paradigm of human beings in the succeeding several decades and there are still new discoveries occurring surrounding the evolution theory, cf. [ChR]. In mathematics in general and in number theory in particular, this is the year when Riemann published his epoch-making paper [Rie2], which is his unique paper on the distribution of primes, with a statement of the most important **Riemann hypothesis**. We are still struggling through inheriting these magnificent legacies. Since Riemann introduced the Riemann zeta-function in that paper, there is an enormous amount of papers and books related to it and generalizations thereof. It is preferable that there be some accessible surveys which make clear the history and relations of all these zeta-functions from a certain higher standpoint: "vom höheren Standpunkte aus"—from part of the title of Klein's book, or with some ever-lasting mnemonics, so that this tradition will be inherited by the coming generations.

Good master, what shall I do to inherit eternal life? — Luke 18:18.

Thus, in this trilogy, we will be the bards with a bird's eye of view.

We take the zeta-symmetry—functional equation—as our guiding principle and will try to make up a chart of the sea of zeta-functions for the coming young navigators to continue the tradition. To take the distribution

1

of prime elements associated with the zeta-functions (or the zeta-functions associated to a certain set of prime elements) into view, we need to introduce the notion of zero-free region, which is furnished by assuming the Euler product. However, we do not assume the Euler product here because it restricts the class of zeta-functions to the one connected with number theory. What we would like to present is a chart to which any sincere researcher in research areas where the appearance of the zeta-symmetry can make a recourse. A chart which shows what you need at a browse, without too much pain.

In our previous "Vista" books [Vista] and [Vi2], we tried to give a fundamental source of special functions and their applications, which, to our great joy, received rather favorable reviews and reactions. In the present case, we would like to present a source book which will be an enlightening and enjoyable handbook of zeta-functions and a.a. equivalent formulas to their symmetry. As you can see on reading through the present book, the history of zeta-functions is, so to say, that of equivalent formulas to functional equations. Here the "every 50 years" phenomena appear to the effect that there appears a repetition of what was obtained 50 years ago, which will be illustrated later. It is like a grandfather telling the theory of zeta-functions to his grandson. This phenomenon being very common in all fields of human activities, mathematics in general, number theory or the theory of zeta-functions in particular, is no exception. The difficulty of such meta-physical subjects is that they are mostly too high-brow and entangled for the public to feel like following. Thus here is a need for bards who can write sagas which are accessible to the public. In the present book, we are not so ambitious to do so and will try to make this a source book useful to researchers, not necessarily mathematicians. In one of the reviews on the Vista books, there is a passage indicating that the books are available for non-mathematicians, too. This is what we would like to achieve, providing something good to scientists and engineers. We are anyway comrades, tovarishch, making a mad pursuit.

We are the weavers who weave a tapestry which can be used as a magic carpet. You sit on it and say the magic words "a là peanut butter sandwiches". Then you would fly provided that you do not think about a monkey with three golden hairs when you say the magic words. To make full use of the book, you need to remember some notations on the H- and G-functions and then the magic will work and you can find what you expect. A tapestry is to have many patterns. In this book they are provided by a number of worked-out examples.

Things done without example in their issue are to be feared.

— Shakespeare, Henry VIII, 1.2.

In this regard, we can claim that the reader will benefit, even by browsing through these examples.

This book is just a starter, the first volume of the planned trilogy— contribution to the theory of zeta-functions. Although we succeeded in making a thorough treatment of the cases of zeta-functions satisfying the functional equation with two gamma factors, we were able to make the slightest touch of the case of more gamma factors, which we have to postpone in later volumes. The planned volumes are

Volume II, modular relation ultimatum

and

Volume III, modular relation niezwyciężony,

which are hoped to be inherited by the coming generations.

In Volume II, we are going to treat what should come after the functional equation—post symmetry, i.e. approximate functional equations and mean values, asymptotic formulas for the coefficient sums, Hamburger type theorems, i.e. characterization of the zeta-function through their functional equations, (Voronoǐ) summation formulas, the Riemann-Siegel formula, explicit formulas–explicit expressions of relevant functions in terms of zeros of the zeta-functions, a generalization of the modular relation in the case where the generating functions have infinitely many poles, etc. The last a few items are intimately related to the Euler product.

It will turn out that our theory of modular relations without Euler product covers a wider class of zeta-functions (e.g. the Hurwitz zeta-functions or Epstein zeta functions) but lacks some delicate details that the class of zeta-functions with the Euler product possesses, notably, those which the

Selberg class $\mathcal{S}$ of zeta-functions have, for which we refer to the work of Kaczorowski and Perelli [KP7] and references therein referred to.

Thus in the coming Volume III, we will include the influence of the Euler product in due course and might be able to step into the forbidden critical strip.

We started thinking of writing this book some 10 years ago. The first author used the G-functions in his earlier apprenticeship paper [Ka] but never thought such fancy goods could be of any use in number theoretic environment in view of the way they are used in statistics and related fields, e.g. [MaSa]. More recently there appeared the application of the G-functions to transcendence theory by Nesterenko, Zudilin [Zu] et al, which came to our attention after we started writing the draft. They seem to be centered around the case of unit argument. Then after a time lapse of 30 years, the second author made a colossal contribution to the theory of zeta-functions by making the effective use of the H-functions. Meanwhile there were many sprouts of this flourishing and we were motivated by a vast amount of literature, and we can mention just part of them. Certainly Lavrik [Lav5] gave a decisive influence because he is the first who mentioned the **extraneous parameter**, which is the variable τ in the upper half-plane (3.100).

As will be expounded at the end of §1.3, we mean by the **Riemann-Hecker-Bochner correspondence** the correspondence between the automorphic forms, dwellers in the upper half-plane $\mathfrak{H}$ (3.100) and the zeta-functions, dwellers in the right half-plane $\operatorname{Re} z > 0$ under the rotation

$$\tau = iz : \text{upper half-plane} \leftrightarrow \text{right half-plane.} \tag{1.1}$$

In the right half-plane, however, there is the principle of analytic continuation holding true and the positive real axis usually prevails. Its multiplicative group structure is in shorthand referred to as the reciprocal relation

$$x \times \frac{1}{x} = 1 \quad x > 0, \tag{1.2}$$

which appears in disguised form subsequently. For example, in (3.82) or in §6.2.1.

In this series of books, we mean by the **modular relation** the relation for the key-function $\mathrm{X}(z,s)$ defined by (3.2) under the zeta-symmetry $s \leftrightarrow r - s$ and the automorphy $z \leftrightarrow \frac{1}{z}$ and it is the one for $z^s \mathrm{X}(z,s)$ that gives the most information in two aspects.

It is arguable whether we should treat the Riemann zeta-function or other zeta-functions satisfying the Riemann functional equation

$$\Gamma\left(\frac{s}{2}\right)\pi^{-\frac{s}{2}}\varphi(s) = \Gamma\left(\frac{r-s}{2}\right)\pi^{-\frac{1-s}{2}}\psi(r-s) \qquad (1.3)$$

as one that satisfies the **Hecke type functional equation**

$$\Gamma(s)\pi^{-s}\varphi(2s) = \Gamma\left(\frac{r}{2}-s\right)\pi^{-\frac{1}{2}+s}\psi(r-2s). \qquad (1.4)$$

Cf. (8.1).

The standpoint of viewing them as Hecke type has been taken by many authors including Bellman [Bel4, (10)], Berndt [BeI, p.356], Berndt [BeII, p.370], [BK, p.121], [KTY7], etc. We note an essential difference between the fundamental sequence in (1.3) and (1.4), i.e. in (1.4), the sequence $\{\lambda_k\}$ (for notation cf. e.g. (1.12)) is squared: $\{\lambda_k^2\}$, i.e., we consider only a subsequence. It would be that in doing so, we may lose some essential **ingredients intrinsic to the original sequence**. E.g. in the case of the Riemann zeta-function, the original sequence is $\{n \in \mathbb{N}\}$, while as a Hecke type zeta, it is $\{n^2 \in \mathbb{N}\}$. This treatment corresponds to viewing the Riemann zeta as an Epstein zeta of the unary quadratic form. However, in most of the formulas that we will obtain, those for the Riemann zeta are those for the Hecke type with the replacement of the variable s by $2s$ (and replacement of the fundamental sequences by their squares is not very much perceivable). The following Exercises 1.1 and 1.2 below help to exhibit this difference.

We introduce the Riemann zeta-function as a special case of the Dedekind zeta-function in §3.4.1. It is defined as the Dirichlet series which is absolutely convergent in the half-plane $\sigma := \operatorname{Re} s > 1$ which is then continued meromorphically over the whole s-plane with a simple pole at $s = 1$ with reside 1. It satisfies the celebrated **Riemann functional equation** (1.3):

$$\pi^{-\frac{s}{2}}\Gamma\left(\frac{s}{2}\right)\zeta(s) = \pi^{-\frac{1-s}{2}}\Gamma\left(\frac{1-s}{2}\right)\zeta(1-s). \qquad (1.5)$$

Let χ be a Dirichlet character to the modulus q and let $\tau(\chi)$ be the normalized Gauss sum:

$$\tau(\chi) = \sum_{a=1}^{q}\chi(a)e^{2\pi i\frac{a}{q}}. \qquad (1.6)$$

The **Dirichlet L-function** $L(s, \chi)$ is defined by

$$L(s, \chi) = \sum_{n=1}^{\infty} \frac{\chi(n)}{n^s}, \quad \sigma > 1. \tag{1.7}$$

It satisfies the functional equation, which takes the following concise form for χ a *primitive* character:

$$L(1 - s, \chi) = q^{s-1}(2\pi)^{-s}\Gamma(s)\left(e^{-\frac{\pi i}{2}s} + \chi(-1)e^{\frac{\pi i}{2}s}\right)\tau(\chi)L(s, \bar{\chi}), \tag{1.8}$$

which is usually stated in the form (see e.g. [Dav, Chapter 9] or [NTA, Chapter 5])

$$\xi(1 - s, \chi) = \frac{i^{\mathfrak{a}}\sqrt{q}}{\tau(\chi)}\xi(s, \bar{\chi}), \tag{1.9}$$

where

$$\xi(s, \chi) = \pi^{-\frac{s+\mathfrak{a}}{2}}\Gamma\left(\frac{s + \mathfrak{a}}{2}\right)L(s, \chi), \tag{1.10}$$

and where $\mathfrak{a}$ is 0 or 1 according as $\chi(-1) = 1$ or $\chi(-1) = -1$ ($\chi(-1) = (-1)^{\mathfrak{a}}$):

$$\mathfrak{a} = \mathfrak{a}(\chi) = \frac{1 - \chi(-1)}{2} = \begin{cases} 0 & \chi \text{ even} \\ 1 & \chi \text{ odd} \end{cases}. \tag{1.11}$$

Conventions

We make some conventions which will be applied throughout in what follows.

(i) Whenever we refer to a functional equation, we concentrate on the gamma factors. In this respect we incorporate the factors of either of the sequences, in the basic sequences or the coefficients, where in the very definition of a Dirichlet series

$$\varphi(s) = \sum_{k=1}^{\infty} \frac{\alpha_k}{\lambda_k^s} \tag{1.12}$$

which we assume is absolutely convergent for $\sigma > \sigma_\varphi$ we call the increasing sequence of positive reals $\{\lambda_k\}$ the basic sequence and the sequence $\{\alpha_k\}$ of complex numbers the coefficients. With another Dirichlet series

$$\psi(s) = \sum_{k=1}^{\infty} \frac{\beta_k}{\mu_k^s}, \tag{1.13}$$

which we assume is absolutely convergent for $\sigma > \sigma_\psi$, we shall discuss various forms of the functional equation and **we treat the two Dirichlet series φ and ψ as given once and for all in the forms** (1.12) and (1.13), referring to "given two Dirichlet series".

For example, in (1.5), we write

$$\varphi(s) = \pi^{-\frac{s}{2}} \zeta(s) = \sum_{k=1}^{\infty} \frac{1}{(\sqrt{\pi}k)^s} \tag{1.14}$$

which is absolutely convergent for $\sigma := \operatorname{Re} s > 1$. Then (1.5) is equi-vocal to

$$\Gamma\left(\frac{s}{2}\right) \varphi(s) = \Gamma\left(\frac{1-s}{2}\right) \varphi(1-s). \tag{1.15}$$

However, it is more far-reaching to consider

$$\Gamma\left(\frac{s}{2}\right) \varphi(s) = \Gamma\left(\frac{r-s}{2}\right) \psi(r-s). \tag{1.16}$$

(1.3) and (1.4) are unified in the single gamma factor case (5.3) or in a slightly more general (4.22). We refer to (1.16) as the **Riemann type functional equation**.

In the cases of (1.9) and (1.10), we put

$$\varphi(s) = \pi^{-\frac{s+a}{2}} L(s, \chi) = \sum_{k=1}^{\infty} \frac{\alpha_k}{(\sqrt{\pi}k)^s}, \quad \alpha_k = \pi^{-\frac{a}{2}} \chi(k) \tag{1.17}$$

and

$$\psi(s) = \pi^{-\frac{s+a}{2}} L(s, \bar{\chi}) = \sum_{k=1}^{\infty} \frac{\beta_k}{(\sqrt{\pi}k)^s}, \quad \beta_k = \frac{i^a \sqrt{q}}{\tau(\chi)} \pi^{-\frac{a}{2}} \bar{\chi}(k). \tag{1.18}$$

Then we think of (1.9) as

$$\Gamma\left(\frac{s+a}{2}\right) \varphi(s) = \Gamma\left(\frac{1-s+a}{2}\right) \psi(1-s). \tag{1.19}$$

In this way, we seal up the additional factors in the Dirchlet series themselves and we will not give a special mention to them but the corresponding results are to be readily read off by specialization of the sequences.

(ii) By abuse of language, by the functional equation for the Dedekind zeta-function (to be developed in §3.2, §3.4.1, §4.4) we mean several objects that have the same form of the functional equation. That is, when we speak of the Dedekind zeta-function of an imaginary quadratic field we also refer to the Epstein zeta-function of a positive definite quadratic form or the one associated to a modular form (which is known as the Hecke correspondence) and when we speak of the Dedekind zeta-function of a real quadratic field we also refer to the Epstein zeta-function of an indefinite quadratic form, the Maass wave zeta-function (§7.1.2) or the one associated to the Dirichlet divisor problem. Here the divisor problem refers to an asymptotic formula for the summatory function of the divisor function and the divisor function is the special case $\sigma_0(n)$ of the sum-of-divisors function in (3.86).

(iii) When we state the modular relations, we often say that they are equivalent to the functional equation to mean that they are all consequences of the functional equation in question and some of them imply the functional equation.

(iv) We usually refer the sum of the residues as the residual function and denote it by $P(x)$, where P refers to capital ρ rather than P as in the tradition of G. H. Hardy [HarAO], where he means by $P(x)$ the error term of the Gauss circle problem and since the number of integers in the circle of radius $\sqrt{x}$ is denoted by $R(x)$, his $P(x)$ is to mean the Greek capital.

Following Bochner [Bo1], we define the **residual function**

$$P(x) = \frac{1}{2\pi i} \int_C \chi(s) x^{-s} ds,$$

where C encircles all the singularities of $\chi(s)$ in $\mathcal{S}$, where χ is the key-function to be defined later. Cf. also [Ahm]. We often express the residual function explicitly as the sum of residues but sometimes we just denote it as $P(x)$.

(v) As the summation variable we almost always use k instead of more familiar n or m so as to avoid the confusion of the indexes in the H-functions $H^{m,n}_{p,q}$. The prime on the summation sign means that the last term is to be halved if the summation variable equals to the bound. This appears especially when we apply the Riesz sum and the partial sum as its special case. Cf. Chapter 6.

σ almost always indicates the real part of the complex variable s save for Koshlyakov's σ-function.

(vi) Although we restate the following in §2.6, we fix the notation and stick to it once and for all.

With the **coefficients array** $(0 \leq n \leq p, 0 \leq m \leq q)$,

$$\Delta = \begin{pmatrix} \{(1 - a_j, A_j)\}_{j=1}^n; & \{(a_j, A_j)\}_{j=n+1}^p \\ \{(b_j, B_j)\}_{j=1}^m & ;\{(1 - b_j, B_j)\}_{j=m+1}^q \end{pmatrix} \tag{1.20}$$

associated is the **processing gamma factor** $(A_j, B_j > 0)$

$$\Gamma(s \mid \Delta) \tag{1.21}$$

$$= \Gamma\left(s \left| \begin{matrix} \{(1 - a_j, A_j)\}_{j=1}^n; & \{(a_j, A_j)\}_{j=n+1}^p \\ \{(b_j, B_j)\}_{j=1}^m & ;\{(1 - b_j, B_j)\}_{j=m+1}^q \end{matrix} \right.\right)$$

$$= \frac{\prod\limits_{j=1}^m \Gamma(b_j + B_j s) \prod\limits_{j=1}^n \Gamma(a_j - A_j s)}{\prod\limits_{j=n+1}^p \Gamma(a_j + A_j s) \prod\limits_{j=m+1}^q \Gamma(b_j - B_j s)} \quad (A_j, B_j > 0)$$

and the H-function

$$H(z \mid \Delta)$$

$$= H_{p,q}^{m,n}\left(z \left| \begin{matrix} (1 - a_1, A_1), \ldots, (1 - a_n, A_n), (a_{n+1}, A_{n+1}), \ldots, (a_p, A_p) \\ (b_1, B_1), \ldots, (b_m, B_m), (1 - b_{m+1}, B_{m+1}), \ldots, (1 - b_q, B_q) \end{matrix} \right.\right)$$

$$= \frac{1}{2\pi i} \int_L \Gamma(s \mid \Delta) \, z^{-s} \, ds$$

$$= \frac{1}{2\pi i} \int_L \frac{\prod\limits_{j=1}^m \Gamma(b_j + B_j s) \prod\limits_{j=1}^n \Gamma(a_j - A_j s)}{\prod\limits_{j=n+1}^p \Gamma(a_j + A_j s) \prod\limits_{m+1}^q \Gamma(b_j - B_j s)} \, z^{-s} \, ds, \tag{1.22}$$

where the path L is subject to the poles separation conditions.

We note

Mnemonics

+	−
m	n
p	q

Here m is an abbreviation for the B-group (b_j, B_j)'s with positive B_j's. Similarly, n is an abbreviation for A-group (a_j, A_j)'s with negative A_j's.

Exercise 1.1 Prove that (1.5) and the following are equivalent.

$$\sum_{n=1}^{\infty} e^{-\pi n^2 x} = x^{-\frac{1}{2}} \sum_{n=1}^{\infty} e^{-\frac{\pi n^2}{x}} + \frac{1}{2}\left(x^{-\frac{1}{2}} - 1\right), \quad \mathrm{Re}(x) > 0, \tag{1.23}$$

$$\pi^{-\frac{s}{2}}\Gamma\left(\frac{s}{2}\right)\zeta(s) = \pi^{-\frac{s}{2}}\sum_{n=1}^{\infty}\frac{1}{n^s}\Gamma\left(\frac{s}{2}, \pi n^2 w\right) \tag{1.24}$$

$$+ \pi^{-\frac{1-s}{2}}\sum_{n=1}^{\infty}\frac{1}{n^{1-s}}\Gamma\left(\frac{1-s}{2}, \frac{\pi n^2}{w}\right) + \frac{w^{\frac{s-1}{2}}}{s-1} - \frac{w^{\frac{s}{2}}}{s}, \quad \mathrm{Re}(w) > 0,$$

$$\pi^{-s}\Gamma(s)\sum_{n=1}^{\infty}\frac{1}{(n^2 + \alpha^2)^s} = 2\alpha^{\frac{1}{2}-s}\sum_{n=1}^{\infty}n^{s-\frac{1}{2}}K_{s-\frac{1}{2}}(2\pi\alpha n) \tag{1.25}$$

$$+ \frac{1}{2}\pi^{-s+\frac{1}{2}}\alpha^{1-2s}\Gamma\left(s - \frac{1}{2}\right) - \frac{1}{2}\pi^{-s}\alpha^{-2s}\Gamma(s),$$

where $\Gamma(s.z)$ and $K_\nu(z)$ indicate the incomplete gamma function (2.12) and the modified K-Bessel function (2.31), resp.

This is essentially Lavrik's Example 1 [Lav5, p.522] that indicates the equivalence of (1.5), (1.23), (1.24), while (1.25) is not mentioned. Note that (1.23) is the celebrated theta transformation formula (1.72). Indeed, Riemann [Rie2] deduced the functional equation (1.5) from the theta-transformation formula (1.23). Cf. §2.2 for more details.

The following Exercise 1.2 is [KTY7, Example 1.1, p.164].

Exercise 1.2 Prove that (1.23) and the following are equivalent

$$\pi^{-s}\Gamma(s)\zeta(2s) = \pi^{-\left(\frac{1}{2}-s\right)}\Gamma\left(\frac{1}{2} - s\right)\zeta(1 - 2s), \tag{1.26}$$

$$\pi^{-s}\Gamma(s)\zeta(2s) = \pi^{-s}\sum_{n=1}^{\infty}\frac{1}{n^{2s}}\Gamma(s, \pi n^2 w) \tag{1.27}$$

$$+ \pi^{s-\frac{1}{2}}\sum_{n=1}^{\infty}\frac{1}{n^{1-2s}}\Gamma\left(\frac{1}{2} - s, \frac{\pi n^2}{w}\right) + \frac{w^{s-\frac{1}{2}}}{2s-1} - \frac{w^s}{2s}, \quad \mathrm{Re}(w) > 0,$$

$$\pi^{-s}\Gamma(s)\sum_{n=-\infty}^{\infty}\frac{1}{(n^2 + \alpha^2)^s} = 2\alpha^{\frac{1}{2}-s}\sum_{n=-\infty}^{\infty}|n|^{s-\frac{1}{2}}K_{s-\frac{1}{2}}\left(2\pi\alpha|n|\right), \tag{1.28}$$

where the term corresponding to $n = 0$ on the right-hand side is to be understood to mean

$$\lim_{u \to 0} u^{s-\frac{1}{2}} K_{s-\frac{1}{2}}(2\pi\alpha u) = \frac{1}{2}(\pi\alpha)^{\frac{1}{2}-s}\Gamma\left(s - \frac{1}{2}\right), \qquad (1.29)$$

which will be restated as (2.35) in §2.3 below.

Solution The equivalence of (1.5) and (1.23) ((1.26) and (1.23)) is the most celebrated modular relation (cf. §2.2 on zeta-functions).

The equivalence of (1.5) and (1.24) ((1.26) and (1.27)) is the Ewald expansion, which will be developed in Chapter 5.

Similarly, the equivalence of (1.5) and (1.25) ((1.26) and (1.28)) is the Fourier-Bessel expansion, which will be developed in Chapter 4. We only indicate the deduction of (1.25) and (1.28) from a general modular relation. In accordance with $\lambda_n = n^2$, we write α^2 for α. Then $\varphi(s, \alpha^2)$ in (4.84) becomes $\sum_{n=1}^{\infty} \frac{1}{(n^2+\alpha^2)^s}$. Since

$$A^{-s} \int_0^{\infty} e^{-\alpha^2 u} u^{s-1} \mathrm{P}\left(\frac{u}{\pi}\right) \mathrm{d}u = \frac{1}{2}\pi^{1/2-s}\alpha^{1-2s}\Gamma\left(s - \frac{1}{2}\right) - \frac{1}{2}\alpha^{-2s}\Gamma(s),$$

we obtain (1.25) from (4.82).

Multiplying (1.25) by 2, moving the last term on the right of (1.25) to the left and adopting the interpretation (1.29) completes the proof of (1.28).

Remark 1.1 (1.28) *or* (1.25) *are known as* **Watson's formula** *[Wa]. Cf. (4.32) below. It is stated as [BK, Theorem 8.4, p.121] and is a starting point for proving [BK, Theorem 4.8] below.*

The special case $s = 1$ *of* (1.25) *gives the* **partial fraction expansion for the hyperbolic cotangent function** *(due to Euler) [Vista, p.100]:*

$$\sum_{k \in \mathbb{Z}} \frac{1}{k^2 + w^2} = \frac{\pi}{w}\left(\frac{2}{e^{2\pi w} - 1} + 1\right) = \frac{\pi}{w} \coth \pi w. \qquad (1.30)$$

(1.30) *is [MKn2, (2.10)] and is the main ingredient in Siegel's proof of the Hamburger theorem (Remark 3.1, (ii)).*

As will be seen steadily, one of the main ingredients that make a distinction between these two types of zeta-functions is the reduction formula (2.93), which necessitates the sequence to be squared (in general raised to the power $\frac{1}{C}$ if we have C for $\frac{1}{2}$).

1.2   Eternal return or every 50 years

1.2.1   *Poincaré recurrence theorem*

The eternal return used to be once a dominating pessimistic view of the 18th century physics and it is remotely but somewhat relevant to what we are going to state, so that we find it appropriate to state and give a proof of the Poincaré recurrence theorem, following Tipler [Tip, pp.418-419].

Theorem 1.1   (Poincaré recurrence theorem) *Let V be a finite and bounded region in the phase space and let f be a continuous, volume preserving 1 : 1 map from V to V: $f : V \to f(V) = V$. Then for any point and in any of its neighborhood U, there exists a point $x \in V$ which returns to U, i.e. there exists an $n \in \mathbb{N}$ such that $f^n x \in U$, where we write fx to mean $f(x)$.*

Proof.   Consider the sequence of actions of f on U:

$$U, fU, f^2U, \ldots .$$

Since f is volume-preserving, all of the terms of this sequence have the same finite volume. If they never intersect, then, all the terms being contained in V, the volume of V would be infinite, a contradiction. Hence there must be $i, j \in \mathbb{N}$, $i > j$ such that $f^iU \cap f^jU \neq \emptyset$, or $f^nU \cap U \neq \emptyset$, on setting $n = i - j$. Hence there exists a point x such that $f^n x$ also belongs to U, whence the theorem. $\square$

1.2.2   *Lerch-Chowla-Selberg formula*

This phenomenon of every 50 years revival of fashion, intrinsic to the human society, has been noticed first by Kurokawa [Kur] in mathematics as far as we know. There must be similar observations made earlier many times, every 50 years or so. More often 60 years period would appear since this is the span of two generations. You and your grandson may not share the same level of intelligence because of the inevitable generation gap. Thus it would look natural that the knowledge obtained two generations ago may be forgotten and not inherited by the offsprings.

Prior to [Kur] by two years, F. Sato [Sa] (there are many Sato's and we are very specific to this Professor Fumihiro Sato), in expounding the theory of pre-homogeneous vector spaces, mentioned Eisenstein's proof of the functional equation, which has been referred to on [Vista, pp.70-72].

Although we referred to [Kur] in [Vi2], there are no comments on that paper in the Chowla-Selberg exposition on [Vi2, pp.122-123], and we shall now provide what could have appeared there.

Matyas Lerch [Ler1], starting from the work of Malmstén [Ma], established **Lerch's formula**

$$\zeta'(0, x) = \log \Gamma(x). \tag{1.31}$$

Using (1.31), he obtained [Ler3] an important relation between the values of the gamma function and the automorphic function (elliptic function), which has been known as the **Chowla-Selberg formula**. Cf. [WeE, Chapter X]. The most enlightening explanations on historic as well as mathematical part can be found in Landau [LanCS] and Deninger [Den], in the latter which the Chowla-Selberg formula for a real quadratic field is also deduced using the R-function.

Theorem 1.2 (Lerch-Chowla-Selberg formula)

$$\frac{w}{2} \sum_{n=1}^{|\Delta|-1} \left(\frac{|\Delta|}{n} \right) \log \Gamma \left(\frac{n}{|\Delta|} \right) = \sum_{(a,b,c)} \log \left[\sqrt{\frac{2|\Delta|\pi}{c}} \eta(\omega_1)\eta(\omega_2) \right], \tag{1.32}$$

where $\Delta < 0$ *is the discriminant of the imaginary quadratic field* $\Omega = \mathbb{Q}(\sqrt{\Delta})$, w *indicates the order of the group of roots of unity in* Ω, $\omega_1 = \frac{-b+i\sqrt{|\Delta|}}{2c}$, $\omega_2 = \frac{-b-i\sqrt{|\Delta|}}{2c}$ *(whence* $\eta(\omega_1)\eta(\omega_2) = |\eta(\omega_1)|^2$*) and on the right* (a, b, c) *runs through the complete set of representatives of classes of quadratic forms of discriminant* $-\Delta$.

For the notation, cf. Section 3.4.

In the times of Kronecker, after the tradition of Gauss, the theory of binary quadratic forms was the main issue of research than the theory of quadratic fields, both being equivalent. For classification of this cf. Davenport [Dav] which in turn depends on Landau [LanZT].

$\left(\frac{|\Delta|}{n} \right) = \chi_\Omega(n)$ on the left of (1.32) indicates the Kronecker-Jacobi symbol associated to Ω and is one of primitive odd characters modulo $|\Delta|$. Berger [Berg] and Lerch [Ler1] independently expressed the values of $L'(1, \chi_\Omega)$ in terms of the gamma values. Both authors appealed to Kummer's Fourier series for the $\log \Gamma$ to deduce the results. Lerch [Ler1] then used the Kronecker limit formula to deduce Theorem 1.2.

Corollary 1.1 *The Lerch-Chowla-Selberg formula* (1.32) *is a consequence of the functional equations for the Dedekind zeta-functions of the* *rational and the imaginary quadratic fields.*

Proof.   The Kronecker limit formula in Corollary 4.3 is a consequence of the Chowla-Selberg type formula (Corollary 4.2), which is equivalent to the functional equation for the Dedekind zeta-functions of the imaginary quadratic field. (4.93) must be divided by the number w of roots of unity and summed over the h classes. Hence the right-hand side of (1.32) is a consequence of the functional equation.

On the other hand, the left-hand side follows from the decomposition

$$\zeta_\Omega(s) = \zeta(s)L(s,\chi_\Omega). \tag{1.33}$$

The evaluation of $L'(1,\chi_\Omega)$ amounts to that of the Fourier series

$$\sum_{n=1}^{\infty} \frac{\log n \sin 2\pi nx}{n} \tag{1.34}$$

which is simply $\log\Gamma(x) - \log\Gamma(1-x)$ and thus the Kummer Fourier series for the loggamma function (1.35) given below gives the result. Hence the left-hand side is a consequence of the functional equation for the Dedekind zeta-function for the rational field.   □

Proposition 1.1   ([Vi2, Chapter 7]) *For $0 < x < 1$ we have*

$$\log\frac{\Gamma(x)}{\sqrt{2\pi}} = \sum_{n=1}^{\infty}\left(\frac{1}{2n}\cos(2\pi nx) + \frac{\gamma + \log(2\pi n)}{\pi n}\sin(2\pi nx)\right), \tag{1.35}$$

which is in a long run a consequence of the functional equation for the Riemann zeta-function.

The decomposition formula (1.33) holds true for an abelian field over the rationals, i.e. for the mth cyclotomic field $\mathbb{Q}(\zeta)$, where $\zeta = \zeta_m$ is a primitive mth root of 1.

Gut [Gut] was the first who expressed the constant term of the Dedekind zeta-function for the $\mathbb{Q}(\zeta)$ in terms of the R-function.

Definition 1.1   Let $\alpha > 0$. Then the **Deninger** $R = R_\alpha$**-function** is defined as the principal solution to the difference equation

$$f_\alpha(x+1) - f_\alpha(x) = \log^\alpha x \quad x > 0. \tag{1.36}$$

Corollary 1.2   *For $\alpha = 1$, $f_1(x) = \log\Gamma(x)$ and for $\alpha = 2$, $f_2(x) = \zeta''(0,x) = \frac{\partial^2}{\partial^2 s}\zeta(s,x)|_{s=0}$, where $\zeta(s,x)$ indicates the Hurwitz zeta-function defined by (2.22).*

The higher-order R_α-function has been applied to the expressing the higher-derivatives of the L-function at $s = 1$ in [CKK].

1.2.3 *Knopp-Hasse-Sondow formula*

The contents of this subsection is mostly due to [XHW]. There are many known integral expressions for the Riemann zeta-function which provides us with a preliminary stage for analytic continuation. Notably, [Vi2, Lemma 2.3, p. 43] ([SC, (30), p. 100]) reads

$$\Gamma(s)\zeta(s) = \int_0^\infty t^s \frac{1}{e^t - 1} \frac{dt}{t} = \int_0^\infty t^s \frac{e^{-t}}{1 - e^{-t}} \frac{dt}{t} \tag{1.37}$$

which is true for $\sigma > 1$.

We may view (1.37) as the gamma transform of the theta function $\vartheta_1(t)$ in the following manner.

Let $\varkappa > 0$ and let $\vartheta_\varkappa(t)$ denote the theta series of order $\varkappa > 0$:

$$\vartheta_\varkappa(t) = \sum_{n=-\infty}^{\infty} e^{-n^\varkappa t} \tag{1.38}$$

in case $\varkappa$ is even and

$$\vartheta_\varkappa(t) = \sum_{n=0}^{\infty} e^{-n^\varkappa t} \tag{1.39}$$

in case $\varkappa$ is odd, for $\mathrm{Re}\, t > 0$. Note that

$$\vartheta_1(t) = \frac{1}{e^t - 1} \tag{1.40}$$

and $\vartheta_2(t)$ is the elliptic theta-function, which satisfies the theta-transformation formula (1.23), (1.72) (cf. Section 1.3).

A possible application of higher-order theta-functions has been already noted much earlier. In [HarDP, p.7] he mentioned possible applications of the higher-order Lambert series

$$F_{\alpha,\beta}(s) = \sum_{n=1}^{\infty} \frac{d(n)}{n^\beta} e^{-sn^\alpha}, \tag{1.41}$$

where $d(n)$ indicates the number of divisors of n, called the **divisor function**, which is the special case $\sigma_0(n)$ of the sum-of-divisors function in (3.86). He also refers to the original papers of Mellin [Mel1], [Mel2] as the main source for a large amount of resources.

The **gamma transform** [Vi2, p.25] reads for $\lambda > 0$ and $\sigma > 0$

$$\Gamma(s)\lambda^{-s} = \int_0^\infty t^s e^{-\lambda t}\, \frac{dt}{t} \tag{1.42}$$

whence in particular,

$$\Gamma(s)\zeta(\varkappa s) = \int_0^\infty t^s \vartheta_\varkappa(t)\, \frac{dt}{t} \tag{1.43}$$

for $\varkappa\sigma > 1$.

In this context, $\vartheta_2(t)$ along with its transformation formula, has been used extensively in literature as a source of the functional equation for the Riemann zeta-function.

Goss [Gs] uses $\vartheta_1(t)$ to provide an analytic continuation of the Riemann zeta-function whose method consists in removing the singularity at $t = 0$ of the integrand $\frac{1}{e^t-1}$ or writing $z = e^{-t}$, that of the fraction $\frac{z}{1-z}$ at $z = 1$. Hence Goss' method can be generalized in the following setting. With a function analytic at $z = 1$ such that $f(1) = 1$, we form the difference

$$\frac{z}{1-z} - \frac{f(z)}{1-z} = \frac{f(z) - z}{z-1}, \tag{1.44}$$

whence it follows that the difference is analytic at $z = 1$.

Goss uses $f(z) = \frac{mz^m}{1+z+\cdots+z^{m-1}}$, i.e. the difference in (1.44) is $\phi_m(z) = \frac{z}{1-z} - \frac{mz^m}{1-z^m}$.

Our aim is to find a class of such functions f which give rise to analytic continuation of the relevant zeta-function.

Theorem 1.3 *Suppose $f(z)$ is an analytic in the circle $|z-1| \le r$ $(r > 1)$ and bounded on the boundary and that $f(0) = 0$ and $f(1) = 1$. Then*

$$F(s) := \Gamma(s)\zeta(s) - \int_0^\infty t^s \left.\frac{f(z)}{1-z}\right|_{z=e^{-t}} \frac{dt}{t} \tag{1.45}$$

$$= \int_0^\infty t^s \left.\frac{f(z) - z}{z-1}\right|_{z=e^{-t}} \frac{dt}{t}$$

is analytic over the whole plane.

Proof. In the above-mentioned region of analyticity of $f(z)$, it can be expressed as a Taylor series at $z = 1$, which is absolutely and uniformly convergent, so that all operations, conducted below, of changing order and term-wise integration are legitimate. By the Taylor expansion of $f(z)$, we obtain

$$\frac{f(z) - z}{z - 1} = \frac{1}{z - 1}\left(1 + f'(1)(z - 1) + \sum_{n=2}^{\infty} \frac{f^{(n)}(1)}{n!}(z - 1)^n - z\right).$$

Substituting $z = e^{-t}$ and using the binomial expansion, we obtain

$$\frac{f(z) - z}{z - 1} = -1 + f'(1) + \sum_{n=1}^{\infty} \frac{f^{(n+1)}(1)}{(n + 1)!} \sum_{k=0}^{n} (-1)^{n-k}\binom{n}{k}e^{-kt}.$$

Separating the term with $k = 0$ from the inner sum, we obtain

$$\text{LHS} = -1 + f'(1) + \sum_{n=1}^{\infty} \frac{f^{(n+1)}(1)}{(n + 1)!}(-1)^n$$

$$+ \sum_{n=1}^{\infty} \frac{f^{(n+1)}(1)}{(n + 1)!} \sum_{k=1}^{n} (-1)^{n-k}\binom{n}{k}e^{-kt}$$

whose first term is $f(0)$, which is 0. Hence

$$\frac{f(z) - z}{z - 1} = \sum_{n=1}^{\infty} \frac{f^{(n+1)}(1)}{(n + 1)!} \sum_{k=1}^{n} (-1)^{n-k}\binom{n}{k}e^{-kt}. \tag{1.46}$$

Substituting (1.46) in (1.45) and using (1.42), we have

$$F(s) = \sum_{n=1}^{\infty} \frac{f^{(n+1)}(1)}{(n + 1)!} \sum_{k=1}^{n} (-1)^{n-k}\binom{n}{k}k^{-s} \tag{1.47}$$

or by changing the order of summation

$$F(s) = \sum_{k=1}^{\infty} \frac{(-1)^k}{k^s} \sum_{n=k}^{\infty} \frac{f^{(n+1)}(1)}{(n + 1)!}(-1)^n. \tag{1.48}$$

By Cauchy's inequality,

$$\frac{f^{(n+1)}(1)}{(n + 1)!} = O\left(\frac{1}{r^{n+1}}\right), \tag{1.49}$$

whence the Majorant series for $F(s)$ is

$$\sum_{k=1}^{\infty} \frac{1}{k^{\sigma}} \sum_{n=k+1}^{\infty} \frac{1}{r^n} = O\left(\sum_{k=1}^{\infty} \frac{1}{k^{\sigma}r^{k+1}}\right) = O(1) \tag{1.50}$$

for all s. Then the Weierstrass Majorant test implies that the series (1.48) is (absolutely and) uniformly convergent over the whole plane, implying the analyticity of $F(s)$. □

Note that the inner sum in (1.48) is

$$f(0) - \sum_{n=1}^{k-1} \frac{f^{(n+1)}(1)}{(n+1)!}(-1)^n = \sum_{n=2}^{k} \frac{f^{(n)}(1)}{n!}(-1)^n. \tag{1.51}$$

Example 1.1 To elucidate Goss' procedure, we choose

$$f(z) = \frac{mz^m(z-1)}{(1+z+\cdots+z^{m-1})(z-1)} = m\frac{z^{m+1} - z^m}{z^m - 1}. \tag{1.52}$$

By the Euclidean division, we have

$$f(z) = m\frac{(z^m - 1)(z-1) + z - 1}{z^m - 1} = m(z-1) + m\frac{z-1}{z^m - 1} = m(z-1) + g(z),$$

say, where $g(z) = m\frac{z-1}{z^m-1}$. Let $\zeta = e^{2\pi i/m}$ be the piervot'ny mth root of 1. Then

$$z^{m-1} + \cdots + z + 1 = \prod_{k=1}^{m-1}(z - \zeta^k)$$

and

$$\prod_{k=1}^{m-1}(1 - \zeta^k) = m.$$

We may easily obtain the partial fraction expansion of $g(z)$:

$$g(z) = m\frac{1}{1 + z + \cdots + z^{m-1}}$$

$$= m\frac{1}{\prod_{k=1}^{m-1}(z - \zeta^k)} = m\left(\sum_{j=1}^{m-1} \frac{a_j}{z - \zeta^j}\right), \tag{1.53}$$

say. Since

$$a_j = \lim_{z \to 1}(z - \zeta^j)g(z) = \frac{1}{\prod_{k \neq j}(1 - \zeta^k)} = \frac{1 - \zeta^j}{m}, \tag{1.54}$$

we have

$$f(z) = m(z-1) + \sum_{j=1}^{m-1} \frac{1 - \zeta^j}{z - \zeta^j}. \tag{1.55}$$

Hence for $\ell \in \mathbb{N}$,

$$f^{(\ell)}(z) = \delta_{\ell,1}m + (-1)^{\ell}\ell! \sum_{j=1}^{m-1} \frac{1 - \zeta^j}{\left(z - \zeta^j\right)^{\ell+1}}, \tag{1.56}$$

where $\delta_{\ell,1} = 1$ or 0 according as $\ell = 1$ or $\ell \geq 2$, whence

$$f^{(\ell)}(1) = \delta_{\ell,1}m + (-1)^{\ell}\ell! \sum_{j=1}^{m-1} \frac{1}{\left(1 - \zeta^j\right)^{\ell}}. \tag{1.57}$$

To secure the analyticity condition in Theorem 1.3, which is given by

$$\left|1 - e^{\frac{2\pi i}{m}}\right| > r(\geq 1),$$

it would be better to have smaller values of m, say, $m = 2$. In this case, $r = 2 - \delta$ $(\delta > 0)$ and the conditions in Theorem 1.3 are satisfied.

But in this case, the function $F(s)$ reduces to the familiar $(1 - 2^{1-s})\zeta(s)$ for which the most famous procedure is that of Euler transformation. [Has], [So]. Cf. Corollary 1.3 below.

We may also take $m = 4$, in which case $r = \sqrt{2} - \delta$ $(\delta > 0)$. The case $m = 6$ does not work since then $|1 - e^{(2\pi i)/6}| = \sqrt{2 - \sqrt{2}} < 1$. For bigger m one has to make some trick and in Goss's case, it is integration by parts.

Corollary 1.3 *The Knopp-Hasse-Sondow difference operator expression*

$$(1 - 2^{1-s})\zeta(s) = \sum_{n=0}^{\infty} \frac{1}{2^{n+1}} \sum_{k=0}^{n} (-1)^k \binom{n}{k}(k+1)^{-s} = \sum_{n=0}^{\infty} \frac{\Delta_n 1^{-s}}{2^{n+1}} \tag{1.58}$$

is a corollary to Theorem 1.3, where the difference operator is defined by the sum in the mid term (cf. (6.3)).

Proof. We consider the case $m = 2$ of Example 1.1. Then $\zeta = -1$ and (1.57) reads

$$f^{(\ell)}(1) = \frac{(-1)^{\ell}\ell!}{2^{\ell}} \tag{1.59}$$

for $\ell \geq 2$. Hence (1.47) reads

$$(1 - 2^{1-s})\zeta(s) = F(s) = \sum_{n=1}^{\infty} \frac{(-1)^{n+1}}{2^{n+1}} \sum_{k=1}^{n} (-1)^{n-k} \binom{n}{k} k^{-s} \tag{1.60}$$

$$= \sum_{n=1}^{\infty} \frac{1}{2^{n+1}} S(n),$$

say, where

$$S(n) := \sum_{k=0}^{n-1} (-1)^k \binom{n}{k+1} (k+1)^{-s}. \tag{1.61}$$

Since $\binom{n+1}{k+1} = \binom{n}{k} + \binom{n}{k+1}$, it follows that

$$S(n) = \sum_{k=0}^{n} (-1)^k \binom{n+1}{k+1} (k+1)^{-s} - (-1)^n (n+1)^{-s} \tag{1.62}$$

$$- \sum_{k=0}^{n-1} (-1)^k \binom{n}{k} (k+1)^{-s}$$

$$= S(n+1) - \sum_{k=0}^{n} (-1)^k \binom{n}{k} (k+1)^{-s},$$

which amounts to the difference equation

$$S(n+1) - S(n) = \sum_{k=0}^{n} (-1)^k \binom{n}{k} (k+1)^{-s}. \tag{1.63}$$

Since

$$\sum_{n=1}^{\infty} \frac{1}{2^{n+1}} S(n+1) = 2 \sum_{n=1}^{\infty} \frac{1}{2^{n+2}} S(n+1)$$

$$= 2 \left(\sum_{n=1}^{\infty} \frac{1}{2^{n+1}} S(n) - \frac{1}{4} S(1) \right) = 2(1 - 2^{1-s}) \zeta(s) - \frac{1}{2} \tag{1.64}$$

by (1.60), it follows from (1.63) that

$$\sum_{n=1}^{\infty} \frac{1}{2^{n+1}} \sum_{k=0}^{n} (-1)^k \binom{n}{k} (k+1)^{-s} \tag{1.65}$$

$$= \sum_{n=1}^{\infty} \frac{1}{2^{n+1}} (S(n+1) - S(n)) = (1 - 2^{1-s}) \zeta(s) - \frac{1}{2},$$

whence (1.58) follows, completing the proof. □

1.3 The theta-transformation formula

We begin with Problem 17 in [WW] which gives a proof of the theta-transformation formula via contour integrals. We denote the elliptic theta function (or series) which is the $\varkappa = 2$ case of (1.38) by ϑ:

$$\vartheta(t) = \vartheta_2(t) = \sum_{n=-\infty}^{\infty} e^{-\pi n^2 t} \qquad (1.66)$$

for $t > 0$ (eventually we may take $\operatorname{Re} t > 0$; (1.75)). Let $\alpha > 0$ and $N > 0$ be a large integer. Let $C = C_N$ denote the rectangle with vertices at $\pm \left(N + \frac{1}{2}\right) \pm \alpha i$. Then integrate the function $f(z) = \frac{e^{-\pi z^2 t}}{e^{2\pi i z} - 1}$ around C_N and letting $N \to \infty$ to prove that

$$\vartheta(t) = \int_{-\infty-\alpha i}^{\infty-\alpha i} \frac{e^{-\pi z^2 t}}{e^{2\pi i z} - 1} \, dz + \int_{-\infty+\alpha i}^{\infty+\alpha i} \frac{e^{-\pi z^2 t}}{1 - e^{2\pi i z}} \, dz = I_1 + I_2, \qquad (1.67)$$

say. Expanding the denominator of the integrand of I_1 into the geometric series

$$\frac{1}{e^{2\pi i z} - 1} = -e^{-2\pi i z} \left(1 - e^{-2\pi i z}\right)^{-1} = -e^{-2\pi i z} \sum_{k=0}^{\infty} e^{-2\pi i k z},$$

which is absolutely convergent, we see that I_1 in (1.67) is

$$I_1 = -e^{-2\pi i z} \sum_{k=0}^{\infty} \int_{-\infty-\alpha i}^{\infty-\alpha i} e^{-\pi z^2 t - 2\pi i k z} \, dz = \sum_{k=1}^{\infty} \int_{-\infty-\alpha i}^{\infty-\alpha i} e^{-\pi z^2 t - 2\pi i k z} \, dz$$

by termwise integration.

By Cauchy's theorem, we may prove that the inner integral is equal to the one with $\alpha = 0$, $\int_{-\infty}^{\infty} e^{-\pi x^2 t - 2\pi i k x} \, dx$.

Similarly,

$$I_2 = \sum_{k=0}^{\infty} \int_{-\infty+\alpha i}^{\infty+\alpha i} e^{-\pi z^2 t + 2\pi i k z} \, dz = \int_{-\infty}^{\infty} e^{-\pi x^2 t + 2\pi i k x} \, dx.$$

Hence it follows that

$$\vartheta(t) = \sum_{k=1}^{\infty} \int_{-\infty}^{\infty} e^{-\pi x^2 t} \left(e^{-2\pi i k x} + e^{2\pi i k x}\right) \, dx + \int_{-\infty}^{\infty} e^{-\pi x^2 t} \, dx \qquad (1.68)$$

$$= 2 \sum_{k=1}^{\infty} \int_{-\infty}^{\infty} e^{-\pi x^2 t} \cos 2\pi k x \, dx + \int_{-\infty}^{\infty} e^{-\pi x^2 t} \, dx$$

$$= \sum_{k=-\infty}^{\infty} \int_{-\infty}^{\infty} e^{-\pi x^2 t} \cos 2\pi k x \, dx.$$

The next step is to apply Example 3 [WW, p.114] which asserts that for $\lambda > 0$,

$$\int_{-\infty}^{\infty} e^{-\lambda x^2} \cos 2\lambda ax \, dx = 2\lambda^{-\frac{1}{2}} e^{-\lambda a^2} \int_{-\infty}^{\infty} e^{-x^2} \, dx. \qquad (1.69)$$

This is also proved by the Cauchy residue theorem.
Substituting (1.69) in (1.68), we obtain

$$\vartheta(t) = \frac{1}{\sqrt{\pi t}} \vartheta\left(\frac{1}{t}\right) \int_{-\infty}^{\infty} e^{-x^2} \, dx. \qquad (1.70)$$

Putting $t = 1$, we may evaluate the probability integral

$$\int_{-\infty}^{\infty} e^{-x^2} \, dx = \sqrt{\pi}. \qquad (1.71)$$

Substituting (1.71) in (1.70), we conclude the **theta-transformation formula**

$$\vartheta(t) = \frac{1}{\sqrt{t}} \vartheta\left(\frac{1}{t}\right) \qquad (1.72)$$

valid for $\operatorname{Re} t > 0$.

This method is stated in [MitK] as one of the methods to evaluate the probability integral.

If we do not reduce the integral to the one along the real axis, then for (1.68), we have

$$\vartheta(t) = e^{\pi \alpha^2 - 2\pi k \alpha} \sum_{k=1}^{\infty} \int_{-\infty}^{\infty} e^{-\pi x^2 t} \left(e^{2\pi i (t\alpha - k) x} + e^{-2\pi i (t\alpha - k) x} \right) dx \qquad (1.73)$$

$$+ \int_{-\infty + \alpha i}^{\infty + \alpha i} e^{-\pi z^2 t} \, dx$$

$$= 2 e^{\pi \alpha^2 - 2\pi k \alpha} \sum_{k=1}^{\infty} \int_{-\infty}^{\infty} e^{-\pi x^2 t} \cos 2\pi (t\alpha - k) x \, dx + \int_{-\infty + \alpha i}^{\infty + \alpha i} e^{-\pi z^2 t} \, dz.$$

We remark that in deriving (1.72), we applied the Cauchy residue theorem three times. We may clarify the situation and why three times applications lead to (1.72) by our theory of modular relation which originates from Riemann (succeeded in a remarkably successful way by Hecke and then by Bochner for a more extensive class of zeta-functions). According to this Riemann-Hecke-Bochner correspondence, the theta-transformation formula is equivalent to the functional equation for the associated zeta-function, which is the Riemann zeta-function in this case. Since the functional equation exhibits the symmetry with respect to the critical line $\sigma = \frac{1}{2}$, the above argument may be thought of as the result of applying the symmetry three times, resulting one time application, i.e. amounting to the equivalence to the functional equation.

To emphasize the ideas of Hecke, it is more appropriate to work with the theta series

$$\Theta(\tau) = \sum_{-\infty}^{\infty} e^{\pi i n^2 \tau} \tag{1.74}$$

for $\tau \in \mathfrak{H}$, the upper half-plane (3.100), i.e. $\operatorname{Im} \tau > 0$.

$$\Theta(iz) = \vartheta(z), \tag{1.75}$$

with $\operatorname{Re} z > 0$. $\Theta(\tau)$ and $\vartheta(z)$ shift to each other under the rotation (1.1) For $\Theta(\tau)$, (1.72) reads

$$\Theta\left(-\frac{1}{\tau}\right) = \left(\frac{\tau}{i}\right)^{\frac{1}{2}} \Theta(\tau), \tag{1.76}$$

the result of action of the Spiegelung (3.102).

1.4 Summation formulas

There are some summation formulas known of number-theoretic nature. We shall refer to just three of them, the Poisson summation formula, the Lipschitz summation formula and (a generalization of) the Plana summation formula. The Lipschitz summation formula is almost equivocal to the functional equation and is expounded in [NTA, Theorem 5.2, p.128]. Cf. also Theorem 4.6 below and [PP] for the latest reference. For the most relevant Voronoĭ summation formula, we refer to [BeV] and [AAW4], and many references given therein. We hope to return to the study on the summation formulas in subsequent volumes including truncated ones.

1.4.1 *Poisson summation formula*

The **Poisson summation formula** has been playing a prominent role in establishing a huge class of identities in scientific disciplines. We state the multi-dimensional version according to [Bak, pp.11-113].

Definition 1.2   For a function $f(t) \in L^1(\mathbb{R}^n)$, its Fourier transform $\hat{f} = \hat{f}(t)$ is defined by

$$\hat{f}(t) = \int_{-\infty}^{\infty} f(t)e(-t \cdot x) \, dx, \tag{1.77}$$

where $t \cdot x$ means the inner product. It is a continuous function on $\mathbb{R}^n$, written $\hat{f} \in C(\mathbb{R}^n)$. If $\hat{f}(t) \in L^1(\mathbb{R}^n)$, then we have the inversion formula

$$f(t) = \int_{-\infty}^{\infty} \hat{f}(t)e(t \cdot x) \, dx \tag{1.78}$$

holds. If $f(t)$ is periodic in each variable t_j and integrable over the n-dimensional unit interval $U^n = [0,1)^n$, then we view f as a function on the n-dimensional torus $\mathbb{T} = \mathbb{R}/\mathbb{Z}$ and then the Fourier transform (1.77) takes the form

$$\hat{f}(t) = \int_{\mathbb{T}^n} f(t)e(-t \cdot x) \, dx. \tag{1.79}$$

The formal series

$$\mathcal{F}(f)(t) = \sum_{n \in \mathbb{Z}^n} \hat{f}(n)e(t \cdot x) \tag{1.80}$$

is the Fourier series for f.

Lemma 1.1   (Poisson summation formula) *If $\sum_{n \in \mathbb{Z}^n} |\hat{f}(n)| < \infty$ and $f \in C(\mathbb{R}^n)$, then*
(i)

$$f(t) = \mathcal{F}(f)(t) = \sum_{n \in \mathbb{Z}^n} \hat{f}(n)e(t \cdot x). \tag{1.81}$$

(ii) *If*

$$\sum_{n \in \mathbb{Z}^n} \sup_{t \in U^n} |f(n + t)| < \infty \tag{1.82}$$

and

$$\sum_{m \in \mathbb{Z}^n} |\hat{f}(m)| < \infty \tag{1.83}$$

are satisfied, then we have the **Poisson summation formula**

$$\sum_{n \in \mathbb{Z}^n} \hat{f}(n) = \sum_{m \in \mathbb{Z}^n} f(m). \tag{1.84}$$

1.4.2 Generalization of the Plana summation formula

The following lemma gives a possibility of continuing the resulting integral in Theorem 1.4 to the left.

Lemma 1.2 *Suppose that $f(x)$ is defined and analytic in the half-plane $\operatorname{Re} x \geq 0$. Then the two representations for $\mathfrak{F}(s)$ (inversion of (6.61)) and*

$$\mathfrak{F}(s) = -\frac{1}{\sin \frac{\pi}{2} s} \int_0^\infty \frac{f(ix) - f(-ix)}{2i} x^s \frac{\mathrm{d}x}{x} \tag{1.85}$$

for $0 < \sigma < 2$, are equivalent. Further, in the case where

$$f(ix) - f(-ix) = o(x), \quad \sigma \to 0+, \tag{1.86}$$

(1.85) holds true for $-1 < \sigma < 0$.

Proof. By rotating the positive real axis by $\frac{\pi}{2}$, we obtain

$$\mathfrak{F}(s) = \int_0^\infty f(ix)(ix)^{s-1} i\mathrm{d}x = e^{\frac{\pi i}{2} s} \int_0^\infty f(ix) x^{s-1} \mathrm{d}x$$

and similarly

$$\mathfrak{F}(s) = e^{-\frac{\pi i}{2} s} \int_0^\infty f(-ix) x^{s-1} \mathrm{d}x.$$

Hence

$$\sin \frac{\pi s}{2} \mathfrak{F}(s) = \frac{e^{\frac{\pi i}{2} s} - e^{-\frac{\pi i}{2} s}}{2i} \mathfrak{F}(s) = -\int_0^\infty \frac{f(ix) - f(-ix)}{2i} x^{s-1} \mathrm{d}x,$$

which is (1.85). $\qquad \square$

The following theorem is a slight generalization of the **Plana summation formula** [KoshII, p. 217] which is the case $z = 1$.

We refer to the following growth condition of Stirling type:

$$\mathfrak{F}(s) = O\left(e^{-\frac{\pi}{2}|t|} |t|^{a\sigma - b}\right) \tag{1.87}$$

uniformly in σ for $\sigma_1 \leq \sigma \leq \sigma_2$, where $a > 0, b > 0$ are subject to

$$\frac{a}{2} + b \geq 1. \tag{1.88}$$

Theorem 1.4 (Generalized Plana summation formula) *Suppose that the Mellin transform $\mathfrak{F}(s)$ of the function $f(x)$ $(x > 0)$ satisfies the Stirling type condition (1.87). Suppose $f(x)$ is defined and analytic in the half-plane $\sigma \geq 0$ in Lemma 1.2, that (1.86) holds, and that $\mathfrak{F}(s)$ is regular in the strip*

$\beta \leq \sigma \leq \alpha$ *($-1 < \beta < 1, 0 < \alpha$) except for a simple pole at $s = 0$. Then we have Plana's summation formula as a modular relation* (Re $z > 0$)

$$\sum_{n=1}^{\infty} F(n)f(nz) = \zeta_{\Omega}(0) \operatorname*{Res}_{s=0} \mathfrak{F}(s) + \mathfrak{F}(1)z^{-1} \tag{1.89}$$

$$- \frac{2}{z} \int_0^{\infty} \frac{f(ix) - f(-ix)}{2i} \sigma\left(\frac{x}{z}\right) \mathrm{d}x,$$

provided that the infinite series and the integral in (1.89) are (absolutely) convergent. The first term on the right may be expressed as $\zeta_{\Omega}(0)f(0)$ under some growth condition on $\mathfrak{F}(s)$ and the second term is $\frac{\rho_{\Omega}}{z} \int_0^{\infty} f(x)\mathrm{d}x$, where $\rho_{\Omega} = -\frac{2^{r+1}\pi^{r_2}\zeta_{\Omega}^{(r)}(s)}{\sqrt{|\Delta|}}$ indicates the residue of the Dedekind zeta-function at $s = 1$ given in (3.152).

Proof of Theorem 1.4. We move the line of integration to $\sigma = -\alpha$ by the Cauchy residue theorem. Then we have for Re $z > 0$ and $\alpha > 1$, using the estimates $\zeta_{\Omega}(s) = O\left(|t|^{\varkappa(1-\sigma)}\right)$ and (1.87)

$$\sum_{n=1}^{\infty} F(n)f(nz) = \frac{1}{2\pi i} \int_{(\alpha)} \mathfrak{F}(s)\zeta_{\Omega}(s)z^{-s}\,\mathrm{d}s \tag{1.90}$$

$$= R_0 + R_1 + \frac{1}{2\pi i} \int_{(1+\alpha)} \mathfrak{F}(1-s)\zeta_{\Omega}(1-s)z^{s-1}\,\mathrm{d}s,$$

by (1.87), where $\alpha > 1$ and we made the change of variable s by $1-s$ in the integral on the right, and where R_j indicates the residue at $s = j, j = 0, 1$ of the integrand, so that

$$R_1 = \rho_{\Omega}\mathfrak{F}(1)z^{-1}, \quad R_0 = -\zeta_{\Omega}(0) \operatorname*{Res}_{s=0} \mathfrak{F}(s).$$

We apply the functional equation in the form

$$\zeta_{\Omega}(1-s) = A^{1-2s}G(1-s)\zeta_{\Omega}(s), \tag{1.91}$$

so that the integrand of the integral, say I_{α}, on the right of (1.90) reduces to

$$I_{\alpha} = \frac{A}{2\pi i z} \int_{(\alpha)} \mathfrak{F}(1-s)G(1-s) \sum_{n=1}^{\infty} F(n)\left(\frac{A^2 n}{z}\right)^{-s}\,\mathrm{d}s. \tag{1.92}$$

Now we use Lemma 1.2 in the form $(-1 < \sigma < 0)$

$$\mathfrak{F}(1-s) = -\frac{1}{\cos\frac{\pi}{2}s} \int_0^{\infty} \frac{f(ix) - f(-ix)}{2i} x^{-s}\,\mathrm{d}x \tag{1.93}$$

and change the order of integration in (1.92) (which is legitimate by the absolute convergence as is proved in [KoshII]), we obtain

$$I_\alpha = \frac{2}{z}\frac{A}{\pi}\int_0^\infty \frac{f(ix) - f(-ix)}{2i}\sum_{n=1}^\infty \sigma_1(x,n)\frac{\mathrm{d}x}{x}, \qquad (1.94)$$

where

$$\sigma_1(x,n) = \frac{1}{2\pi i}\int_{(c)}\frac{\pi}{2\cos\left(\frac{\pi}{2}s\right)}G(1-s)(A^2 nx/z)^{-s}\mathrm{d}s = K(A^2 nx/z), \quad (1.95)$$

whence $\frac{A}{\pi}\sum_{n=1}^\infty \sigma_1(x,n) = \sigma\left(\frac{x}{z}\right)$ and (1.90) follows, completing the proof.

Corollary 1.4 ([SC, p. 90, (9)]). *Under the conditions of Theorem 1.4, we have Plana's summation formula*

$$\begin{aligned}
\sum_{n=1}^\infty f(n) &= -\frac{1}{2}f(0) + \int_0^\infty f(x)\mathrm{d}x \\
&- 2\int_0^\infty \frac{f(ix) - f(-ix)}{2i}\frac{1}{e^{2\pi x} - 1}\mathrm{d}x,
\end{aligned} \qquad (1.96)$$

provided that the domain of $f(x)$ is extended to

$$x = |x|e^{i\theta}, \quad -\frac{\pi}{2} < \theta < \frac{\pi}{2}$$

and $f(x)$ is analytic there, and that the infinite series and the integral in (1.96) are (absolutely) convergent.

Our expression in the following Theorem 1.5 for the Hurwitz-Lerch zeta function $\Phi(z,s,a)$ ((5.38)) has its genesis in the confluent hypergeometric function and Plana's summation formula, which in the long run, amounts to the functional equation for the Riemann zeta-function.

Theorem 1.5 *For $\mathrm{Re}\, a > 0$, we have*

$$\begin{aligned}
\Phi(z,s,a) &= \frac{1}{2}a^{-s} + \int_0^\infty \frac{z^x}{(x+a)^s}\mathrm{d}x \\
&- 2\int_0^\infty \left(x^2 + a^2\right)^{-\frac{s}{2}}\sin\left(x\log z - s\arctan\frac{x}{a}\right)\frac{\mathrm{d}x}{e^{2\pi x} - 1}.
\end{aligned} \qquad (1.97)$$

Corollary 1.5   (Hermite's formula [SC, p. 91, (12)]). *For* $\mathrm{Re}\, a > 0$ *we have*

$$
\zeta(s,a) = \frac{1}{2}a^{-s} + \frac{a^{1-s}}{s-1}
$$
$$
+ 2\int_0^\infty \left(a^2 + x^2\right)^{-\frac{s}{2}} \sin\left(s \arctan \frac{x}{a}\right) \frac{\mathrm{d}x}{e^{2\pi x} - 1}.
$$

(1.98)

Corollary 1.6   (Binet's second expression [SC, p. 17, (30)]). *For* $\mathrm{Re}\, a > 0$ *we have*

$$
\log \frac{\Gamma(a)}{\sqrt{2\pi}} = \left(a - \frac{1}{2}\right) \log a + 2 \int_0^\infty \frac{\arctan \frac{x}{a}}{e^{2\pi x} - 1}\,\mathrm{d}x.
$$

(1.99)

Fig. 1.1   Joseph Fourier

Chapter 2

Grocery of Special Functions

2.1 Formulas for the gamma function and their use

We refer to [Vista] for a concise theory of the Euler gamma function. It is introduced most commonly as the principal solution to the **difference equation**

$$\Gamma(s+1) = s\Gamma(s), \quad \log\Gamma(s+1) - \log\Gamma(s) = \log s \qquad (2.1)$$

with the normalization

$$\Gamma(1) = 1. \qquad (2.2)$$

The following two well-known formulas for the gamma function are often used subsequently without further mentioning.

The reciprocity relation $\qquad \Gamma(s)\Gamma(1-s) = \dfrac{\pi}{\sin(\pi s)}, \qquad (2.3)$

The duplication formula $\qquad \Gamma(2s) = \pi^{-\frac{1}{2}} 2^{2s-1}\Gamma(s)\,\Gamma\left(s + \dfrac{1}{2}\right). \qquad (2.4)$

More generally, we have the **multiplication formula** for the gamma function [ErdT, I,p.4]: For integer $m \geq 2$

$$\Gamma(ms) = (2\pi)^{-\frac{m-1}{2}} m^{ms-1/2} \prod_{r=0}^{m-1} \Gamma\left(s + \frac{r}{m}\right). \qquad (2.5)$$

This is used e.g. by Bochner [Bo2] to reduce the gamma factor $\Gamma\left(\frac{n}{m}s + a\right)$ in terms of $\Gamma(s+b)$. As will be seen, this procedure is the reduction of the H-function to the G-function.

Exercise 2.1 Prove by (2.1) and (2.2) that

$$\left(-\frac{1}{s}\frac{d}{ds}\right)^{\rho} s^{2\nu-2z} = \frac{2^{\rho}\Gamma(z-\nu+\rho+1)}{\Gamma(z-\nu)} s^{2\nu-2z-2\rho}. \tag{2.6}$$

Exercise 2.2 By (2.4) prove that

$$\left(-\frac{1}{s}\frac{d}{ds}\right)^{\rho} s^{-2z-1} = 2^{-\rho}\frac{\Gamma(2z+2\rho+1)\Gamma(z+1)}{\Gamma(z+\rho+1)\Gamma(2z+1)} s^{-2z-2\rho-1}. \tag{2.7}$$

Below we will see that (2.3), resp. (2.4) will be magnified to (2.101), resp. (2.97). At present we shall see that (2.3) is used in the derivation of (2.8) in the following context.

Example 2.1 We presuppose basics for G- and allied functions and prove

$$G_{1,3}^{3,1}\left(z\,\middle|\,\begin{matrix}a\\b,b+\frac{1}{2},a\end{matrix}\right) \tag{2.8}$$

$$= 2^{-b}\sqrt{\pi}\,\Gamma(1-2a+2b)\Big(\left(-2i\sqrt{z}\right)^{2b} U\big(2b-2a+1,2b-2a+1,-2\sqrt{z}\,i\big)$$

$$+ \left(2i\sqrt{z}\right)^{2b} U\big(2b-2a+1,2b-2a+1,2\sqrt{z}\,i\big)\Big),$$

which entails

$$G_{1,3}^{3,1}\left(z\,\middle|\,\begin{matrix}a\\0,\frac{1}{2},a\end{matrix}\right) \tag{2.9}$$

$$= \sqrt{\pi}\,\Gamma(1-2a)\Big(e^{-a\pi i+2\sqrt{z}\,i}\,\Gamma\big(2a,2\sqrt{z}\,i\big) + e^{a\pi i-2\sqrt{z}\,i}\,\Gamma\big(2a,-2\sqrt{z}\,i\big)\Big).$$

The workable procedure is always the same.

Make the denominators of coefficients coincide. Cancel common factors. Extract the sines and cosines by the reciprocity relation of the gamma function, therewith shifting the gamma factors in the denominator to the numerator. Use Euler's identity. Reduce to the G-functions.

By definition

$$LHS = \frac{1}{2\pi i}\int_L \Gamma(b+w)\Gamma\left(b+\frac{1}{2}w\right)\Gamma(a+w)\Gamma(1-a-w)\,z^{-w}dw.$$

Applying (2.4) three times, we transform the integrand into

$$2^{2-b-w}\pi^{3/2}\Gamma(2b+2w)\Gamma(2a+2w)\Gamma(1-2a-2w)\frac{\cos(\pi(a+w))}{\pi}$$

$$= 2^{1-b}\sqrt{\pi}\,\Gamma(2b+2w)\Gamma(2a+2w)\Gamma(1-2a-2w)$$

$$\times\left(e^{\pi ia}(-2i\sqrt{z})^{-2w}+e^{-\pi ia}(2i\sqrt{z})^{-2w}\right).$$

Hence by the change of variable formula (2.93),

$$LHS = 2^{-b}\sqrt{\pi}\left(G_{1,2}^{2,1}\left(-2i\sqrt{z}\,\middle|\,\begin{matrix}2a\\2b,2a\end{matrix}\right)+G_{1,2}^{2,1}\left(2i\sqrt{z}\,\middle|\,\begin{matrix}2a\\2b,2a\end{matrix}\right)\right)$$

$$= 2^{-b}\sqrt{\pi}\,\Gamma(1-2a+2b)\Big((-2i\sqrt{z})^{2b}U(2b-2a+1,2b-2a+1,-2i\sqrt{z})$$

$$+(2i\sqrt{z})^{2b}U(2b-2a+1,2b-2a+1,2i\sqrt{z})\Big),$$

which is (2.8) and (2.8) in turn leads to (2.9) in view of (2.119).

Exercise 2.3   Prove the identities

$$H_{1,3}^{2,1}\left(z\,\middle|\,\begin{matrix}(1-s,1)\\(\frac{1-s}{2},\frac{1}{2}),(0,1),(1-\frac{s}{2},\frac{1}{2})\end{matrix}\right)\tag{2.10}$$

$$=\frac{\Gamma(s)}{2^{1-s}\sqrt{\pi}}\left(e^{\frac{1-s}{2}\pi i-2zi}\,\Gamma(1-s,-2zi)+e^{-\frac{1-s}{2}\pi i+2zi}\,\Gamma(1-s,2zi)\right)$$

and

$$H_{1,3}^{2,1}\left(z\,\middle|\,\begin{matrix}(1-2s,2)\\(\frac{1}{2}-s,1),(0,2),(1-s,1)\end{matrix}\right)\tag{2.11}$$

$$=\frac{\Gamma(2s)}{2^{2-2s}\sqrt{\pi}}\left(e^{-\frac{1-2s}{2}\pi i+2\sqrt{z}\,i}\,\Gamma(1-s,2\sqrt{z}\,i)\right.$$

$$\left.+e^{\frac{1-2s}{2}\pi i-2\sqrt{z}\,i}\,\Gamma(1-s,-2\sqrt{z}\,i)\right).$$

Also work out another case:

$$H_{2,4}^{2,2}\left(z\,\middle|\,\begin{matrix}(1-(1+\frac{s}{2}),\frac{1}{2}),(1-s,1)\\(-\frac{s}{2},\frac{1}{2}),(0,1),(1-\frac{s}{2},1),(1-\frac{s+1}{2},\frac{1}{2})\end{matrix}\right)$$

$$=\frac{1}{2^{1-s}\pi}G_{1,3}^{3,1}\left(z^2\,\middle|\,\begin{matrix}-\frac{s}{2}\\-\frac{s}{2},0,\frac{1}{2}\end{matrix}\right).$$

(2.10) is used in [KTT1] and (2.11) will be used in §8.9.

Solution By (2.4), the left-hand side of (2.10) reduces to

$$\frac{1}{2^{1-s}\pi} G_{1,3}^{3,1}\left(z^2 \left|\begin{array}{c} \frac{1-s}{2} \\ \frac{1-s}{2}, 0, \frac{1}{2} \end{array}\right.\right),$$

whence the result follows by (2.8). The second formula amounts to the first by (2.93).

Applying (2.4) twice, we find that the LHS of the third equality amounts to

$$\frac{1}{2^{2-s}\pi} H_{1,3}^{2,1}\left(z \left|\begin{array}{c} (-\frac{s}{2}, \frac{1}{2}) \\ (-\frac{s}{2}, \frac{1}{2}), (0, \frac{1}{2}), (\frac{1}{2}, \frac{1}{2}) \end{array}\right.\right)$$

to which we apply (2.93) and then (2.8).

Definition 2.1   The **incomplete gamma function** which will play a central role in Ewald expansion (Chapter 5) is defined by

$$\Gamma(s, z) = \int_z^\infty t^s e^{-t} \frac{dt}{t}, \quad \sigma > 0. \tag{2.12}$$

For generalizations of the incomplete gamma function, we refer to [CZ].

Exercise 2.4   (i) Prove that

$$\Gamma(1, z) = e^{-z}, \quad \Gamma\left(\frac{1}{2}, z\right) = \sqrt{\pi}\, \mathrm{erfc}(\sqrt{z}), \tag{2.13}$$

where erfc is the error function.

(ii) Prove the formula for the Laplace transform

$$\mathcal{L}\left[\Gamma\left(\nu, \frac{a}{t}\right)\right](p) = 2a^{\frac{\nu}{2}} p^{\frac{\nu}{2}-1} K_\nu(2\sqrt{ap}), \tag{2.14}$$

whence deduce

$$\mathcal{L}\left[\mathrm{erfc}\left(at^{-1/2}\right)\right](p) = \frac{1}{p} e^{-2a\sqrt{p}} \tag{2.15}$$

for $\nu = \frac{1}{2}$. Also deduce

$$\mathcal{L}[\Gamma(\nu, at)](p) = \frac{\Gamma(\nu)}{p}\left(1 - \frac{1}{(1 + \frac{p}{a})^\nu}\right). \tag{2.16}$$

(iii) Check the formula [ErdH, II, p.149]

$$C(z, a) = \int_z^\infty t^{a-1} \cos t \, dt \tag{2.17}$$

$$= \frac{1}{2} e^{\frac{\pi i}{2} a} \Gamma(a, -iz) + \frac{1}{2} e^{-\frac{\pi i}{2} a} \Gamma(a, iz)$$

and its counterpart

$$S(z, a) = \int_z^\infty t^{a-1} \sin t \, dt \tag{2.18}$$

$$= \frac{1}{2i} e^{\frac{\pi i}{2} a} \Gamma(a, -iz) - \frac{1}{2i} e^{-\frac{\pi i}{2} a} \Gamma(a, iz)$$

for $\operatorname{Re} a > 0$.

(iv) Check the formulas [ErdH, II, p.143]

$$E_1(z) = -\operatorname{Ei}(-z) = \int_z^\infty e^{-t} t^{-1} \, dt \tag{2.19}$$

$$= \Gamma(0, z) = e^{-z} \Psi(1, 1, z)$$

and

$$\operatorname{li}(x) = \operatorname{Ei}(\log x) = \int_0^x \frac{1}{\log t} \, dt = -E_1(-\log x). \tag{2.20}$$

Remark 2.1 *The limit of the integral in* (2.19) *is called the* **logarithm integral** *and constitutes the main term of the* **prime number theorem**:

$$\pi(x) := \sum_{\substack{p \leq x \\ p:prime}} 1 = \operatorname{Li}(x) + E(x), \tag{2.21}$$

where $\operatorname{Li}(x) = \int_0^\infty \frac{1}{\log t} \, dt$. *Some equivalent forms of* (2.21) *appear in Chapter 6,* (6.121) *and Theorem 6.13.*

2.2 Zeta-functions

There are two excellent surveys on the theory of zeta-functions in general. [SpH] is a very comprehensive account of the theory of zeta-functions, restricted mainly to the Riemann and allied zeta-functions, although [SpH, Vol. II] contains a more extensive class of zeta-functions. [Kaw] is quite a complete survey on zeta-functions up to 70's in Japanese, though. [Joy] is, its title notwithstanding, quite an excellent survey of various zeta-functions including modern ones. Also [WeIII] contains important material for the theory of zeta-functions, esp. [We1], [We3], [We4]. [We2], [Weds] and other papers in [SpH] will be discussed in our coming Volume II. There are many pther classes of zeta-functions whose complete theory have not fully appeared in book form. There exists a vast literature on zeta-functions arising from representation theory cf. e.g. [Bu1] or from the theory of pre-homogeneous vector spaces (p.v.), cf. [MS] and many others. Hopefully, a comprehensive book will appear on zeta-functions in due course.

Of course, regarding a specific zeta-function, there are more literature specified to that zeta-function. [Titr] is a masterpiece work on the Riemann zeta-function. [Pra] is a rather extensive survey on Dirichlet L-functions. So is Suetuna [Sue] which is also devoted to analytic number theory in algebraic number fields. For a modern authoritative account of analytic number theory, we refer to [IwKo]. The recent [AIK] contains interesting expositions on the Bernoulli numbers and zeta-functions.

We follow [Sa] to give a table of zeta-functions whose functional equations follow from the theta-transformation formula. Although his presentation is from the view point of prehomogeneous vector spaces and his table contains the p.v. and the relevant invariant, we refer to the zeta-function part only. For recent results, we refer e.g. to Ohno et al [Oh], [OTW].

Table 2.1   Zeta-table

zeta-functions	generalizations
Riemann [Rie2]	Hecke L-functions with Grössen characters[1]
Dedekind	Hecke L-functions with Grössen characters
Epstein (positive definite)	Epstein zeta-function with harmonic polynomials
Siegal (indefinite) [Sie3]	Maass [Maa3], Hejhal
Hey (Simple algebra) [Hey]	Godement-Jacquet [GJ]
Koecher [Koc]	Maass [Maa1], [Maa2], [Maa4]
Eisenstein series of $\mathrm{SL}_n(\mathbb{Z})$	E-series of $\mathrm{SL}_n(\mathbb{Z})$ with representation of SO_n
Shintani	?

[1] means [Heck1], [Heck2].

There is a large class of perturbed zeta-functions including the generalized Hurwitz-Lerch zeta-function in §5.2.1. It reduces to the Hurwitz zeta-function and the Lipschitz-Lerch transcendent, which then reduce to the Hurwitz zeta-function

$$\zeta(s, x) = \sum_{n=0}^{\infty} \frac{1}{(n+w)^s}, \quad \sigma > 1. \tag{2.22}$$

Table 2.2 Functional equation-table

	Landau-Walfisz	Bochner-Chandrasekharan-Narasimhan
	$Z(s) = \sum_{k=1}^{\infty} \alpha_k \lambda_k^{-s}$, $\sigma_Z \geq r$	$\varphi(s) = \sum_{k=1}^{\infty} \alpha_k \lambda_k^{-s}$, σ_φ
	$W(s) = \sum_{k=1}^{\infty} \beta_k \mu_k^{-s}$, $\sigma_W \geq r$	$\psi(s) = \sum_{k=1}^{\infty} \beta_k \mu_k^{-s}$, σ_ψ
	W is regular at $s = r$, $r > 0$	$r \in \mathbb{R}$
	$\beta_1, \cdots, \beta_M > 0$, $\delta_1, \cdots, \delta_N > 0$	$\alpha_1, \cdots, \alpha_M > 0$
	$\alpha_1, \cdots, \alpha_M \in \mathbb{R}$, $\gamma_1, \cdots, \gamma_N \in \mathbb{R}$	$\beta_1, \cdots, \beta_M \in \mathbb{C}$
	$G(s) = \frac{\Delta_2(r-s)}{\Delta_1(s)} = \frac{\prod_{\nu=1}^{N} \Gamma(\gamma_\nu - \delta_\nu s)}{\prod_{\nu=1}^{N} \Gamma(\alpha_\mu + \beta_\mu s)}$	$G(s) = \frac{\Delta(r-s)}{\Delta(s)} = \prod_{\nu=1}^{N} \frac{\Gamma(\alpha_\nu r + \beta_\nu - \alpha_\nu s)}{\Gamma(\beta_\nu + \alpha_\nu s)}$
	$G(0) = 0$, has a pole at $s = r$	-
	$\alpha_\mu + r\beta_\mu > 0$, $1 \leq \mu \leq M$	-
	$\sum_{\nu=1}^{N} \beta_\nu = \sum_{\nu=1}^{N} \delta_\nu = \frac{H}{2}$	$\sum_{\nu=1}^{N} \alpha_\nu = A$
$H = \frac{2\eta}{r}$, $\eta := \sum_{\nu=1}^{N} \gamma_\nu - \sum_{\mu=1}^{N} \alpha_\mu + \frac{M-N}{2}$		$\eta = Ar$
	$\frac{1}{2} < \eta < \frac{3}{2}$, $\frac{5}{2} - \eta < H$	-
	$Z(s) = G(s)W(r - s)$	$\varphi(s) = G(s)\psi(s)$
	$Z(s) = O(e^{\gamma|t|})$	$\chi(s) = O(e^{c|t|^\alpha})$, $c > 0, \alpha > 0$

Here the last row specifies the growth conditions in the strip $\sigma_1 \leq \sigma \leq \sigma_2$ and the constants $\gamma = \gamma(\sigma_1, \sigma_2)$.

Since the functional equation appears in the form of $\varphi(s) = G(s)\psi(s)$ in Table 2.2, we keep this notation $G(s)$ to indicate the **gamma quotient**. Some aspects of this form of functional equations have been developed in [KTT2] and we refer to it for details. The gamma quotient typically appears in §4.6 and (3.177) prior to it; cf. the remark at the end of §4.6.

2.3 Bessel functions

In this section we assemble definitions and properties of well-known Bessel functions, which will be touched upon again in the context of *H*-or *G*-functions. We note *a priori* that the added provisos to the complex integral representations (Mellin-Barnes integrals) come from the separation of poles requirement of the relevant gamma factors (cf. definition (1.22) of the *H*-function).

The *J*-Bessel function $J_s(z)$ is a solution to the Bessel differential equation

$$x^2 \frac{d^2 w}{dx^2} + x \frac{dw}{dx} + (x^2 - \nu^2)w = x \frac{d}{dx}\left(x \frac{dw}{dx}\right) + (x^2 - \nu^2)w = 0 \quad (2.23)$$

and is given as a power series (2.46) ([ErdH, III, p.4]) below. What is relevant to us is the integral representation [ErdH, III,(34) p.21]

$$J_\nu(x) = \frac{1}{2}\frac{1}{2\pi i}\int_{(c)}\frac{\Gamma(\frac{1}{2}\nu + \frac{1}{2}z)}{\Gamma(1 + \frac{1}{2}\nu - \frac{1}{2}z)}\left(\frac{x}{2}\right)^{-z}\mathrm{d}z \qquad (2.24)$$

$$= \frac{1}{2}H_{0,2}^{1,0}\left(\frac{x}{2}\,\middle|\,\begin{array}{c} - \\ (\frac{\nu}{2},\frac{1}{2}),(-\frac{\nu}{2},\frac{1}{2}) \end{array}\right)$$

for $-\operatorname{Re}\nu < c < 1$ ($x > 0$, which may read $\operatorname{Re} x > 0$). We note that on the left-side of [ErdH, III,(34) p.21] should read $4\pi i J_\nu(x)$ and that without the restriction $-\operatorname{Re}\nu < c < 1$, (2.24) still makes sense, but it need not represent the Bessel function.

Exercise 2.5 Check that (2.24) reduces to [PBM, 8.4.19.1, p. 651]

$$J_\nu(2\sqrt{x}) = G_{0,2}^{1,0}\left(\frac{x}{2}\,\middle|\,\begin{array}{c} - \\ \frac{\nu}{2},-\frac{\nu}{2} \end{array}\right), \qquad (2.25)$$

which in turn amounts to [ErdH, III,(36) p.83]

$$J_\nu(x) = \frac{1}{2\pi i}\int_{(0)}\frac{\Gamma(-z)}{\Gamma(1 + \nu + z)}\left(\frac{x}{2}\right)^{\nu+2z}\mathrm{d}z \qquad (2.26)$$

$$= \frac{1}{2}G_{2,0}^{0,1}\left(\frac{x}{2}\,\middle|\,\begin{array}{c} 1,1+\nu \\ - \end{array}\right)$$

as well as (2.110).

Solution All are immediate consequences of formulas in Proposition 2.3.

We note that in general the Bessel functions, say F_s reduce in the case of s being a half-integer $\nu + \frac{1}{2}$, to an elementary function ([ErdT, II]); cf. (2.27), (2.32). $J_s(z)$ reduces to

$$J_{\nu+1/2}(z) = (-1)^\nu\frac{1}{\sqrt{\frac{\pi}{2}z}}z^{\nu+1}\left(\frac{\mathrm{d}}{z\mathrm{d}z}\right)^\nu\frac{\sin z}{z} \qquad (2.27)$$

$$= \frac{1}{\sqrt{\frac{\pi}{2}z}}\left(\sin\left(z - \frac{\pi}{2}n\right)\sum_{m=0}^{[\frac{n}{2}]}(-1)^m\binom{n+1/2}{2m}(2z)^{-2m}\right.$$

$$\left. + \cos\left(z - \frac{\pi}{2}n\right)\sum_{m=0}^{[\frac{n-1}{2}]}(-1)^m\binom{n+1/2}{2m+1}(2z)^{-2m-1}\right),$$

where $[x]$ indicates the integer part of x, i.e. the greatest integer not exceeding x.

Associated to the J-Bessel function is its complex version ($\nu \in \mathbb{Z}$)

$$I_\nu(z) = e^{-\frac{\pi}{2}i\nu} J_\nu(ze^{-\frac{\pi}{2}i\nu}) \tag{2.28}$$

called the modified Bessel function of the first kind. It is a solution to

$$x^2 \frac{\mathrm{d}^2 w}{\mathrm{d}x^2} + x\frac{\mathrm{d}w}{\mathrm{d}x} - (x^2 + \nu^2)w = 0 \tag{2.29}$$

and is given by the power series (2.45).

$$-\pi Y_s(x) = \frac{1}{2\pi i} \int_{(c)} \Gamma(z)\Gamma(z-s) \cos \pi(z-s) \left(\frac{x}{2}\right)^{s-2z} \mathrm{d}z, \tag{2.30}$$

for $0 \le \mathrm{Re}(s) < c < \mathrm{Re}(s) + \frac{3}{4} < \frac{3}{2}$ [ErdH].

The K-Bessel function $K_s(z)$ may be defined by ([Vi2, (6.14), p.107])

$$K_\nu(z) = \frac{1}{2}\left(\frac{z}{2}\right)^\nu \int_0^\infty e^{-t-\frac{z^2}{4t}} t^{-\nu-1}\mathrm{d}t,$$
$$\mathrm{Re}(\nu) > -\frac{1}{2}, \quad |\arg z| < \frac{\pi}{4}. \tag{2.31}$$

$K_s(z)$ reduces ([ErdH, (40), p.10]) to

$$K_{\nu+\frac{1}{2}}(z) = \left(\frac{\pi}{2z}\right)^{1/2} e^{-z} \sum_{m=0}^\nu (2z)^{-m} \frac{\Gamma(\nu+m+1)}{m!\,\Gamma(\nu+1+m)}. \tag{2.32}$$

A special case of (2.32) which is often used without notice is

$$K_{\pm\frac{1}{2}}(z) = \sqrt{\frac{\pi}{2z}}\, e^{-z}. \tag{2.33}$$

The following differential operator ([Wat, p.79]) is often used

$$\left(-\frac{1}{s}\frac{d}{ds}\right)^\rho \left\{s^\nu K_\nu(s)\right\} = s^{\nu-\rho} K_{\nu-\rho}(s), \tag{2.34}$$

e.g. cf. Exercises 2.1, 2.2.

We recall the following limit relation ([Vi2, (6.22),p. 110]) which is already referred to as (1.29) as well as the symmetry property which will turn out to be relevant to the functional equation:

$$\lim_{z\to 0} z^\nu K_\nu(2az) = \frac{1}{2}a^{-\nu}\Gamma(\nu), \quad \mathrm{Re}\,\nu \ge 0, \nu \ne 0. \tag{2.35}$$

$$K_{-\nu}(z) = K_\nu(z). \tag{2.36}$$

For an interesting interpretation of (2.35), cf. [CZ, p.9].

2.4 Ψ-functions

Definition of the Ψ-function. For $p, q = 0, 1, 2, \ldots$, we call

$$
{}_p\Psi_q\left(\begin{array}{l}(a_1, A_1), \ldots, (a_p, A_p) \\ (b_1, B_1), \ldots, (b_q, B_q)\end{array}; z\right)
$$

$$
= \sum_{k=0}^{\infty} \frac{\displaystyle\prod_{j=1}^{p} \Gamma(a_j + A_j k)}{\displaystyle\prod_{j=1}^{q} \Gamma(b_j + B_j k)} \frac{z^k}{k!} \tag{2.37}
$$

and its normalization

$$
{}_p\Psi_q^*\left(\begin{array}{l}(a_1, A_1), \ldots, (a_p, A_p) \\ (b_1, B_1), \ldots, (b_q, B_q)\end{array}; z\right)
$$

$$
= \frac{\displaystyle\prod_{j=1}^{q} \Gamma(b_j)}{\displaystyle\prod_{j=1}^{p} \Gamma(a_j)} \,{}_p\Psi_q\left(\begin{array}{l}(a_1, A_1), \ldots, (a_p, A_p) \\ (b_1, B_1), \ldots, (b_q, B_q)\end{array}; z\right) \tag{2.38}
$$

the Ψ-series (the Ψ-function). By the very definition (2.42) in the next section, the Ψ-functions are generalizations of hypergeometric functions.

Relation between H and Ψ-functions.

The following relation between H and Ψ-function is well known. If either

$$
\sum_{j=1}^{p} A_j < \sum_{j=1}^{q} B_j
$$

or

$$
\sum_{j=1}^{p} A_j = \sum_{j=1}^{q} B_j \quad \text{and} \quad |z| < \frac{B_1^{B_1} \cdots B_q^{B_q}}{A_1^{A_1} \cdots A_p^{A_p}},
$$

then

$$H_{p,q+1}^{1,p}\left(z\left|\begin{array}{c}(1-a_1,A_1),\ldots,(1-a_p,A_p)\\(0,1),(1-b_1,B_1),\ldots,(1-b_q,B_q)\end{array}\right.\right)$$
$$={}_p\Psi_q\left(\begin{array}{c}(a_1,A_1),\ldots,(a_p,A_p)\\(b_1,B_1),\ldots,(b_q,B_q)\end{array};-z\right). \tag{2.39}$$

In Chapter 5, we shall apply the following complementary formula (cf. [MaSa, p.11, (1.7.8)])

$$H_{1,2}^{1,1}\left(z\left|\begin{array}{c}(1-a,A)\\(c,1),(1-b,B)\end{array}\right.\right)=z^c\,{}_1\Psi_1\left(\begin{array}{c}(a+Ac,A)\\(b+Bc,B)\end{array};-z\right), \tag{2.40}$$

where ${}_1\Psi_1$ is one of the Ψ-functions defined by (2.37):

$${}_1\Psi_1\left(\begin{array}{c}(a,A)\\(b,B)\end{array};z\right)=\sum_{k=0}^{\infty}\frac{\Gamma(a+Ak)}{\Gamma(b+Bk)}\frac{z^k}{k!}.$$

2.5 Generalized hypergeometric functions

Generalized hypergeometric function. We recall the Pochhammer symbol ([Vi2, p.27])

$$(a)_k=\prod_{j=0}^{k-1}(a+j)=\overbrace{a(a+1)\cdots(a+k-1)}^{k}=\frac{\Gamma(a+k)}{\Gamma(a)} \tag{2.41}$$

for $k\in\mathbb{N}$ and $(a)_0=1$ by (2.1). For $p,q=0,1,2,\ldots,$

$$\begin{aligned}&{}_pF_q\left(\begin{array}{c}a_1,a_2,\ldots,a_p\\b_1,b_2,\ldots,b_q\end{array};z\right)\\&={}_p\Psi_q^*\left(\begin{array}{c}(a_1,1),\ldots,(a_p,1)\\(b_1,1),\ldots,(b_q,1)\end{array};z\right)\\&=\sum_{k=0}^{\infty}\frac{(a_1)_k(a_2)_k\cdots(a_p)_k}{(b_1)_k(b_2)_k\cdots(b_q)_k}\frac{z^k}{k!},\end{aligned} \tag{2.42}$$

the second equality being due to (2.38), are called the generalized hypergeometric series (as functions, they are called the **generalized hypergeometric functions**).

In the case, where $a_1, a_2, \ldots, a_p, b_1, b_2, \ldots, b_q \neq 0, -1, -2, \ldots$, the radius of convergence is

$$\lim_{k \to \infty} \left| \frac{(a_1)_k(a_2)_k \cdots (a_p)_k}{(b_1)_k(b_2)_k \cdots (b_q)_k} \frac{(b_1)_{k+1}(b_2)_{k+1} \cdots (b_q)_{k+1}}{(a_1)_{k+1}(a_2)_{k+1} \cdots (a_p)_{k+1}} \frac{(k+1)!}{k!} \right|$$

$$= \lim_{k \to \infty} \left| \frac{(k+b_1)(k+b_2) \cdots (k+b_q)}{(k+a_1)(k+a_2) \cdots (k+a_p)} (k+1) \right|$$

$$= \begin{cases} \infty, & (p < q+1) \\ 1, & (p = q+1) \\ 0, & (p > q+1) \end{cases}.$$

Example 2.2

$$_0F_0\left(\begin{matrix} - \\ - \end{matrix}; z\right) = \sum_{k=0}^{\infty} \frac{z^k}{k!} = e^z, \tag{2.43}$$

$$_1F_0\left(\begin{matrix} a \\ - \end{matrix}; z\right) = \sum_{k=0}^{\infty} (a)_k \frac{z^k}{k!} = (1-z)^{-a}, \tag{2.44}$$

$$_0F_1\left(\begin{matrix} - \\ b \end{matrix}; z\right) = \sum_{k=0}^{\infty} \frac{1}{(b)_k} \frac{z^k}{k!} = \frac{\Gamma(b)}{\sqrt{z}^{b-1}} I_{b-1}\left(2\sqrt{z}\right), \tag{2.45}$$

$$_0F_1\left(\begin{matrix} - \\ b \end{matrix}; -z\right) = \sum_{k=0}^{\infty} \frac{1}{(b)_k} \frac{(-z)^k}{k!} = \frac{\Gamma(b)}{\sqrt{z}^{b-1}} J_{b-1}\left(2\sqrt{z}\right), \tag{2.46}$$

where in (2.45), I_ν is defined by (2.28).

Example 2.3 (Confluent hypergeometric functions) We note the following notation ([ErdH, p.248])

$$_1F_1\left(\begin{matrix} a \\ c \end{matrix}; z\right) \equiv \Phi(a, c; z), \tag{2.47}$$

with Φ Humbert's symbol, where $_1F_1$ is defined by (2.42). Its counterpart $U = \Psi$ (sometimes referred to as the Tricomi function) is defined by ([ErdH, (2), p.255], [Tem, p.175])

$$U(a, c; z) \equiv \Psi(a, c; z)$$

$$= \frac{1}{\Gamma(a)} \int_0^\infty e^{-zt} t^{a-1}(1+t)^{c-1-a} \, dt, \ \operatorname{Re} a > 0. \tag{2.48}$$

These functions are the solutions (the latter in Re $x > 0$) of the confluent hypergeometric equation

$$x\frac{\mathrm{d}^2 y}{\mathrm{d}x^2} + (c - x)\frac{\mathrm{d}y}{\mathrm{d}x} - ay = 0 \tag{2.49}$$

and are referred to as the **confluent hypergeometric function** of the first and the second kind, respectively. In the case of the confluent hypergeometric function of the second kind, we shall use the U notation predominantly over Ψ notation so as to avoid confusion with the Ψ-function while for the first kind, we use both notation Φ and $_1F_1$ interchangeably. The following relation holds true:

$$U(a, c; z) = \Psi(a, c; z) \tag{2.50}$$
$$= \frac{\Gamma(1 - c)}{\Gamma(1 - c + a)}\Phi(a, c; z) + \frac{\Gamma(c - 1)}{\Gamma(a)}x^{1-c}\Phi(1 - c + a, 2 - c; z),$$

[ErdH, (7), p.257], which is sometimes used as a definition of the Tricomi function [PBM, p.799].

We shall therefore use several symbols to denote the confluent hypergeometric functions of the first kind

$$\Phi(a, c; z) = {}_1F_1\left(\begin{matrix} a \\ c \end{matrix}; z\right) = {}_1\Psi_1^*\left(\begin{matrix} (a, 1) \\ (c, 1) \end{matrix}; z\right) = \frac{\Gamma(c)}{\Gamma(a)}{}_1\Psi_1\left(\begin{matrix} (a, 1) \\ (c, 1) \end{matrix}; z\right). \tag{2.51}$$

Example 2.4 (Whittaker functions) The Whittaker function $W_{\kappa,\mu}$ is one of two independent solutions to the Whittaker differential equation (cf. [ErdH, p.248, (4)]).

$$\frac{\mathrm{d}^2 w}{\mathrm{d}x^2} + \left(-\frac{1}{4} + \frac{\kappa}{x} + \frac{\frac{1}{4} - \mu}{x^2}\right)w = 0. \tag{2.52}$$

Integral representations for $_{p+1}F_{q+1}$ in terms of $_pF_q$.

Proposition 2.1 *We have a general integral expression* ([PBM, 9, p.438])

$$_{p+1}F_{q+1}\left(\begin{matrix} a_1, a_2, \ldots, a_p, a_{p+1} \\ b_1, b_2, \ldots, b_q, b_{q+1} \end{matrix}; z\right) = \frac{\Gamma(b_{q+1})}{\Gamma(a_{p+1})\Gamma(b_{q+1} - a_{p+1})} \tag{2.53}$$

$$\times \int_0^1 t^{a_{p+1}-1}(1 - t)^{b_{q+1}-a_{p+1}-1}{}_pF_q\left(\begin{matrix} a_1, a_2, \ldots, a_p \\ b_1, b_2, \ldots, b_q \end{matrix}; tz\right)\mathrm{d}t.$$

Proof.   We perform the termwise integration in

$$\int_0^1 t^{b-1}(1-t)^{c-b-1}\left(\sum_{k=0}^\infty a_k(tz)^k\right)dt$$

which is permissible in our immediate application to the hypergeometric function (absolutely convergent in the unit circle). Then the resulting integral $\int_0^1 t^{b+k-1}(1-t)^{c-b-1}dt$ being the beta-function, we deduce that

$$\sum_{k=0}^\infty a_k \frac{(b)_k}{(c)_k}z^k = \frac{\Gamma(c)}{\Gamma(b)\Gamma(c-b)}\int_0^1 t^{b-1}(1-t)^{c-b-1}\left(\sum_{k=0}^\infty a_k(tz)^k\right)dt. \quad (2.54)$$

Viewing

$$_{p+1}F_{q+1}\left(\begin{matrix} a_1, a_2, \ldots, a_p, a_{p+1} \\ b_1, b_2, \ldots, b_q, b_{q+1} \end{matrix}; z\right)$$

as the power series $\sum_{k=0}^\infty a_k \frac{(b)_k}{(c)_k}z^k$ with $a_k = \frac{(a_1)_k(a_2)_k\cdots(a_p)_k}{(b_1)_k(b_2)_k\cdots(b_q)_k}$ and $\frac{(b)_k}{(c)_k} = \frac{(a_{p+1})_k}{(b_{q+1})_k}$, we apply (2.54) ro conclude (2.53).   □

Example 2.5   (i) Using (2.43) and (2.44) in (2.53), respectively, we have

$$\Phi(a,b;z) = {}_1F_1\left(\begin{matrix} a \\ b \end{matrix};z\right)$$

$$= \frac{\Gamma(b)}{\Gamma(a)\Gamma(b-a)}\int_0^1 t^{a-1}(1-t)^{b-a-1}e^{tz}\,dt, \quad \operatorname{Re}a > 0 \quad (2.55)$$

$$_2F_1\left(\begin{matrix} a, b \\ c \end{matrix};z\right) = \frac{\Gamma(c)}{\Gamma(b)\Gamma(c-b)}\int_0^1 t^{b-1}(1-t)^{c-b-1}(1-tz)^{-a}\,dt. \quad (2.56)$$

(ii) We now dwell on novel applications of hypergeometric functions in [KTT3]. Rewrite (2.56) in the form of [GR, p.314, 3.197.1] (with $\mu = \rho = s$):

$$\int_0^\infty x^{\nu-1}(x+\alpha)^{-s}(x+\beta)^{-s}\,dx \quad (2.57)$$

$$= \beta^{-s}\alpha^{\nu-s}B(\nu, 2s-\nu)\,{}_2F_1\left(\begin{matrix} s, \nu \\ 2s \end{matrix}; 1 - \frac{\alpha}{\beta}\right),$$

where $B(\alpha,\beta)$ is the beta-function. Formula (2.57) is valid only if

$$|\arg\alpha - \arg\beta| < \pi \quad (2.58)$$

(note that the argument is taken in $[-\pi, \pi)$), though in [GR] the condition on the argument of α and β is stated wrongly as

$$|\arg \alpha| < \pi, \quad |\arg \beta| < \pi.$$

Exercise 2.6 Under condition (2.58), deduce (2.57). Similarly, deduce

$$\int_0^\infty x^{\nu-1}(x+\alpha)^{-s}(x+\beta)^{-s}\,dx \tag{2.59}$$

$$= \alpha^{-s}\beta^{\nu-s}B(\nu, 2s - \nu)_2F_1\left(\begin{matrix} s,\, \nu \\ 2s \end{matrix}; 1 - \frac{\alpha}{\beta}\right)$$

and confirm the consistency of (2.57) and (2.59) by Kummer's relation ([ErdH, p.105, (1), (3)]).

Solution E.g. to prove (2.57) we rotate the line of integration from $\mathbb{R}_+$ to $\mathbb{R}_+ e^{i\arg\alpha}$ and apply [ErdH, p.115, (5)].

Exercise 2.7 Deduce the Mellin inversion formulas

$$(x+\alpha)^{-s}(x+\beta)^{-s} \tag{2.60}$$

$$= \frac{\alpha^{-s}\beta^{-s}}{2\pi i}\int_{(c)} \alpha^\nu B(\nu, 2s - \nu)_2F_1\left(\begin{matrix} s,\, \nu \\ 2s \end{matrix}; 1 - \frac{\alpha}{\beta}\right) x^{-\nu}\,d\nu$$

and

$$(x+\alpha)^{-s}(x+\beta)^{-s} \tag{2.61}$$

$$= \frac{\alpha^{-s}\beta^{-s}}{2\pi i}\int_{(c)} \beta^\nu B(\nu, 2s - \nu)_2F_1\left(\begin{matrix} s,\, \nu \\ 2s \end{matrix}; 1 - \frac{\beta}{\alpha}\right) x^{-\nu}\,d\nu,$$

where $1 < c < 2\sigma$.

Remark 2.2 *We may refer to (2.60) and (2.61) as the* **hypergeometric transform** *in the same context as with the beta transform (4.2) and the gamma transform ((1.42), [Vi2, p.25]). To apply them care must be taken, see Proposition 2.2.*

Exercise 2.8 Recall the following formula for the Gauss hypergeometric function:

$$_2F_1\left(\begin{matrix} a,\, b \\ c \end{matrix}; z\right) = (1-z)^{-a}\frac{\Gamma(c)\Gamma(b-a)}{\Gamma(c-a)\Gamma(b)}\,_2F_1\left(\begin{matrix} a,\, c-b \\ 1+a-b \end{matrix}; \frac{1}{1-z}\right) \tag{2.62}$$

$$+ (1-z)^{-b}\frac{\Gamma(c)\Gamma(a-b)}{\Gamma(c-b)\Gamma(a)}\,_2F_1\left(\begin{matrix} c-a,\, b \\ 1-a+b \end{matrix}; \frac{1}{1-z}\right)$$

valid for $|\arg(-z)| < \pi$, $|\arg(1 - z)| < \pi$, $a - b \notin \mathbb{Z}$ ([ErdH, p.109 (3)]). Start from the initial domain of α and β lying in the first and the fourth quadrant, respectively, and continue (2.60) into another region

$$\frac{\pi}{2} \leq \arg \alpha < \pi, \quad -\pi < \arg \beta \leq -\frac{\pi}{2}. \tag{2.63}$$

Proposition 2.2 (Hypergeometric transform) *To apply the hypergeometric transforms* (2.60), (2.61) *to the series whose terms satisfy* (2.58), *we are to use*

$$(x + \alpha)^{-s}(x + \beta)^{-s} \tag{2.64}$$
$$= \frac{\beta^{-2s}}{2\pi i} \int_{(c)} \beta^\nu \frac{\Gamma(2s - \nu)\Gamma(\nu - s)}{\Gamma(s)} {}_2F_1\left(\begin{array}{c} s, 2s - \nu \\ 1 + s - \nu \end{array} ; \frac{\alpha}{\beta} \right) x^{-\nu} d\nu$$
$$+ \frac{\alpha^{-s}\beta^{-s}}{2\pi i} \int_{(c)} \alpha^\nu \frac{\Gamma(\nu)\Gamma(s - \nu)}{\Gamma(s)} {}_2F_1\left(\begin{array}{c} s, \nu \\ 1 - s + \nu \end{array} ; \frac{\alpha}{\beta} \right) x^{-\nu} d\nu$$

and

$$(x + \alpha)^{-s}(x + \beta)^{-s} \tag{2.65}$$
$$= \frac{\alpha^{-2s}}{2\pi i} \int_{(c)} \alpha^\nu \frac{\Gamma(2s - \nu)\Gamma(\nu - s)}{\Gamma(s)} {}_2F_1\left(\begin{array}{c} s, 2s - \nu \\ 1 + s - \nu \end{array} ; \frac{\beta}{\alpha} \right)$$
$$+ \frac{\alpha^{-s}\beta^{-s}}{2\pi i} \int_{(c)} \beta^\nu \frac{\Gamma(\nu)\Gamma(s - \nu)}{\Gamma(s)} {}_2F_1\left(\begin{array}{c} s, \nu \\ 1 - s + \nu \end{array} ; \frac{\beta}{\alpha} \right),$$

valid for $0 < c < 2\sigma$, in place of them, respectively.

Proof. It suffices to note that the true form of analytic continuation of $B(\nu, 2s - \nu) {}_2F_1(s, \nu; 1 + s - \nu; 1 - \frac{\alpha}{\beta})$ (α and β are interchanged) into the region (2.63) is given by

$$B(\nu, 2s - \nu) {}_2F_1\left(s, \nu; 1 + s - \nu; 1 - \frac{\alpha}{\beta} \right) \tag{2.66}$$
$$= \alpha^{-s}\beta^s \frac{\Gamma(2s - \nu)\Gamma(\nu - s)}{\Gamma(s)} {}_2F_1\left(s, 2s - \nu; 1 + s - \nu; \frac{\beta}{\alpha} \right)$$
$$+ \alpha^{-\nu}\beta^\nu \frac{\Gamma(\nu)\Gamma(s - \nu)}{\Gamma(s)} {}_2F_1\left(s, \nu; 1 - s + \nu; \frac{\beta}{\alpha} \right).$$

$\square$

Let $Z(s)$ denote the Epstein zeta-function of the positive definite binary quadratic form given in (4.88). In [KTT2] the integral formula is proved.

Theorem 2.1 *We have for $\sigma > 0$*

$$Z(s) = \frac{2}{a^s}\zeta(2s) + \frac{2^{2s}a^{s-1}\sqrt{\pi}}{\Delta^{s-\frac{1}{2}}}\frac{\Gamma(s-\frac{1}{2})}{\Gamma(s)}\zeta(2s-1) + \frac{2}{a^s}\{T_1(s) + T_2(s)\}, \quad (2.67)$$

where $T_1(s)$ and $T_2(s)$ are given by

$$T_1(s) = \frac{\bar{\omega}^{-2s}}{2\pi i}\int_{(-\varepsilon)}\left(1 + e^{\pi i(\nu - 2s)}\right)\bar{\omega}^\nu\frac{\Gamma(2s-\nu)\Gamma(\nu-s)}{\Gamma(s)} \quad (2.68)$$

$$\times\,_2F_1\left(s, 2s-\nu; 1+s-\nu; \frac{\omega}{\bar{\omega}}\right)\zeta(2s-\nu)\zeta(\nu)d\nu$$

and

$$T_2(s) = \frac{\bar{\omega}^{-s}\omega^{-s}}{2\pi i}\int_{(-\varepsilon)}\left(1 + e^{-\pi i\nu}\right)\omega^\nu\frac{\Gamma(\nu)\Gamma(s-\nu)}{\Gamma(s)} \quad (2.69)$$

$$\times\,_2F_1\left(s, \nu; 1-s+\nu; \frac{\omega}{\bar{\omega}}\right)\zeta(2s-\nu)\zeta(\nu)d\nu.$$

The Gauss hypergeometric function is often used in other number-theoretic settings including Example 5.1 in Chapter 5, too. In [Sar], it is used with (2.62). The following exercise will be used in solving Exercise 5.2.

Exercise 2.9

(i) Deduce from (2.56) that

$$_2F_1\left(\begin{matrix}\nu, s\\s+1\end{matrix}; -z\right) = \int_0^z x^{s-1}(x+1)^{-\nu}\,dx. \quad (2.70)$$

([ErdT, p.310]).

(ii) Let $P(x) = \sum_{k=1}^L \rho_k x^{-s_k}$ be the residual function. Transform the integral of $P(x)$ as follows.

$$\int_0^1 x^{s-1}P(x+\delta)\,dx = \sum_{k=1}^L \rho_k\delta^{s-s_k}\int_0^{1/\delta} x^{s-1}(x+1)^{-s_k}\,dx$$

$$= s^{-1}\sum_{k=1}^L \rho_k\delta^{-s_k}{}_2F_1\left(\begin{matrix}s_k, s\\s+1\end{matrix}; -\frac{1}{\delta}\right). \quad (2.71)$$

(iii) Check the following ([PBM, 8, p.431]).

$$_2F_1\left(\begin{matrix}0, b\\c\end{matrix}; z\right) = \lim_{z\to 0}{}_2F_1\left(\begin{matrix}a, b\\c\end{matrix}; z\right) = 1. \quad (2.72)$$

(iv) Check the following ([PBM, p.455, 28, p.797]).

$$
{}_2F_1\left({\nu, s \atop s+1} : -z\right) = s(-z)^{-s}B_{-s}(s, 1 - \nu), \qquad (2.73)
$$

where

$$
B_z(\alpha, \beta) = \int_0^z t^{\alpha-1}(1 - t)^{\beta-1}\,\mathrm{d}t
$$

indicates the **incomplete beta-function**.

The relation between the G-function and generalized hypergeometric functions. We suppose that all the poles of

$$
\prod_{j=1}^m \Gamma(b_j + s) \prod_{j=1}^n \Gamma(a_j - s)
$$

are of degree 1. Then we have

$$
G_{p,q}^{m,n}\left(z \left|\begin{array}{l} 1 - a_1, \ldots, 1 - a_n, a_{n+1}, \ldots, a_p \\ b_1, \ldots, b_m, 1 - b_{m+1}, \ldots, 1 - b_q \end{array}\right.\right) \qquad (2.74)
$$

$$
= \begin{cases}
\displaystyle\sum_{k=1}^m \frac{\displaystyle\prod_{\substack{j=1 \\ j \neq k}}^m \Gamma(b_j - b_k) \prod_{j=1}^n \Gamma(a_j + b_k)}{\displaystyle\prod_{j=n+1}^p \Gamma(a_j - b_k) \prod_{j=m+1}^q \Gamma(b_j + b_k)} z^{b_k} \\
\qquad \times {}_pF_{q-1}\left(\begin{array}{l} \{a_j + b_k\}_{j=1}^n, \{1 - a_j + b_k\}_{j=n+1}^p \\ \{1 - b_j + b_k\}_{\substack{j=1 \\ j \neq k}}^m, \{b_j + b_k\}_{j=m+1}^q \end{array}; (-1)^{p+m+n}\, z\right), \\
\hfill \text{if } p < q \text{ or } (p = q, |z| < 1) \\[2em]
\displaystyle\sum_{k=1}^n \frac{\displaystyle\prod_{\substack{j=1 \\ j \neq k}}^n \Gamma(a_j - a_k) \prod_{j=1}^m \Gamma(b_j + a_k)}{\displaystyle\prod_{j=m+1}^q \Gamma(b_j - a_k) \prod_{j=n+1}^p \Gamma(a_j + a_k)} \frac{1}{z^{a_k}} \\
\qquad \times {}_qF_{p-1}\left(\begin{array}{l} \{b_j + a_k\}_{j=1}^m, \{1 - b_j + a_k\}_{j=m+1}^q \\ \{1 - a_j + a_k\}_{\substack{j=1 \\ j \neq k}}^n, \{a_j + a_k\}_{j=n+1}^p \end{array}; (-1)^{q+m+n}\, \frac{1}{z}\right), \\
\hfill \text{if } p > q \text{ or } (p = q, |z| > 1).
\end{cases}
$$

In particular, we have

$$G_{p,q+1}^{1,n}\left(z \left| \begin{array}{l} 1 - a_1, \ldots, 1 - a_n, a_{n+1}, \ldots, a_p \\ 0, 1 - b_1, \ldots, 1 - b_q \end{array} \right.\right) \hspace{2cm} (2.75)$$

$$= \begin{cases} \dfrac{\displaystyle\prod_{j=1}^{n} \Gamma(a_j)}{\displaystyle\prod_{j=n+1}^{p} \Gamma(a_j) \prod_{j=1}^{q} \Gamma(b_j)} \, {}_pF_q\left(\begin{array}{l} \{a_j\}_{j=1}^{n}, \{1 - a_j\}_{j=n+1}^{p} \\ \{b_j\}_{j=1}^{q} \end{array} ; -(-1)^{p+n} \, z \right), \\ \hspace{4cm} \text{if } p < q + 1 \text{ or } (p = q + 1, |z| < 1) \\[1em] \displaystyle\sum_{k=1}^{n} \dfrac{\Gamma(a_k) \displaystyle\prod_{\substack{j=1 \\ j \neq k}}^{n} \Gamma(a_j - a_k)}{\displaystyle\prod_{j=1}^{q} \Gamma(b_j - a_k) \prod_{j=n+1}^{p} \Gamma(a_j + a_k)} \dfrac{1}{z^{a_k}} \\ \hspace{1cm} \times {}_{q+1}F_{p-1}\left(\begin{array}{l} a_k, \{1 - b_j + a_k\}_{j=1}^{q} \\ \{1 - a_j + a_k\}_{\substack{j=1 \\ j \neq k}}^{n}, \{a_j + a_k\}_{j=n+1}^{p} \end{array} ; (-1)^{q+n} \, \dfrac{1}{z} \right), \\ \hspace{4cm} \text{if } p > q + 1 \text{ or } (p = q + 1, |z| > 1) \end{cases}$$

and

$$G_{p,q+1}^{1,p}\left(z \left| \begin{array}{l} 1 - a_1, \ldots, 1 - a_p \\ 0, 1 - b_1, \ldots, 1 - b_q \end{array} \right.\right) \hspace{2cm} (2.76)$$

$$= \begin{cases} \dfrac{\displaystyle\prod_{j=1}^{p} \Gamma(a_j)}{\displaystyle\prod_{j=1}^{q} \Gamma(b_j)} \, {}_pF_q\left(\begin{array}{l} a_1, \ldots, a_p \\ b_1, \ldots, b_q \end{array} ; -z \right), \\ \hspace{4cm} \text{if } p < q + 1 \text{ or } (p = q + 1, |z| < 1) \\[1em] \displaystyle\sum_{k=1}^{p} \dfrac{\Gamma(a_k) \displaystyle\prod_{\substack{j=1 \\ j \neq k}}^{p} \Gamma(a_j - a_k)}{\displaystyle\prod_{j=1}^{q} \Gamma(b_j - a_k)} \dfrac{1}{z^{a_k}} \\ \hspace{1cm} \times {}_{q+1}F_{p-1}\left(\begin{array}{l} a_k, 1 - b_1 + a_k, \ldots, 1 - b_q + a_k \\ 1 - a_1 + a_k, \overset{k}{\vee}., 1 - a_p + a_k \end{array} ; (-1)^{p+q} \, \dfrac{1}{z} \right), \\ \hspace{4cm} \text{if } p > q + 1 \text{ or } (p = q + 1, |z| > 1) \end{cases}$$

where the symbol $\overset{k}{\vee}$ means that the k-th term should be omitted.

In particular, in the case $p = q + 1$, the right-hand side of the following identity gives an analytic continuation to $|z| > 1$ of the left-hand side.

$$
\begin{aligned}
&{}_{q+1}F_q\left(\begin{matrix} a_1, \ldots, a_{q+1} \\ b_1, \ldots, b_q \end{matrix} ; z \right) \\
&= \frac{\displaystyle\prod_{j=1}^{q} \Gamma(b_j)}{\displaystyle\prod_{j=1}^{q+1} \Gamma(a_j)} \, G_{q+1,q+1}^{1,q+1}\left(-z \left|\begin{matrix} 1 - a_1, \ldots, 1 - a_{q+1} \\ 0, 1 - b_1, \ldots, 1 - b_q \end{matrix}\right.\right) \quad (z \notin [0, \infty)),
\end{aligned}
\tag{2.77}
$$

and we have

$$
\begin{aligned}
&{}_{q+1}F_q\left(\begin{matrix} a_1, \ldots, a_{q+1} \\ b_1, \ldots, b_q \end{matrix} ; -z \right) \\
&= \frac{\displaystyle\prod_{j=1}^{q} \Gamma(b_j)}{\displaystyle\prod_{j=1}^{q+1} \Gamma(a_j)} \sum_{k=1}^{q+1} \frac{\Gamma(a_k) \displaystyle\prod_{\substack{j=1 \\ j \neq k}}^{q+1} \Gamma(a_j - a_k)}{\displaystyle\prod_{j=1}^{q} \Gamma(b_j - a_k)} \frac{1}{z^{a_k}} \\
&\quad \times {}_{q+1}F_q\left(\begin{matrix} a_k, 1 - b_1 + a_k, \ldots, 1 - b_q + a_k \\ 1 - a_1 + a_k, \overset{k}{\ldots}, 1 - a_{q+1} + a_k \end{matrix} ; -\frac{1}{z} \right).
\end{aligned}
\tag{2.78}
$$

Example 2.6

$$
{}_1F_0\left(\begin{matrix} a \\ - \end{matrix} ; -z \right) = \frac{1}{z^a} \, {}_1F_0\left(\begin{matrix} a \\ - \end{matrix} ; -\frac{1}{z} \right),
\tag{2.79}
$$

$$
\begin{aligned}
{}_2F_1\left(\begin{matrix} a, b \\ c \end{matrix} ; -z \right) &= \frac{\Gamma(c)\,\Gamma(b-a)}{\Gamma(b)\,\Gamma(c-a)} \frac{1}{z^a} \, {}_2F_1\left(\begin{matrix} a, 1 - c + a \\ 1 - b + a \end{matrix} ; -\frac{1}{z} \right) \\
&\quad + \frac{\Gamma(c)\,\Gamma(a-b)}{\Gamma(a)\,\Gamma(c-b)} \frac{1}{z^b} \, {}_2F_1\left(\begin{matrix} b, 1 - c + b \\ 1 - a + b \end{matrix} ; -\frac{1}{z} \right).
\end{aligned}
\tag{2.80}
$$

Exercise 2.10   Check the following special case of (2.80):

$$
\begin{aligned}
&{}_2F_1\left(\begin{matrix} \nu, s \\ s + 1 \end{matrix} ; -\frac{1}{x} \right) \\
&= \frac{s}{s - \nu} x^\nu \, {}_2F_1\left(\begin{matrix} \nu, \nu - s \\ \nu - s + 1 \end{matrix} ; -x \right) + \frac{\Gamma(s+1)\Gamma(\nu - s)}{\Gamma(\nu)} x^s.
\end{aligned}
\tag{2.81}
$$

([GR, p.1043]).

Solution Indeed, the left-hand side becomes

$$\frac{s}{s-\nu} x^\nu {}_2F_1\left(\begin{array}{c} \nu, \nu - s \\ \nu - s + 1 \end{array}; -x\right)$$
$$+ \frac{\Gamma(s+1)\Gamma(\nu-s)}{\Gamma(\nu)} x^s {}_2F_1\left(\begin{array}{c} s, 0 \\ 1 - \nu + s \end{array}; -x\right), \tag{2.82}$$

whence (2.81) follows by (2.72).

Exercise 2.11 Check that (2.74) with $m = q = 2, n = 0$ and $p = 1$:

$$G_{1,2}^{2,0}\left(z \left| \begin{array}{c} b \\ s, a \end{array} \right.\right) = \frac{\Gamma(a-s)}{\Gamma(b-s)} z^s {}_1F_1\left(\begin{array}{c} 1 - b + s \\ 1 - a + s \end{array}; -z\right)$$
$$+ \frac{\Gamma(s-a)}{\Gamma(b-a)} z^a {}_1F_1\left(\begin{array}{c} 1 - b + a \\ 1 - s + a \end{array}; -z\right)$$

leads to (2.50) in view of

$$U(a, c; z) = z^{1-c} U(a + 1 - c, 2 - c; z). \tag{2.83}$$

2.6   Fox *H*-functions

Definition of the (Fox) *H*-function

In what follows we shall use the power function $z^{-s} = e^{-s \log z}$ either by considering $\log z$ ($z \neq 0$) on its Riemann surface or by taking its principal values, with $z \notin (-\infty, 0]$.

We choose the path L which separates the poles of the integrand in such a way that the poles of the gamma factor

$$\frac{\prod\limits_{j=1}^{n} \Gamma(a_j - A_j s)}{\prod\limits_{j=n+1}^{p} \Gamma(a_j - A_j s) \prod\limits_{m+1}^{q} \Gamma(b_j - B_j s)}$$

lie to the right of L and the poles of the gamma factor

$$\frac{\prod\limits_{j=1}^{m} \Gamma(b_j + B_j s)}{\prod\limits_{j=n+1}^{p} \Gamma(a_j - A_j s) \prod\limits_{j=m+1}^{q} \Gamma(b_j - B_j s)}$$

lie to the left of L. We call this the **poles separation condition** and will tacitly assume this in almost all cases. For the **coefficients array** $(0 \le n \le p, 0 \le m \le q)$

$$\Delta = \begin{pmatrix} \{(1 - a_j, A_j)\}_{j=1}^n; & \{(a_j, A_j)\}_{j=n+1}^p \\ \{(b_j, B_j)\}_{j=1}^m & ;\{(1 - b_j, B_j)\}_{j=m+1}^q \end{pmatrix} \tag{1.20}$$

we associate the **processing gamma factor** $(A_j, B_j > 0)$

$$\Gamma(s \,|\, \Delta) \tag{1.21}$$

$$= \Gamma\left(s \left| \begin{array}{l} \{(1 - a_j, A_j)\}_{j=1}^n; & \{(a_j, A_j)\}_{j=n+1}^p \\ \{(b_j, B_j)\}_{j=1}^m & ;\{(1 - b_j, B_j)\}_{j=m+1}^q \end{array} \right.\right)$$

$$= \frac{\prod\limits_{j=1}^m \Gamma(b_j + B_j s) \prod\limits_{j=1}^n \Gamma(a_j - A_j s)}{\prod\limits_{j=n+1}^p \Gamma(a_j + A_j s) \prod\limits_{j=m+1}^q \Gamma(b_j - B_j s)} \quad (A_j, B_j > 0)$$

and define the **Fox H-function** by

$$H(z \,|\, \Delta)$$

$$= H_{p,q}^{m,n}\left(z \left| \begin{array}{l} (1 - a_1, A_1), \ldots, (1 - a_n, A_n), (a_{n+1}, A_{n+1}), \ldots, (a_p, A_p) \\ (b_1, B_1), \ldots, (b_m, B_m), (1 - b_{m+1}, B_{m+1}), \ldots, (1 - b_q, B_q) \end{array} \right.\right)$$

$$= \frac{1}{2\pi i} \int_L \Gamma(s \,|\, \Delta) \, z^{-s} \, ds$$

$$= \frac{1}{2\pi i} \int_L \frac{\prod\limits_{j=1}^m \Gamma(b_j + B_j s) \prod\limits_{j=1}^n \Gamma(a_j - A_j s)}{\prod\limits_{j=n+1}^p \Gamma(a_j + A_j s) \prod\limits_{m+1}^q \Gamma(b_j - B_j s)} z^{-s} ds, \tag{1.22}$$

where the path L is subject to the poles separation conditions given above.

For L we principally use the following.

(1) L is a deformed Bromwich contour, i.e. an infinite vertical line extending from $\gamma - i\infty$ to $\gamma + i\infty$ with indentation in finite part. We denote an integral along such an L by

$$\int_L = \lim_{X \to \infty} \int_{\gamma - iX}^{\gamma + iX},$$

which we often denote by $\int_{(\gamma)}$. In the case

$$c = \sum_{j=1}^{n} A_j + \sum_{j=1}^{m} B_j - \sum_{j=n+1}^{p} A_j - \sum_{j=m+1}^{q} B_j > 0, \quad \text{and} \quad |\arg(z)| < \frac{c}{2}\pi$$

(2.84)

the integral converges absolutely (absolute convergence sometimes occurs in the case $c = 0$ and $z > 0$, too).

In what follows we sometimes use the symbol to denote the coefficients array, which simply indicates that the coefficients in capitals are positive:

$$\Omega = \left(\coprod_{n=0}^{\infty} (\mathbb{C} \times \mathbb{R}^+)^n \right)^4.$$

(2.85)

2.7 Formulas for Fox and Meijer functions

The following formulas play the most prominent role in transforming the H-functions appearing in various contexts.

The following inversion formula is used to transform the H-function into the form which is listed in the table and will be used without notice.

$$H_{q,p}^{n,m} \left(z \left| \begin{array}{l} \{(1 - b_j, B_j)\}_{j=1}^m, \{(b_j, B_j)\}_{j=m+1}^q \\ \{(a_j, A_j)\}_{j=1}^n, \{(1 - a_j, A_j)\}_{j=n+1}^p \end{array} \right. \right)$$
$$= H_{p,q}^{m,n} \left(\frac{1}{z} \left| \begin{array}{l} \{(1 - a_j, A_j)\}_{j=1}^n, \{(a_j, A_j)\}_{j=n+1}^p \\ \{(b_j, B_j)\}_{j=1}^m, \{(1 - b_j, B_j)\}_{j=m+1}^q \end{array} \right. \right).$$

(2.86)

Proposition 2.3 *We have the following relations between H-functions:*

$$H(z \,|\, \Lambda + c) = z^c \, H(z \,|\, \Lambda) \quad (c \in \mathbb{C}),$$

(2.87)

$$H\left(z \left| \frac{1}{C} \Lambda \right. \right) = C \, H(z^C \,|\, \Lambda) \quad (C \in \mathbb{R}^+),$$

(2.88)

$$H(z \,|\, \Lambda^*) = H\left(\frac{1}{z} \,\middle|\, \Lambda \right),$$

(2.89)

where for

$$\Lambda = \left(\begin{array}{l} \{(a_j, A_j)\}_{j=1}^n ; \{(a_j, A_j)\}_{j=n+1}^p \\ \{(b_j, B_j)\}_{j=1}^m ; \{(b_j, B_j)\}_{j=m+1}^q \end{array} \right) \in \Omega$$

and $c \in \mathbb{C}$,

$$\Lambda + c = \begin{pmatrix} \{(a_j + A_j c, A_j)\}_{j=1}^{n}; \{(a_j + A_j c, A_j)\}_{j=n+1}^{p} \\ \{(b_j + B_j c, B_j)\}_{j=1}^{m}; \{(b_j + B_j c, B_j)\}_{j=m+1}^{q} \end{pmatrix} \in \Omega, \qquad (2.90)$$

and for $C \in \mathbb{R}^+$,

$$C\Lambda = \begin{pmatrix} \{(a_j, CA_j)\}_{j=1}^{n}; \{(a_j, CA_j)\}_{j=n+1}^{p} \\ \{(b_j, CB_j)\}_{j=1}^{m}; \{(b_j, CB_j)\}_{j=m+1}^{q} \end{pmatrix} \in \Omega, \qquad (2.91)$$

and

$$\Lambda^* = \begin{pmatrix} \{(1 - b_j, B_j)\}_{j=1}^{m}; \{(1 - b_j, B_j)\}_{j=m+1}^{q} \\ \{(1 - a_j, A_j)\}_{j=1}^{n}; \{(1 - a_j, A_j)\}_{j=n+1}^{p} \end{pmatrix} \in \Omega. \qquad (2.92)$$

In particular, Formula (2.88) *entails the* $H \to G$ *formula:*

$$H_{p,q}^{m,n}\left(z \, \middle| \, \begin{matrix} (a_1, \frac{1}{C}), \ldots, (a_n, \frac{1}{C}), (a_{n+1}, \frac{1}{C}), \ldots, (a_p, \frac{1}{C}) \\ (b_1, \frac{1}{C}), \ldots, (b_m, \frac{1}{C}), (b_{m+1}, \frac{1}{C}), \ldots, (b_q, \frac{1}{C}) \end{matrix} \right) \qquad (2.93)$$

$$= C \, G_{p,q}^{m,n}\left(z^C \, \middle| \, \begin{matrix} a_1, \ldots, a_n, a_{n+1}, \ldots, a_p \\ b_1, \ldots, b_m, b_{m+1}, \ldots, b_q \end{matrix} \right) \qquad (C > 0).$$

Proof follows from the following relations between the Gamma factors:

$$\Gamma(s \, | \, \Lambda + c) = \Gamma(s + c \, | \, \Lambda) \quad (c \in \mathbb{C}),$$

$$\Gamma(s \, | \, C\Lambda) = \Gamma(Cs \, | \, \Lambda) \quad (C \in \mathbb{R}^+),$$

$$\Gamma(s \, | \, \Lambda^*) = \Gamma(-s \, | \, \Lambda).$$

For another element

$$\Lambda' = \begin{pmatrix} \{(a_j', A_j')\}_{j=1}^{n'}; \{(a_j', A_j')\}_{j=n'+1}^{p'} \\ \{(b_j', B_j')\}_{j=1}^{m'}; \{(b_j', B_j')\}_{j=m'+1}^{q'} \end{pmatrix}$$

of Ω, we define

$$\Lambda \oplus \Lambda' \qquad (2.94)$$

$$= \begin{pmatrix} \{(a_j, A_j)\}_{j=1}^{n}, \{(a_j', A_j')\}_{j=1}^{n'}; \{(a_j, A_j)\}_{j=n+1}^{p}, \{(a_j', A_j')\}_{j=n'+1}^{p'} \\ \{(b_j, B_j)\}_{j=1}^{m}, \{(b_j', B_j')\}_{j=1}^{m'}; \{(b_j, B_j)\}_{j=m+1}^{q}, \{(b_j', B_j')\}_{j=m'+1}^{q'} \end{pmatrix}.$$

Proposition 2.4 *Suppose that there exists a path L parallel to the imaginary axis, such that*

$$H(z \,|\, \Lambda) = \frac{1}{2\pi i} \int_L \Gamma(s \,|\, \Lambda) \, z^{-s} \, \mathrm{d}s,$$

$$H(z \,|\, \Lambda') = \frac{1}{2\pi i} \int_L \Gamma(s \,|\, \Lambda') \, z^{-s} \, \mathrm{d}s,$$

$$H(z \,|\, \Lambda \oplus \Lambda') = \frac{1}{2\pi i} \int_L \Gamma(s \,|\, \Lambda \oplus \Lambda') \, z^{-s} \, \mathrm{d}s.$$

Then we have

$$\int_0^\infty H\big(tz \,|\, \Lambda\big) \, H\!\left(\frac{1}{tz'} \,\middle|\, \Lambda'\right) \frac{\mathrm{d}t}{t} = H\!\left(\frac{z}{z'} \,\middle|\, \Lambda \oplus \Lambda'\right). \tag{2.95}$$

Proof. As we have

$$\int_0^\infty\!\left(\int_0^\infty H\big(tz \,|\, \Lambda\big) \, H\!\left(\frac{1}{tz'} \,\middle|\, \Lambda'\right) \frac{\mathrm{d}t}{t}\right) z^{s-1} \mathrm{d}z$$

$$= \int_0^\infty\!\left(\int_0^\infty H\big(tz \,|\, \Lambda\big) z^{s-1} \mathrm{d}z\right) H\!\left(\frac{1}{tz'} \,\middle|\, \Lambda'\right) \frac{\mathrm{d}t}{t}$$

$$= \int_0^\infty H\big(x \,|\, \Lambda\big) x^{s-1} \mathrm{d}x \cdot \int_0^\infty H\big(y \,|\, \Lambda'\big) y^{s-1} \mathrm{d}y \cdot z'^s$$

$$= \Gamma\big(s \,|\, \Lambda\big) \Gamma\big(s \,|\, \Lambda'\big) z'^s = \Gamma\big(s \,|\, \Lambda \oplus \Lambda'\big) z'^s,$$

the proposition follows from the Mellin inversion theorem. $\square$

Theorem 2.2 *The* **reduction-augmentation formula** *reads*

$$H\!\left(z \,\middle|\, \Lambda \oplus \begin{pmatrix} (c,C); & - \\ - & ;(c,C) \end{pmatrix}\right)$$

$$= H\!\left(z \,\middle|\, \Lambda \oplus \begin{pmatrix} - & ;(c,C) \\ (c,C); & - \end{pmatrix}\right) \tag{2.96}$$

$$= H\big(z \,|\, \Lambda\big).$$

We now state the duplication formula and the reciprocity formula which are most often used in practice.

Theorem 2.3 (Duplication formula)
For

$$\Lambda = \begin{pmatrix} \{(a_j, A_j)\}_{j=1}^n ; \{(a_j, A_j)\}_{j=n+1}^p \\ \{(b_j, B_j)\}_{j=1}^m ; \{(b_j, B_j)\}_{j=m+1}^q \end{pmatrix} \in \Omega$$

and

$$\Lambda' = \begin{pmatrix} \{(\frac{a_j}{2}, A_j), (\frac{a_j+1}{2}, A_j)\}_{j=1}^n ; \{(\frac{a_j}{2}, A_j), (\frac{a_j+1}{2}, A_j)\}_{j=n+1}^p \\ \{(\frac{b_j}{2}, B_j), (\frac{b_j+1}{2}, B_j)\}_{j=1}^m ; \{(\frac{b_j}{2}, B_j), (\frac{b_j+1}{2}, B_j)\}_{j=m+1}^q \end{pmatrix} \in \Omega$$

and for any $\Xi \in \Omega$, *we have*

$$H\left(z \mid \Lambda' \oplus \Xi\right) = (\sqrt{\pi})^{m+n-(q-m)-(p-n)} 2^{\sum\limits_{j=1}^{p} a_j - \sum\limits_{j=1}^{q} b_j + m - (p-n)}$$

$$\times H\left(4^{-\sum\limits_{j=1}^{p} A_j + \sum\limits_{j=1}^{q} B_j} z \,\middle|\, 2\Lambda \oplus \Xi\right). \tag{2.97}$$

In particular, we have

$$H\left(z \mid \Lambda'\right)$$

$$= (2\pi)^{m+n-\frac{p+q}{2}} 2^{\sum\limits_{j=1}^{p} a_j - \sum\limits_{j=1}^{q} b_j - \frac{p-q}{2}} H\left(4^{-\sum\limits_{j=1}^{p} A_j + \sum\limits_{j=1}^{q} B_j} z \,\middle|\, 2\Lambda\right)$$

$$= (2\pi)^{m+n-\frac{p+q}{2}} 2^{\sum\limits_{j=1}^{p} a_j - \sum\limits_{j=1}^{q} b_j - \frac{p-q}{2} - 1} H\left(2^{-\sum\limits_{j=1}^{p} A_j + \sum\limits_{j=1}^{q} B_j} \sqrt{z} \,\middle|\, \Lambda\right). \tag{2.98}$$

In the traditional notation, (2.97) reads

$$H^{2m+m',2n+n'}_{2p+p',2q+q'}\left(z \,\middle|\, \begin{matrix} \{(\frac{a_j}{2}, A_j), (\frac{a_j+1}{2}, A_j)\}_{j=1}^n, \{(a'_j, A'_j)\}_{j=1}^{n'}, \\ \{(\frac{b_j}{2}, B_j), (\frac{b_j+1}{2}, B_j)\}_{j=1}^m, \{(b'_j, B'_j)\}_{j=1}^{m'}, \end{matrix}\right.$$

$$\left.\begin{matrix} \{(\frac{a_j}{2}, A_j), (\frac{a_j+1}{2}, A_j)\}_{j=n+1}^p, \{(a'_j, A'_j)\}_{j=n'+1}^{p'} \\ \{(\frac{b_j}{2}, B_j), (\frac{b_j+1}{2}, B_j)\}_{j=m+1}^q, \{(b'_j, B'_j)\}_{j=m'+1}^{q'} \end{matrix}\right)$$

$$= (\sqrt{\pi})^{m+n-(q-m)-(p-n)} 2^{\sum\limits_{j=1}^{p} a_j - \sum\limits_{j=1}^{q} b_j + m - (p-n)} \tag{2.99}$$

$$\times H^{m+m',n+n'}_{p+p',q+q'}\left(4^{-\sum\limits_{j=1}^{p} A_j + \sum\limits_{j=1}^{q} B_j} z \,\middle|\, \begin{matrix} \{(a_j, 2A_j)\}_{j=1}^n, \{(a'_j, A'_j)\}_{j=1}^{n'}, \\ \{(b_j, 2B_j)\}_{j=1}^m, \{(b'_j, B'_j)\}_{j=1}^{m'}, \end{matrix}\right.$$

$$\left.\begin{matrix} \{(a_j, 2A_j)\}_{j=n+1}^p, \{(a'_j, A'_j)\}_{j=n'+1}^{p'} \\ \{(b_j, 2B_j)\}_{j=m+1}^q, \{(b'_j, B'_j)\}_{j=m'+1}^{q'} \end{matrix}\right).$$

Remark 2.3 *In practice, we use a concise version of the bulky formula in Theorem 2.3 concentrating on one group, say B(m) in Mnemonics. I.e., if the elements* $\left(\frac{b_j}{2}, B_j\right), \left(\frac{b_j+1}{2}, B_j\right)$ *contract to* $(b_j, 2B_j)$ *with the new factor*

$2^{m-\sum_{j=1}^{m} b_j} \pi^{m/2}$, *then*

$$H_{p',q'}^{2m+m',n'}\left(z \left| \begin{array}{c} -,\qquad\qquad -,\{(a'_j,A'_j)\}_{j=1}^{n'},\{(a'_j,A'_j)\}_{j=n'+1}^{p'} \\ \{(\frac{b_j}{2},B_j),(\frac{b_j+1}{2},B_j)\}_{j=1}^{m},-,\{(b'_j,B'_j)\}_{j=1}^{m'},\{(b'_j,B'_j)\}_{j=m'+1}^{q'} \end{array}\right.\right)$$

$$= (\sqrt{\pi})^m \, 2^{m-\sum\limits_{j=1}^{m} b_j} \tag{2.100}$$

$$\times H_{p',q'}^{m+m',n'}\left(4^{\sum\limits_{j=1}^{m} B_j} z \left| \begin{array}{c} -,\qquad\qquad -,\{(a'_j,A'_j)\}_{j=1}^{n'},\{(a'_j,A'_j)\}_{j=n'+1}^{p'} \\ \{(b_j,2B_j)\}_{j=1}^{m},-,\{(b'_j,B'_j)\}_{j=1}^{m'},\{(b'_j,B'_j)\}_{j=m'+1}^{q'} \end{array}\right.\right).$$

Formulas for other groups are similar.

Theorem 2.4 (Reciprocity formula) *The reciprocity formula for the gamma function and Euler's identity lead to*

$$H\left(z \left| \Lambda \oplus \begin{pmatrix} -;(c,C) \\ -;(c,C) \end{pmatrix}\right.\right) \tag{2.101}$$

$$= \frac{1}{2\pi i}\left\{e^{c\pi i}\, H\left(e^{-C\pi i}z \middle| \Lambda\right) - e^{-c\pi i}\, H\left(e^{C\pi i}z \middle| \Lambda\right)\right\}.$$

In the traditional notation, (2.101) reads

$$H_{p+1,q+1}^{m,n}\left(z \left| \begin{array}{c} \{(a_j,A_j)\}_{j=1}^{n},\{(a_j,A_j)\}_{j=n+1}^{p},(c,C) \\ \{(b_j,B_j)\}_{j=1}^{m},\{(b_j,B_j)\}_{j=m+1}^{q},(c,C) \end{array}\right.\right)$$

$$= \frac{1}{2\pi i}\left\{e^{c\pi i}\, H_{p,q}^{m,n}\left(e^{-C\pi i}z \left| \begin{array}{c} \{(a_j,A_j)\}_{j=1}^{n},\{(a_j,A_j)\}_{j=n+1}^{p} \\ \{(b_j,B_j)\}_{j=1}^{m},\{(b_j,B_j)\}_{j=m+1}^{q} \end{array}\right.\right)\right. \tag{2.102}$$

$$\left. - e^{-c\pi i}\, H_{p,q}^{m,n}\left(e^{C\pi i}z \left| \begin{array}{c} \{(a_j,A_j)\}_{j=1}^{n},\{(a_j,A_j)\}_{j=n+1}^{p} \\ \{(b_j,B_j)\}_{j=1}^{m},\{(b_j,B_j)\}_{j=m+1}^{q} \end{array}\right.\right)\right\},$$

which entails the following formula for G-functions:

$$G_{p+1,q+1}^{m,n}\left(z \left| \begin{array}{c} a_1,\ldots,a_n,a_{n+1},\ldots,a_p,c \\ b_1,\ldots,b_m,b_{m+1},\ldots,b_q,c \end{array}\right.\right)$$

$$= \frac{1}{2\pi i}\left\{e^{c\pi i}\, G_{p,q}^{m,n}\left(e^{-\pi i}z \left| \begin{array}{c} a_1,\ldots,a_n,a_{n+1},\ldots,a_p \\ b_1,\ldots,b_m,b_{m+1},\ldots,b_q \end{array}\right.\right)\right. \tag{2.103}$$

$$\left. - e^{-c\pi i}\, G_{p,q}^{m,n}\left(e^{\pi i}z \left| \begin{array}{c} a_1,\ldots,a_n,a_{n+1},\ldots,a_p \\ b_1,\ldots,b_m,b_{m+1},\ldots,b_q \end{array}\right.\right)\right\}.$$

Corollary 2.1 *We have*

$$H\left(z \left| \Lambda \oplus \begin{pmatrix} -;(c+1,C) \\ -;(c+1,C) \end{pmatrix}\right.\right) = -H\left(z \left| \Lambda \oplus \begin{pmatrix} -;(c,C) \\ -;(c,C) \end{pmatrix}\right.\right), \tag{2.104}$$

which in the traditional notation, reads

$$
\begin{aligned}
H^{m,n}_{p+1,q+1}&\left(z \,\middle|\, \begin{array}{l} \{(a_j, A_j)\}_{j=1}^n, \{(a_j, A_j)\}_{j=n+1}^p, (c+1, C) \\ \{(b_j, B_j)\}_{j=1}^m, \{(b_j, B_j)\}_{j=m+1}^q, (c+1, C) \end{array} \right) \\
&= -H^{m,n}_{p+1,q+1}\left(z \,\middle|\, \begin{array}{l} \{(a_j, A_j)\}_{j=1}^n, \{(a_j, A_j)\}_{j=n+1}^p, (c, C) \\ \{(b_j, B_j)\}_{j=1}^m, \{(b_j, B_j)\}_{j=m+1}^q, (c, C) \end{array} \right),
\end{aligned}
\tag{2.105}
$$

which entails the G-function version.

2.8 Special cases of *G*-functions

The gamma transform (or the Euler gamma integral) (1.42) is a special case of

$$
\Gamma(a+s) : G^{1,0}_{0,1}\left(z \,\middle|\, \begin{array}{l} - \\ a \end{array}\right) = z^a\, e^{-z}.
\tag{2.106}
$$

The Riesz kernel which produces the Riesz sum is defined by

$$
G^{1,0}_{1,1}\left(z \,\middle|\, \begin{array}{l} a \\ b \end{array}\right) =
\begin{cases}
\dfrac{1}{\Gamma(a-b)}\, z^b (1-z)^{a-b-1}, & |z| < 1 \\
\dfrac{1}{2} & z = 1, a = b+1 \\
0, & |z| > 1.
\end{cases}
\tag{2.107}
$$

Cf. Remark 6.1.

The beta transform (or the beta integral) (4.2) is a special case of

$$
\Gamma(a-s)\Gamma(b+s) :
\tag{2.108}
$$

$$
G^{1,1}_{1,1}\left(z \,\middle|\, \begin{array}{l} a \\ b \end{array}\right) = \Gamma(1-a+b)\, z^b\, (z+1)^{a-b-1}.
$$

The **inverse Heaviside integral** formula ([Vista, p.108], [Vi2, p.98]), a companion to (2.31), reads

$$
\Gamma(a+s)\Gamma(b+s) :
\tag{2.109}
$$

$$
G^{2,0}_{0,2}\left(z \,\middle|\, \begin{array}{l} - \\ a, b \end{array}\right) = 2\, z^{\frac{1}{2}(a+b)} K_{a-b}\!\left(2\sqrt{z}\right).
$$

Its special case is (cf. (2.33))

$$G_{0,2}^{2,0}\left(z \left|\begin{array}{c} - \\ a, a + \frac{1}{2} \end{array}\right.\right) = \sqrt{\pi}\, z^a e^{-2\sqrt{z}}.$$

(2.108) and (2.109) are the only cases of two gamma factors that have been essentially used in converting the integrals into the Riesz sum and the modified K-Bessel function, respectively.

A companion to (2.24) is

$$G_{0,2}^{1,0}\left(z \left|\begin{array}{c} - \\ a, b \end{array}\right.\right) = z^{\frac{1}{2}(a+b)} J_{a-b}\left(2\sqrt{z}\right) \tag{2.110}$$

whose special cases read (cf. [PBM, p.635, 3])

$$G_{0,2}^{1,0}\left(z \left|\begin{array}{c} - \\ a, a + \frac{1}{2} \end{array}\right.\right) = \frac{z^a}{\sqrt{\pi}} \cos\left(2\sqrt{z}\right),$$

$$G_{0,2}^{1,0}\left(z \left|\begin{array}{c} - \\ a + \frac{1}{2}, a \end{array}\right.\right) = \frac{z^a}{\sqrt{\pi}} \sin\left(2\sqrt{z}\right).$$

For a slight generalization, cf. (9.11).

[ErdH, (4), p.256] (or [PBM, 8.4.45.1, p.715]) reads

$$G_{1,2}^{1,1}\left(z \left|\begin{array}{c} 1 - a \\ 0, 1 - b \end{array}\right.\right) = \frac{\Gamma(b)}{\Gamma(a)}\, {}_1F_1\left(\begin{array}{c} a \\ b \end{array}; -z\right) = \frac{\Gamma(b)}{\Gamma(a)}\, \Phi(a, b; -z). \tag{2.111}$$

We note that neither (2.111) or its complement (2.118) appears in the table of formulas for G-functions in [ErdH, pp.216-222] but they do appear as integral representations (Mellin-Barnes integrals, or the G-functions), [ErdH, (4), p.256] and [ErdH, (6), p.256], respectively.

(2.111) as an integral representation is essentially used in the proof of Theorem 4.11.

Exercise 2.12   Show that (2.111) may be put into another form

$$G_{1,2}^{1,1}\left(z \left|\begin{array}{c} a \\ b, c \end{array}\right.\right) = z^b \frac{\Gamma(1 - a + b)}{\Gamma(1 - c + b)}\, {}_1F_1\left(\begin{array}{c} 1 - a + b \\ 1 - c + b \end{array}; -z\right). \tag{2.112}$$

Solution Writing $\alpha = 1 - a + b$, $\beta = 1 - c + b$, we may write the left-hand side as $G_{1,2}^{1,1}\left(z \left|\begin{array}{c} 1 - \alpha + c \\ 1 - \beta + c, c \end{array}\right.\right)$. Hence by (2.87), (2.112) follows.

(2.112) reduces, in view of (2.122), to

$$G_{1,2}^{1,1}\left(z\left|\begin{array}{c}1\\a,0\end{array}\right.\right) = \Gamma(a) - \Gamma(a,z). \tag{2.113}$$

[ErdH, (6), p.216] reads

$$G_{1,2}^{2,0}\left(z\left|\begin{array}{c}a\\b,c\end{array}\right.\right) = z^{\frac{1}{2}(b+c-1)}e^{-\frac{1}{2}z}W_{\kappa,\mu}(z), \tag{2.114}$$

where $W_{\kappa,\mu}$ is the **Whittaker function** with indexes $\kappa = \frac{1}{2}(b+c+1) - a$, $\mu = \frac{1}{2}(b-c)$ in Example 2.4 and [ErdH, (8), p.216] reads

$$G_{1,2}^{2,1}\left(z\left|\begin{array}{c}a\\b,c\end{array}\right.\right) = \Gamma(1-a+b)\,\Gamma(1-a+c)\,z^{\frac{1}{2}(b+c-1)}e^{\frac{1}{2}z}W_{\kappa,\mu}(z), \tag{2.115}$$

where the indexes of the Whittaker function are $\kappa = a - \frac{1}{2}(b+c+1), \mu = \frac{1}{2}(b-c)$.

In view of [ErdH, (2), p.264]

$$W_{\kappa,\mu}(z) = e^{-\frac{1}{2}z}z^{\frac{1}{2}c}U(a,c;z), \quad a = \frac{1}{2} - \kappa + \mu,\ c = 2\mu + 1, \tag{2.116}$$

(2.114) is equi-vocal to

$$G_{1,2}^{2,0}\left(z\left|\begin{array}{c}a\\b,c\end{array}\right.\right) = e^{-z}z^b\,U\left(a-c,b-c+1,z\right), \tag{2.117}$$

where $U(a,c;z)$ is the confluent hypergeometric function of the second kind introduced in Example 2.3 while (2.115), to

$$G_{1,2}^{2,1}\left(z\left|\begin{array}{c}a\\b,c\end{array}\right.\right) \tag{2.118}$$
$$= \Gamma(1-a+b)\,\Gamma(1-a+c)\,z^b\,U\left(b-a+1,b-c+1,z\right).$$

In view of

$$\Gamma(a,z) = e^{-z}U(1-a,1-a;z), \tag{2.119}$$

which is [ErdH, (21), p.266], (2.117) with $a = 1, c = 0$ and a for b reduces to

$$G_{1,2}^{2,0}\left(z\left|\begin{array}{c}1\\a,0\end{array}\right.\right) = \Gamma(a,z), \tag{2.120}$$

while (2.118) with $b = a$ reduces to

$$G^{2,1}_{1,2}\left(z \left| \begin{matrix} a \\ a, b \end{matrix} \right.\right) = \Gamma(1 - a + b)\,\Gamma\left(a - b, z\right) e^z z^b. \qquad (2.121)$$

As noted above, (2.118) is complementary to (2.111) and so are (2.119) and (2.122) in their specifications:

$$\gamma(a, z) = \Gamma(a) - \Gamma(a, z) = a^{-1} z^a \Phi(a, a + 1; z), \qquad (2.122)$$

which is [ErdH, (22), p.266]. We also note

$$\Gamma(a, z) = e^{-z} z^a \Psi(1, a + 1; z), \qquad (2.123)$$

which is [PBM, 7.11.4.11, p.584].

(2.115) also reduces when $b = a$ to

$$G^{2,1}_{1,2}\left(2z \left| \begin{matrix} 1 - a \\ 0, 1 - 2a \end{matrix} \right.\right) = \frac{1}{\sqrt{\pi}}\,\Gamma(a)\Gamma(1 - a)e^z(2z)^{-\left(a - \frac{1}{2}\right)} K_{a - \frac{1}{2}}(z) \qquad (2.124)$$

in view of the relation [ErdH, (13), p.265]

$$W_{0,\mu}(2z) = \left(\frac{2z}{\pi}\right)^{\frac{1}{2}} K_\mu(z). \qquad (2.125)$$

For $G^{3,1}_{1,3}$, cf. (2.8) above. Other formulas will be given at their occurrences, cf. especially, Chapters 6, 8 and 9.

Fig. 2.1   Charles C. Fox

Chapter 3

Unprocessed Modular Relations

3.1 The $H_{0,M}^{M,0} \leftrightarrow H_{0,\tilde{N}}^{\tilde{N},0}$ formula

We suppose the existence of the meromorphic function χ satisfying the functional equation with r a real number,

$$
\chi(s) = \begin{cases} \displaystyle\prod_{j=1}^{M} \Gamma(d_j + D_j s)\, \varphi(s), & \mathrm{Re}(s) > \sigma_\varphi \\[2em] \displaystyle\prod_{j=1}^{\tilde{N}} \Gamma(e_j + E_j(r-s))\, \psi(r-s), & \mathrm{Re}(s) < r - \sigma_\psi \end{cases} \tag{3.1}
$$

and having a finite number of poles s_k $(1 \le k \le L)$, where

$$
D_j, E_j > 0.
$$

In the w-plane we take two deformed Bromwich paths

$$
L_1(s) : \gamma_1 - i\infty \to \gamma_1 + i\infty, \quad L_2(s) : \gamma_2 - i\infty \to \gamma_2 + i\infty \quad (\gamma_2 < \gamma_1)
$$

such that they squeeze a compact set $\mathcal{S}$ with boundary $\mathcal{C}$ for which $s_k \in \mathcal{S}$ $(1 \le k \le L)$

Under these conditions the χ-function, **key-function**, $\mathrm{X}(s, z \,|\, \Delta)$ is usually defined in our theory as

$$
\mathrm{X}(s, z \,|\, \Delta) = \frac{1}{2\pi i} \int_{L_1(s)} \Gamma(w - s \,|\, \Delta)\chi(w)\, z^{-(w-s)} \mathrm{d}w \tag{3.2}
$$

where $\Gamma(s \,|\, \Delta)$ is the processing gamma factor (1.21). In this chapter we consider the case where there is no processing gamma factor, $\Gamma(s \,|\, \Delta) = 1$ and we write $\mathrm{X}(s, z) = \mathrm{X}(s, z \,|\, \emptyset)$.

Then we have the following modular relation.

Theorem 3.1

$$X(s, z) \tag{3.3}$$

$$= \begin{cases} \displaystyle\sum_{k=1}^{\infty} \frac{\alpha_k}{\lambda_k{}^s} H_{0,M}^{M,0}\left(z\lambda_k \,\middle|\, \begin{array}{c} - \\ \{(d_j + D_j s, D_j)\}_{j=1}^M \end{array} \right) \\ \qquad\qquad \text{if } L_1(s) \text{ can be taken to the right of } \sigma_\varphi, \\[2em] \displaystyle\sum_{k=1}^{\infty} \frac{\beta_k}{\mu_k{}^{r-s}} H_{0,\tilde{N}}^{\tilde{N},0}\left(\frac{\mu_k}{z} \,\middle|\, \begin{array}{c} - \\ \{(e_j + E_j(r-s), E_j)\}_{j=1}^{\tilde{N}} \end{array} \right) \\ \qquad + \displaystyle\sum_{k=1}^{L} \mathrm{Res}\Big(\chi(w)\, z^{s-w}, w = s_k\Big) \\ \qquad\qquad \text{if } L_2(s) \text{ can be taken to the left of } r - \sigma_\psi \end{cases}$$

is equivalent to the functional equation (3.1).

Both [BeI] and [BeII] treat the functional equation

$$\Gamma^m(s)\varphi(s) = \Gamma^m(r-s)\psi(r-s) \tag{3.4}$$

for m a positive integer, extending Hecke's functional equation. In [BeI] the equivalence is proved of (3.4) to the modular relation ([BeI, Theorem 1]),

$$\sum_{k=1}^{\infty} \alpha_k E_m(\lambda_k x) = x^{-r} \sum_{k=1}^{\infty} \beta_k E_m(\mu_k/x) + \mathrm{P}(x), \tag{3.5}$$

where $E_m(x)$ indicates **Voronoĭ's function**

$$E_m(x) = \frac{1}{2\pi i} \int_{(c)} \Gamma^m(s) x^{-s}\, \mathrm{d}s = G_{0,m}^{m,0}\left(z \,\middle|\, \begin{array}{c} - \\ \underbrace{0, \ldots, 0}_{m} \end{array} \right) \quad (x > 0), \tag{3.6}$$

and to the Riesz sum of order $\varkappa$ (cf. (6.1)):

$$\frac{1}{\Gamma(\varkappa+1)} \sideset{}{'}\sum_{\lambda_k \leq x} \alpha_k (x - \lambda_k)^{\varkappa}$$

involving the generalized Bessel function $K_\nu(x; \mu; m)$ ([BeI, Definition 4])

$$K_\varkappa(z; r, m) = \frac{z^\varkappa}{2\pi i} \int_L \frac{\Gamma(s)}{\Gamma(\varkappa + 1 - s)} \frac{\Gamma^{m-1}(s)}{\Gamma^{m-1}(r - s)} \left(\frac{z^2}{2^{2m}}\right)^{-s} \mathrm{d}s \qquad (3.7)$$

$$= z^\varkappa H_{0,m}^{m,0}\left(\left(\frac{z}{2^m}\right)^2 \middle| \underbrace{(0,1), \ldots, (0,1)}_{m}, (-\varkappa, 1), \underbrace{(1 - r, 1), \ldots, (1 - r, 1)}_{m}\right),$$

which is slightly different from other definitions of generalized Bessel functions given below. Cf. (9.54) in Chapter 10.

We remark that Bellman had already introduced the Voronoĭ function in 1949 in [Bel2], [Bel3]. In [Bel4, (1b)], he refers to the Siegel integral, which has been introduced by Wishart [Wis] (cf. [Sa] for more details). In [Bel5] he further refers to the most general **Steen's function** $V = V(x; a_1, \ldots, a_n)$ ([St]) as:

$$\frac{1}{2\pi i} \int_0^\infty x^s V(x; a_1, \ldots, a_n) \frac{\mathrm{d}x}{x} = \Gamma(s + a_1) \cdots \Gamma(s + a_n). \qquad (3.8)$$

This is nothing but the G-function formula

$$V(x; a_1, \ldots, a_n) = G_{0,n}^{n,0}\left(x \middle| \begin{array}{c} - \\ a_1, \ldots, a_n \end{array}\right) \qquad (3.9)$$

and will be applied in §8.16 as well as in the following exercise.

Exercise 3.1 Prove the following as an example of (9.45)

$$\sum_{k=1}^\infty \frac{\alpha_k}{\lambda_k^s} G_{0,m}^{m,0}\left(z\lambda_k \middle| \underbrace{s, \ldots, s}_{m}\right) \qquad (3.10)$$

$$= \sum_{k=1}^\infty \frac{\beta_k}{\mu_k^{r-s}} G_{0,m}^{m,0}\left(\frac{\mu_k}{z} \middle| \underbrace{r - s, \ldots, r - s}_{m}\right) + \sum_{k=1}^L \mathrm{Res}\left(\chi(w) z^{s-w}, w = s_k\right),$$

which is a generalization (3.5) above.

For curiosity's sake, we remark that in Chaudhry and Zubair [CZ] the most far-reaching formula for G-functions seems to be [CZ, (4.14), p.197] to the effect that

$$\gamma_\nu(\alpha, x; b) + \Gamma_\nu(\alpha, x; b)$$
$$= 2^{\alpha-2}\pi^{-1}b^{\frac{1}{2}}$$
$$\times G_{0,4}^{4,0}\left(\frac{b^2}{16} \middle| \frac{1}{2}\left(\nu+\frac{1}{2}\right), -\frac{1}{2}\left(\nu+\frac{1}{2}\right), \frac{1}{2}\left(\alpha+\frac{1}{2}\right), \frac{1}{2}\left(\alpha-\frac{1}{2}\right)\right),$$

where

$$\gamma_\nu(\alpha, x; b) = \left(\frac{2b}{\pi}\right)^{1/2}\int_0^x t^{\alpha-\frac{3}{2}}\exp(-t)K_{\nu+\frac{1}{2}}\left(\frac{1}{t}\right)\,dt.$$

We now refer to various **generalized Bessel functions** which have hitherto been introduced by many authors and are integrals of the gamma quotient $G(s)$ (cf. the passage after Table 2.2). Most of them contains the Riesz sum factor $\frac{\Gamma(s)}{\Gamma(s+\varkappa+1)}$ for the sake of enhancing the convergence (cf. §6.1.6 in this regard) as well as the anticipated application to the Riesz sums (§6.1.1).

Lavrik [Lav5] introduces (with $\varkappa = 0$)

$$J_N(\omega, z, \varkappa)$$
$$= H_{1,N+1}^{N+1,0}\left(z \middle| \begin{array}{c} (\varkappa+1, 1) \\ (0,1), (\alpha_1\omega+\beta_1, \alpha_1), \ldots, (\alpha_N\omega+\beta_N, \alpha_N) \end{array}\right) \qquad (3.11)$$
$$= \frac{1}{2\pi i}\int_L \frac{\Gamma(s)}{\Gamma(\varkappa+1+s)}\prod_{j=1}^N \Gamma(\alpha_j s + \alpha_j\omega+\beta_j)\,z^{-s}ds,$$

while Hafner [Haf] introduces the most general

$$f_\varkappa(z) = f_{\varkappa,G}(z) = \frac{z^{\varkappa+r}}{2\pi i}\int_L \frac{\Gamma(r-s)G(r-s)}{\Gamma(\varkappa+1+r-s)}z^{-s}ds$$
$$= \frac{z^{\varkappa+r}}{2\pi i}\int_L \frac{\Gamma(r-s)\prod_{j=1}^N \Gamma(\alpha_j s+\beta_j)}{\Gamma(\varkappa+1+r-s)\prod_{j=1}^N \Gamma(\alpha_j r+\beta_j-\alpha_j s)}z^{-s}ds \qquad (3.12)$$
$$= z^{\varkappa+r}H_{1,N+1}^{N,1}\left(z \middle| \begin{array}{c} (1-r, 1) \\ (\beta_1, \alpha_1), \ldots, (-\varkappa-r, 1), (1-(\alpha_1 r+\beta_1), \alpha_1), \ldots \end{array}\right),$$

where

$$G(s) = \frac{\Gamma(\beta_1+\alpha_1 r-\alpha_1 s)\cdots\Gamma(\beta_N+\alpha_N r-\alpha_N s)}{\Gamma(\beta_1+\alpha_1 s)\cdots\Gamma(\beta_N+\alpha_N s)} \qquad (3.13)$$

and $L = \mathcal{C}_{a,b}$ is a special contour described on [Haf, pp.151-152] with indentation in finite part and may be thought of as a Bromwich path satisfying the poles separation condition. He shows that the function $f_\varkappa$ shares analogous properties with the J Bessel function.

A. Z. Walfisz, in connection with the m-dimensional Piltz divisor problem in algebraic number fields ([AZW2]) introduced the function

$$J_{\rho_1,\rho_2,m}(z) = \frac{1}{2\pi i} \int_{\left(\frac{3}{4}\right)} \frac{\Gamma(s)}{\Gamma(s+1)} G(s)^m \left(\frac{z}{m\varkappa 2^{r_1/\varkappa}}\right)^{m\varkappa s - 1} ds \qquad (3.14)$$

$$= H^{1,\rho_1+\rho_2}_{\rho_1+\rho_2+1,1} \left(\frac{z}{m\varkappa 2^{r_1/\varkappa}} \left| \begin{array}{c} \left(1 - \frac{1}{2}, \frac{1}{2}\right), \ldots, (1-r,1), (1,1), \ldots, \left(0, \frac{1}{2}\right) \\ (0,1) \end{array} \right.\right)$$

with $G(s)$ the gamma quotient of the Dedekind zeta-function (which is a special case of (3.13) with $\alpha_j = \frac{1}{2}, \beta_j = \frac{r}{2}, 1 \le j \le r_1$, etc.). Noting that Hafner's generalized Bessel function may be written as

$$f_\varkappa(z) \qquad (3.15)$$

$$= z^\varkappa H^{1,N}_{N+1,1} \left(\frac{1}{z} \left| \begin{array}{c} (1 - (\alpha_1 r + \beta_1), \alpha_1), (\varkappa + 1, 1), \ldots (\beta_1, \alpha_1), \ldots \\ (0,1) \end{array} \right.\right),$$

Walfisz's function is a special case of that of Hafner's. Walfisz's function $J_{r_1,r_2}(z) = J_{\rho_1,\rho_2,1}(z)$ with $m = 1$ appears in the study of the Idealfunktion which was studied by him in [AZW1] mainly as a generalization of, as well as a sequel to, Hardy's Ω-results paper [HarDP]. We record Walfisz's function [AZW1, (3.12), p.31]

$$J_{r_1,r_2}(z) = \frac{1}{2\pi i} \int_{\left(\frac{3}{4}\right)} \frac{\Gamma(s)}{\Gamma(s+1)} G(s) \left(\frac{z}{\tau}\right)^{\varkappa s - 1} ds \qquad (3.16)$$

with $\tau = \varkappa 2^{r_1/\varkappa}$. We state the special case $J_{2,0}$ which appears in the case of a real quadratic field and is a relative of Koshlyakov's L-function (cf. (3.178)):

$$J_{2,0}(z) = \frac{1}{2\pi i} \int_{\left(\frac{3}{4}\right)} \frac{1}{s} G(s) \left(\frac{z}{4}\right)^{2s-1} ds$$

$$= \frac{2}{z} H^{0,2}_{4,0} \left(\left(\frac{4}{z}\right)^2 \left| \begin{array}{c} \left(1 - \frac{1}{2}, \frac{1}{2}\right), \left(1 - \frac{1}{2}, \frac{1}{2}\right), \left(1, \frac{1}{2}\right), \left(0, \frac{1}{2}\right) \\ - \end{array} \right.\right). \qquad (3.17)$$

The remaining argument being the same as in Exercise 3.8, we obtain

$$J_{2,0}(z) = -\left(\frac{\pi}{2} K_1(z) + Y_1(z)\right) \qquad (3.18)$$

which corresponds to [AZW2, (10.39)] in slightly different form. We note that in exactly the same way, we may deduce A. A. Walfisz's result [AAW4, (17)] $(z > 0)$

$$\tilde{J}_{2,0}(z) = -\sqrt{z}\left(\frac{\pi}{2}K_1(4\sqrt{z}) + Y_1(4\sqrt{z})\right), \qquad (3.19)$$

where

$$\tilde{J}_{2,0}(z) = \frac{1}{2\pi i}\int_{(c)}\frac{1}{s}G(s)\,z^s\mathrm{d}s \qquad (3.20)$$

with $c = \frac{\eta - 5/2}{H}$. Since $J_{2,0}(z) = \frac{4}{z}\tilde{J}_{2,0}\left(\left(\frac{z}{4}\right)^2\right)$, (3.18) follows again.

3.1.1 The $H_{0,1}^{1,0} \leftrightarrow H_{0,1}^{1,0}$ formula

The special case of Theorem 3.1 with $M = \tilde{N} = 1$ includes the rational and imaginary quadratic cases of Theorem 3.12:

Theorem 3.2 *The functional equation*

$$\Gamma(d_1 + D_1 s)\varphi(s) = \Gamma(e_1 + E_1(r - s))\psi(r - s), \qquad (3.21)$$

is equivalent to the modular relation

$$X(s, z \,|\, \Delta) \qquad\qquad\qquad\qquad\qquad\qquad\qquad\qquad (3.22)$$

$$= \begin{cases} \displaystyle\sum_{k=1}^{\infty}\frac{\alpha_k}{\lambda_k^s}H_{0,1}^{1,0}\left(z\lambda_k \,\middle|\, \begin{matrix} - \\ (d_1 + D_1 s, D_1) \end{matrix}\right) \\ \qquad\qquad \text{if } L_1(s) \text{ can be taken to the right of } \max_{1 \le h \le H}(\sigma_{\varphi_h}) \\[2em] \displaystyle\sum_{k=1}^{\infty}\frac{\beta_k}{\mu_k^{r-s}}H_{0,1}^{1,0}\left(\frac{\mu_k}{z} \,\middle|\, \begin{matrix} - \\ (e_1 + E_1(r - s), E_1) \end{matrix}\right) \\ \qquad + \displaystyle\sum_{k=1}^{L}\mathrm{Res}\left(\chi(w)\,z^{s-w}, w = s_k\right) \\ \qquad\qquad \text{if } L_2(s) \text{ can be taken to the left of } \min_{1 \le i \le I}(r - \sigma_{\psi_i}) \end{cases}$$

or

$$\frac{z^{\frac{d_1+D_1 s}{D_1}}}{D_1} \sum_{k=1}^{\infty} \alpha_k \lambda_k^{\frac{d_1}{D_1}} e^{-(z\lambda_k)^{\frac{1}{D_1}}} \tag{3.23}$$

$$= \frac{(z^{-1})^{\frac{e_1+E_1(r-s)}{E_1}}}{E_1} \sum_{k=1}^{\infty} \beta_k \mu_k^{\frac{e_1}{E_1}} e^{-\left(\frac{\mu_k}{z}\right)^{\frac{1}{E_1}}} + \sum_{k=1}^{L} \mathrm{Res}\Big(\chi(w)\, z^{s-w}, w = s_k\Big).$$

Proof follows from

$$\frac{1}{D_1} \sum_{k=1}^{\infty} \frac{\alpha_k}{\lambda_k^s} (z\lambda_k)^{\frac{d_1+D_1 s}{D_1}} e^{-(z\lambda_k)^{\frac{1}{D_1}}} = \frac{1}{E_1} \sum_{k=1}^{\infty} \frac{\beta_k}{\mu_k^{r-s}} \left(\frac{\mu_k}{z}\right)^{\frac{e_1+E_1(r-s)}{E_1}} e^{-\left(\frac{\mu_k}{z}\right)^{\frac{1}{E_1}}}$$

$$+ \sum_{k=1}^{L} \mathrm{Res}\Big(\chi(w)\, z^{s-w}, w = s_k\Big). \tag{3.24}$$

Theorem 3.2 with no perturbation $d_1 = e_1 = 0$ is a typical example of the use of the gamma transform (1.42).

Remark 3.1 *It may be remarked that the K-Bessel function $K_s(z)$ defined by (2.31) and its reduction to exponential functions (2.32) and (2.33) appeared in various contexts in relation to the lattice point problem as will be enunciated in §4.7. It was [HarMB] who first used the integral (2.31), without noticing that it is a K-Bessel function, in his research on the Epstein zeta-function. Bellman [Bel3, (3)] already notices the relevance of Hardy's paper in the context of Maass forms.*

(i) K-Bessel-allied functions in Bellman's papers. Indeed, [Bel3, (3)] is (2.33) with plus sign while [Sieg1] is (2.33) with minus sign. We note the following chain of different symbols for the K-Bessel function, the first equality being [Bel5, (4.1)]

$$V_a(x,y) = x^{2a-1}V_a(|xy|), \quad V_a(|xy|) = \pi^a W_{2-a}((\pi xy)^2) \tag{3.25}$$

and

$$W_a(z) = \int_0^{\infty} e^{-zv - \frac{1}{v}} v^{-a}\, dv = 2z^{\frac{a-1}{2}} K_{a-1}(2\sqrt{z}), \quad \mathrm{Re}\, z > 0. \tag{3.26}$$

(ii) Siegel's proof of Hamburger's theorem. We notice that the main ingredient of Siegel's proof of Hamburger's theorem is indeed the partial fraction expansion for the cotangent function in Example 4.2:

The Riemann type functional equation

$$\Gamma\left(\frac{1}{2}s\right)\varphi(s) = \Gamma\left(\frac{1}{2}(r-s)\right)\psi(r-s) \qquad (3.27)$$

with a simple pole at $s = r > 0$ with residue ρ is equivalent to

$$2\Gamma\left(\frac{1}{2}s\right)\sum_{k=1}^{\infty}\frac{\alpha_k}{(\lambda_k^2 + z^2)^{\frac{1}{2}s}}$$

$$= 4z^{\frac{1}{2}(r-s)}\sum_{k=1}^{\infty}\frac{\beta_k}{\mu_k^{\frac{1}{2}(r-s)}}K_{\frac{1}{2}(r-s)}(2\mu_k z) + 2\Gamma\left(\frac{1}{2}s\right)\varphi(0)z^{-s}$$

$$+ \rho\Gamma\left(\frac{1}{2}(s-r)\right)z^{r-s}. \qquad (3.28)$$

This reduces for $s = 2, r = 1$ to the partial fraction expansion for a generalized cotangent function

$$2\sum_{k=1}^{\infty}\frac{\alpha_k}{\lambda_k^2 + z^2} = 2\sqrt{\pi}z^{-1}\sum_{k=1}^{\infty}\beta_k e^{-2\mu_k z} + 2\varphi(0)z^{-2} + \rho\sqrt{\pi}z^{-1}, \quad (3.29)$$

which amounts to [Sieg1, (10)].
Remarkably enough, Siegel deduces (3.29) not as the Fourier-Bessel expansion but as a consequence of the modular relation, a special case of (3.23):

$$2z^2\sum_{k=1}^{\infty}\alpha_k e^{-(z\lambda_k)^2} = 2z^{-(r-s)}\sum_{k=1}^{\infty}\beta_k e^{-\left(\frac{\mu_k}{z}\right)^2} + \mathrm{P}(z). \qquad (3.30)$$

By a trivial change of variables and specifications, (3.30) amounts to [Sieg1, l.2, p.156]

$$2z^2\sum_{k=1}^{\infty}\alpha_k e^{-\pi n^2 x} = \frac{2}{\sqrt{x}}\sum_{k=1}^{\infty}\beta_k e^{-\frac{\pi n^2}{x}} + \mathrm{P}(x), \quad x > 0. \qquad (3.31)$$

Then multiplying (3.31) by $e^{-\pi t^x}$ and integrating in x over $(0,\infty)$, thereby incorporating (2.33) with minus sign $K_{-\frac{1}{2}}$, we uplift $H_{0,1}^{1,0} \leftrightarrow H_{0,1}^{1,0}$ to $H_{1,1}^{1,1} \leftrightarrow H_{0,2}^{2,0}$.

(iii) *Chowla and Selberg leave the function undefined, denoting it e.g. by*

$$I_a = \int_0^{\infty} e^{-\frac{\pi\sqrt{\Delta}n}{2a}\left(y+y^{-1}\right)}y^{s-\frac{3}{2}}\,\mathrm{d}y = 2K_{s-\frac{1}{2}}\left(\frac{\pi\sqrt{\Delta}n}{a}\right). \qquad (3.32)$$

3.1.2   The $G_{0,2}^{2,0} \leftrightarrow G_{0,2}^{2,0}$ formula

We keep Convention in §9.1, i.e. $\varphi(s) = \psi(s) = \pi^{-s}\zeta(2s + \nu)\zeta(2s - \nu)$. Anticipating an application to (3.114) (where φ is defined by (3.110)) and (3.141) we state the special case of (9.6) as

$$\sum_{k=1}^{\infty} \frac{\alpha_k}{\lambda_k^s} \, G_{0,2}^{2,0}\left(z\lambda_k \, \middle| \, \begin{matrix} - \\ s + \frac{\nu}{2}, s - \frac{\nu}{2} \end{matrix} \right)$$

$$= \sum_{k=1}^{\infty} \frac{\beta_k}{\mu_k^{r-s}} \, G_{0,2}^{2,0}\left(\frac{\mu_k}{z} \, \middle| \, \begin{matrix} - \\ r - s + \frac{\nu}{2}, r - s - \frac{\nu}{2} \end{matrix} \right) \qquad (3.33)$$

$$+ \sum_{k=1}^{L} \mathrm{Res}\Big(\chi(w) \, z^{s-w}, w = s_k \Big).$$

By (2.109), (3.33) reduces to the K-Bessel formula:

$$2 \sum_{k=1}^{\infty} \alpha_k \, K_\nu\left(2\sqrt{z\lambda_k} \right)$$

$$= 2z^{-r} \sum_{k=1}^{\infty} \beta_k \, K_\nu\left(2\sqrt{\frac{\mu_k}{z}} \right) + \sum_{k=1}^{L} \mathrm{Res}\Big(\chi(w) \, z^{-w}, w = s_k \Big). \qquad (3.34)$$

As a parenthetical remark on [BeIII, p.324], Berndt notes that (3.34) and the functional equation (9.1) are equivalent, which is the most relevant to our work. This also follows from (3.5) in view of

$$E_2(z) = G_{0,2}^{2,0}\left(z \, \middle| \, \begin{matrix} - \\ 0, 0 \end{matrix} \right) = 2 \, K_0\big(2\sqrt{z} \big).$$

Remark 3.2   *The above reasoning can be perceived already by (3.89). There are two ways leading to the rephrased Ramanujan formula (3.87). In the beginning all appealed to the use of the special difference of the shift of the argument, i.e. the odd integer value shift, which brings in mind the reduction of the product of two gamma factors into a single gamma factor, thus amounting to the modular relation of Bochner. It seems Guinand who suggested a more general reasoning based on the inverse Heaviside integral. By this, the functional equation is transformed into the K-Bessel series and vice versa. Then taking into account the special feature that the shift is a half-integer, we are led to the degeneracy (2.32) of the Bessel function into the exponential function. Cf. §3.3.3 below.*

3.2 Dedekind zeta-function I

We consider the Dedekind zeta-function of a number field Ω with degree $[\Omega : \mathbb{Q}] = \varkappa = r_1 + 2r_2$ (in standard notation; r_1 and $2r_2$ indicates the number of real and imaginary conjugates of Ω, respectively) and discriminant Δ. For $\varkappa \leq 2$, cf. Let

$$A = \frac{2^{r_2} \pi^{\varkappa/2}}{\sqrt{|\Delta|}} \tag{3.35}$$

and let $r = r_1 + r_2 - 1$ denote the rank of the unit group. The Dedekind zeta-function $\zeta_\Omega(s)$ of the number field Ω is defined by

$$\zeta_\Omega(s) = \sum_{0 \neq \mathfrak{a} \subset \Omega} \frac{1}{(\mathrm{N}\mathfrak{a})^s} = \sum_{k=1}^{\infty} \frac{F(k)}{k^s} \tag{3.36}$$

for $\sigma = \mathrm{Re}\, s > 1$, where $\mathfrak{a}$ runs through all non-zero ideals of Ω and $F(k) = F_\Omega(k)$ indicates the number of ideals of norm k, sometimes called "Idealfunction". Hence $\alpha_k = \beta_k = F(k)$ and $\lambda_k = \mu_k = Ak$, $r = 1$. $\zeta_\Omega(s)$ satisfies the **functional equation**

$$A^{-s} \Gamma^{r_1}\left(\frac{s}{2}\right) \Gamma^{r_2}(s) \zeta_\Omega(s)$$
$$= A^{-(1-s)} \Gamma^{r_1}\left(\frac{1-s}{2}\right) \Gamma^{r_2}(1-s) \zeta_\Omega(1-s), \tag{3.37}$$

and is continued to a meromorphic function with a pole at $s = 1$ with residue (cf. (3.152) for $\varkappa \leq 2$)

$$\rho = \lambda h = \frac{2^{r+1} \pi^{r_2} Rh}{w \sqrt{|\Delta|}}. \tag{3.38}$$

Here w, h and R are the **number of roots of unity** in Ω, the **class number** and the **regulator**, respectively.

Table 3.1   Invariants of a number field

symbol	meaning
$\varkappa = r_1 + 2r_2$	degree of Ω
r_1	number of real conjugates of Ω
$2r_2$	number of imaginary conjugates of Ω
$r = r_1 + r_2 - 1$	rank of the unit group of Ω
Δ	discriminant of Ω
h	class number
R	regulator
w	number of roots of unity in Ω
$\zeta_\Omega(s)$	Dedekind zeta-function of Ω
$F(k) = F_\Omega(k)$	number of ideals whose norm is k
$A = \dfrac{2^{r_2}\pi^{\varkappa/2}}{\sqrt{|\Delta|}}$	coefficient of functional equation for $\zeta_\Omega(s)$
$\lambda h = \dfrac{2^{r+1}\pi^{r_2}Rh}{w\sqrt{|\Delta|}}$	residue at $s = 1$ of $\zeta_\Omega(s)$

We define the meromorphic function in (3.1) as

$$\chi(s) = \Gamma^{r_1}\!\left(\frac{s}{2}\right)\Gamma^{r_2}(s)\varphi(s), \tag{3.39}$$

where

$$\varphi(s) = A^{-s}\zeta_\Omega(s) = \sum_{k=1}^{\infty}\frac{\alpha(k)}{(Ak)^s}. \tag{3.40}$$

The functional equation (3.1) in this case reads

$$\begin{aligned}
\chi(s) &= \Gamma^{r_1}\!\left(\frac{1}{2}s\right)\Gamma^{r_2}(s)\varphi(s) \\
&= \Gamma^{r_1}\!\left(\frac{1}{2} - \frac{1}{2}s\right)\Gamma^{r_2}(1-s)\varphi(1-s) \\
&= \chi(1-s)
\end{aligned} \tag{3.41}$$

and $\chi(s)$ has poles at $s = 1, 0$ (and possibly at negative integer points, which will not be relevant to us).

(3.41) is a special case of (3.1) with $d_j = 0, e_j = 0, D_j = 1\,(1 \le j \le r_1)$, $D_j = \frac{1}{2}\,(r_1 + 1 \le j \le r_1 + r_2 = M)$ and $D_j = 1\,(1 \le j \le r_1)$, $E_j = \frac{1}{2}\,(r_1 + 1 \le j \le r_1 + r_2 = \tilde{N} = M)$ and $r = 1$.

Hence Theorem 3.1 reads

Theorem 3.3

$$\sum_{k=1}^{\infty} \frac{\alpha_k}{(Ak)^s} H_{0,M}^{M,0}\left(z\lambda_k \,\middle|\, \begin{array}{c} - \\ \{(\frac{s}{2},\frac{1}{2})\}_{j=1}^{r_1}, \{(s,1)\}_{j=r_1+1}^{r_2} \end{array} \right)$$

$$= \sum_{k=1}^{\infty} \frac{\alpha_k}{(Ak)^{s\,1-s}} H_{0,\tilde{N}}^{\tilde{N},0}\left(\frac{\mu_k}{z} \,\middle|\, \begin{array}{c} - \\ \{(\frac{1-s}{2},\frac{1}{2})\}_{j=1}^{r_1}, \{(1-s,1)\}_{j=r_1+1}^{r_2} \end{array} \right)$$

$$+ \sum_{k=1}^{L} \operatorname{Res}\!\left(\chi(w)\, z^{s-w}, w = s_k \right). \tag{3.42}$$

Following Koshlyakov, we define the X-function, $X(x) = X_{r_1,r_2}(x)$, to be $H_{0,\varkappa}^{\varkappa,0}\left(x \,\middle|\, \begin{array}{c} - \\ \{(0,\frac{1}{2})\}_{j=1}^{r_1}, \{(0,1)\}_{j=r_1+1}^{r_2} \end{array} \right)$, i.e.

$$X_{r_1,r_2}(x) = \frac{1}{2\pi i} \int_{(c)} \Gamma^{r_1}\!\left(\frac{s}{2}\right) \Gamma^{r_2}(s) x^{-s}\, ds, \quad c > 0, \tag{3.43}$$

for $x > 0$ (eventually $\operatorname{Re}(x) > 0$ being allowed), which proviso will remain true in what follows.

Corollary 3.1 *Theorem 3.3 asserts that*

$$\sum_{k=1}^{\infty} F_\Omega(k) X_{r_1,r_2}(Ak\rho) = \mathrm{P}(x) + \sum_{k=1}^{\infty} F_\Omega(k) X_{r_1,r_2}(Ak\rho^{-1}), \tag{3.44}$$

where $\mathrm{P}(x) = R_0 + R_1$ *is the residual function and*

$$R_j = \operatorname{Res}\left(\chi(s) z^{-s}, s = j \right), \quad j = 0, 1. \tag{3.45}$$

We note that $R_1 = \rho z^{-1}$, where ρ is defined by (3.38).

Remark 3.3 *In Koshlyakov's case which we will study in §3.4, $\varkappa \leq 2$, whence Ω is either the rational or a quadratic field. Corollary 3.1 will be made more explicit as Theorem 3.12 below.*

3.3   Transformation formulas for Lambert series

In what follows we shall state special cases of the following theorem, which in turn is a special case of Theorem 3.2. Cf. (3.89), (3.129) and (3.129) below.

Theorem 3.4 *For Dirichlet series* φ, ψ *that satisfy the functional equation*

$$\Gamma(c + Cs)\varphi(s) = \Gamma(c + C(r - s))\psi(r - s), \tag{3.46}$$

we have the Bochner modular relation $H_{0,1}^{1,0} \leftrightarrow H_{0,1}^{1,0}$

$$z^{\frac{c}{C}} \sum_{k=1}^{\infty} \alpha_k \lambda_k^{\frac{c}{C}} e^{-(z\lambda_k)^{\frac{1}{C}}} = \left(\frac{1}{z}\right)^{r+\frac{c}{C}} \sum_{k=1}^{\infty} \beta_k \mu_k^{\frac{c}{C}} e^{-\left(\frac{\mu_k}{z}\right)^{\frac{1}{C}}}$$

$$+ C \sum_{k=1}^{L} \mathrm{Res}\left(\chi(w) z^{-w}, w = s_k\right). \tag{3.47}$$

3.3.1 *Lambert series*

First we follow [Win2] to introduce the Lambert series. Let $\{a_n\} \subset \mathbb{C}$ be such that

$$\limsup |a_n|^{1/n} \leq 1, \tag{3.48}$$

i.e. such that the power series $\sum_{n=1}^{\infty} a_n z^n$ is absolutely convergent in $|z| < 1$. Then the **Lambert series**

$$f(z) = \sum_{n=1}^{\infty} a_n \frac{z^n}{1 - z^n} \tag{3.49}$$

is absolutely convergent in $|z| < 1$ and represents an analytic function and moreover the power series of this function can be obtained by formal rearrangement of (3.49), i.e.

$$f(z) = \sum_{n=1}^{\infty} b_n z^n \quad (|z| < 1), \tag{3.50}$$

where

$$b_n = \sum_{d|n} a_d. \tag{3.51}$$

Along with (3.49), we may consider

$$f_+(z) := \sum_{n=1}^{\infty} a_n \frac{z^n}{1 + z^n}. \tag{3.52}$$

Riemann's posthumous Fragment [Rie1], based on Jacobi's Fundamenta Nova, §40 [Ja], consists of two parts, Fragment **I** and Fragment **II**, the latter of which contains only formulas and almost no text. Dedekind, entrusted with editing and publishing the paper, succeeded in elucidating the genesis of all the formulas in Fragment **II** by introducing the most celebrated Dedekind eta-function defined by (3.92) below. All the results in Fragment **II** deal with the asymptotic behavior of those modular functions from Jacobi's Fundamenta Nova, for which the variable tends to rational points on the unit circle. After Dedekind, several authors including Smith [Sm], Hardy [HarLam], Rademacher [Ra1] made some more incorporations of Fragment **II**. In 2004, Arias-de-Reyna [AdR] analyzed all the formulas in Fragment **II** again by applying ad hoc methods to each of the formulas. N. Wang [NLW, Theorem 3], referring to the more informative paper of Wintner [Win2] published in 1941, which already gave a close analysis of Fragment **I** and some far-reaching comments on Fragment **II** but was not mentioned by Arias-de-Reyna, chose a suitable R-function in taking the radial limits and applied a form of Dirichlet-Abel Theorem, whereby he almost immediately got the expression in terms of the (differences of) polylogarithm function of order 1, without singularity. This is precisely what Riemann intended to do, i.e. to eliminate the singular part, which turns out to be the Clausen function, by taking the odd part, but could not because of lack of time, and thought it could have been within reach of Riemann (at least in his mind, Riemann must have felt the truth of this type of Dirichlet-Abel theorem), those formulas were simply noted for record by Riemann.

For the moment we shall dwell on Wintner's paper [Win2], elucidating Riemann's procedure in Fragment **I**. Riemann divides the Lambert series by z and integrate the result from 0 to $z = re^{i\theta}$, $r = |z| < 1$, and then puts $r = 1$ to obtain his results, i.e. "Fourier expansions", in a formal way. Wintner's legitimation reads as follows. Slightly modifying Riemann's procedure by the trivial factor $e^{i\theta}$, which makes the function $F = F(r, \theta)$ a function in z, one obtains

$$F(re^{i\theta}) = \int_0^r \frac{f(re^{i\theta})}{r}\,\mathrm{d}r = \sum_{n=0}^{\infty} \frac{b_n}{n}(re^{i\theta})^n,$$

or

$$F(z) = \int_0^r \frac{f(re^{i\theta})}{r}\,\mathrm{d}r = \sum_{n=0}^{\infty} c_n z^n, \tag{3.53}$$

with

$$c_n = \frac{1}{n} \sum_{d|n} a_n. \tag{3.54}$$

In the case of (3.49), we should have

$$c_n = \frac{1}{n} \sum_{d|n} (-1)^n a_n. \tag{3.55}$$

Then Wintner asks if $F(z)$ tends to a measurable boundary function $F(e^{i\theta})$ as $r \to 1$ (within the Stolz path) and if so, then whether or not the boundary function is of class L^p, so that Riemann's formal trigonometric series actually is the Fourier series of the boundary function.

Use is made not only of the celebrated Riesz-Fischer condition [Vi2, Theorem 7.1, p. 139]

$$\sum_{n=-\infty}^{\infty} |c_n|^2 < \infty \tag{3.56}$$

but also a weaker condition

$$\sum_{n=-\infty}^{\infty} |n|^\varepsilon |c_n|^2 < \infty, \tag{3.57}$$

which lies between (3.56) and the L^p-condition

$$\sum_{n=-\infty}^{\infty} |c_n|^{p/(p-1)} < \infty \qquad (p \geq 2) \tag{3.58}$$

for an f to be of class L^p for some $p > 2$. Condition (3.57) implies that the trigonometric series $\sum_{n=-\infty}^{\infty} c_n e^{in\theta}$ is convergent a.e. (almost everywhere) and represents the values of the function a.e. We refer to this as the **Hausdorff-Paley extension of Fischer-Riesz theorem** to be used again in §6.3.4. The contents of this subsection has a close relationship to that of §6.3.4.

Theorem 3.5 (Wintner) *Suppose*

$$a_n = O(n^{\lambda-\delta}) \tag{3.59}$$

for some $\delta > 0$ and a fixed $0 < \lambda \leq \frac{1}{2}$. Then the boundary function $F(e^{i\theta})$ exists and is measurable such that

$$F(re^{i\theta}) \to F(e^{i\theta}) \quad a.e., \ as \ r \to 1 \tag{3.60}$$

along the Stoltz path. If $\lambda < \frac{1}{2}$, then $F(e^{i\theta})$ is of class $L^{1/\lambda}$ and if in (3.59), the exponent can be taken arbitrarily small, then it is of class L^{∞}.

Proof.   Recall that the divisor function, which is the special case of the sum-of-divisors function (3.86) satisfies the estimate

$$d(n) = \sum_{d|n} 1 = O(n^{\varepsilon}) \tag{3.61}$$

for every $\varepsilon > 0$ (e.g. [HaWr]). Hence it follows that

$$c_n = O(n^{\lambda-1-\delta}) \tag{3.62}$$

for some $\delta > 0$. Hence if $\lambda < \frac{1}{2}$, then the L^p-condition (3.58) is satisfied and if $\lambda = \frac{1}{2}$, then the series for $F(re^{i\theta})$ is Cauchy in L^2 and so there exists a function $F(e^{i\theta})$ of class L^2 such that

$$F(e^{i\theta}) \sim \sum_{n=-\infty}^{\infty} c_n e^{in\theta}. \tag{3.63}$$

This together with the condition (3.57) implies that

$$F(e^{i\theta}) = \sum_{n=-\infty}^{\infty} c_n e^{in\theta} \quad \text{a.e.} \tag{3.64}$$

Hence (3.60) follows by Abel's continuity theorem.   □

Example 3.1

(i) In the case $a_{2n} = 0, a_{2n+1} = 4(-1)^{n+1}$, we obtain the Lambert series

$$f(z) = \frac{2K}{\pi} = 4 \sum_{n=1}^{\infty} \frac{(-1)^{n+1}z^{2n+1}}{1 - z^{2n+1}} + 1, \tag{3.65}$$

which in the notation of (3.50) reads

$$f(z) = \frac{2K}{\pi} = 4 \sum_{\ell,m,n=0}^{\infty} d_{4,1}(n)z^{2^{\ell}(4m-1)^2 n} + 1, \tag{3.66}$$

with $d_{4,1}(n)$ denoting the number of divisors of n of the form $4k + 1$. Hence

$$F(z) = \sum_{n=0}^{\infty} \frac{d_{4,1}(n)}{n} z^n, \quad |z| < 1. \tag{3.67}$$

Hence by Theorem 3.5, the boundary function $F(e^{i\theta})$ exists and

$$F(e^{i\theta}) = \sum_{n=1}^{\infty} \frac{d_{4,1}(n)}{n} e^{in\theta} \quad \text{a.e.} \tag{3.68}$$

(ii) In the case $a_n = 1$, we obtain the Lambert series considered by Lambert himself ([Lam])

$$f(z) = \sum_{n=1}^{\infty} \frac{z^n}{1 - z^n} = \sum_{n=1}^{\infty} d(n) z^n, \quad |z| < 1. \tag{3.69}$$

Hence

$$F(z) = \sum_{n=1}^{\infty} \frac{d(n)}{n} z^n, \quad |z| < 1. \tag{3.70}$$

Hence by Theorem 3.5, the boundary function $F(e^{i\theta})$ exists and

$$F(e^{i\theta}) = \sum_{n=0}^{\infty} \frac{d(n)}{n} e^{in\theta} \quad \text{a.e.} \tag{3.71}$$

In connection with (3.71), we introduce the notation and terminology. Let

$$\bar{B}_1(x) = x - [x] - \frac{1}{2} \tag{3.72}$$

be the first **periodic Bernoulli polynomial**, where $[x]$ indicates the greatest integer function. As is well-known, we have the Fourier expansion for $x \notin \mathbb{Z}$

$$\bar{B}_1(x) = -\frac{1}{\pi} \sum_{m=1}^{\infty} \frac{\sin 2\pi m x}{m}, \tag{3.73}$$

while for $x \in \mathbb{Z}$, the right-hand side series of (3.73) converges to 0. Following Wintner [Win1], [Win2], we denote the Fourier series (3.73) by $\psi(x)$. Then

$$\psi(x) = \begin{cases} -\dfrac{1}{\pi} \displaystyle\sum_{m=1}^{\infty} \dfrac{\sin 2\pi m x}{m}, & x \notin \mathbb{Z} \\ 0, & x \in \mathbb{Z} \end{cases} \tag{3.74}$$

is the **sawtooth Fourier series**. Indeed, this was also established as one of the earliest instances of the modular relation by A. Z. Walfisz (6.113).

Exercise 3.2 Prove that the imaginary part of (3.71) gives the expansion treated in [Win1], which in turn gives rise to the series considered by Riemann [Rie1].

Indeed, this was proved by Chowla and Walfisz [CW]. Cf. (6.122) and (6.123) and §6.3.4.

We note that the estimate (3.61) gives rise to one for the sums over all divisors including the estimate for the sum-of-divisors function in the proof of Corollary 6.1 and (6.164).

3.3.2 *Lambert series and short character sums*

For a Dirichlet character χ modulo M, let $L_\chi(t)$ define the Lambert series associated to χ:

$$L_\chi(t) = \sum_{n=1}^{\infty} \chi(n)e^{-nt}, \quad \operatorname{Re} t > 0, \tag{3.75}$$

the series being absolutely convergent in the right half-plane.

Let N be a multiple of M, say $N = uM$ and let r be a positive integer prime to N. The essential case is $u < r$, which we so assume. Then Szmidt, Urbanowicz and Zagier [SUZ, p. 275] deduced the elegant formula

$$\sum_{0<n<\frac{N}{r}} \chi(n)e^{-rnt} = L_\chi(rt) - \frac{\bar{\chi}(r)}{\varphi(r)}e^{-Nt} \sum_{\psi \bmod r} \overline{\psi}(-N)L_{\chi\psi}(t), \tag{3.76}$$

with φ being the Euler φ-function.

Exercise 3.3 Deduce (3.76).

Solution We divide the sum (3.75) into two parts: $S_1 = \sum_{0<n<N/r}$ and $S_2 = \sum_{n \geq N/r}$ and write $rn = m$ in the latter:

$$S_1 = \sum_{0<n<N/r} \chi(n)e^{-rnt} = L_\chi(t) - S_2.$$

Then the second sum becomes $S_2 = \sum_{\substack{m \geq N \\ r|m}} \chi(mr^{-1})e^{-mt}$. Writing $m = n + N$, we obtain $S_2 = \sum_{\substack{m=1 \\ r|n+N}}^{\infty} \bar{\chi}(r)\chi(m)e^{-(n+N)t}$.

Finally, we use the orthogonality relation

$$\frac{1}{\varphi(r)} \sum_{\psi \bmod r} \psi(n)\psi(-N) \tag{3.77}$$

to express the congruence $n \equiv -N \pmod{r}$ to deduce (3.76).

It is rather remarkable that (3.76), though it is just a division of the sum into two parts, can be used to generate the identity as its Taylor coefficients:

$$
\begin{aligned}
& mr^{m-1} \sum_{0<n<N/r} \chi(n)n^{m-1} \\
& = -B_{m,\chi}r^{m-1} + \frac{\bar{\chi}(r)}{\varphi(r)} \sum_{\psi \bmod r} \bar{\psi}(-N)B_{m,\chi\psi}(N),
\end{aligned}
\tag{3.78}
$$

which is then effectively used to obtain non-trivial congruences, cf. [SUZ], [KUW], [Ur1] etc.

(3.78) constitutes a counterpart of Y. Yamamoto's formula, cf. [Ya].

We prove Theorem 3.6 below, forming Riemann-Hecke-Bochner correspondence (cf. Knopp [MKn2]) with Szmidt, Urbanowicz and Zagier's formula (3.76), by applying the Mellin transform technique. This aspect is already stated in [SUZ, p. 275, l. 8] to deduce their formula [SUZ, (5)], in the spirit of Riemann, splitting up the range of integral.

Theorem 3.6 *Let χ be a Dirichlet character* $\bmod M$, *not necessarily primitive. Let N be a multiple of M and let $r > \frac{N}{M}$ be a positive integer prime to N. Then we have the modular relation*

$$
\sum_{0<n<\frac{N}{r}} \frac{\chi(n)}{n^s} = L(s,\chi) - \frac{\overline{\chi}(r)}{\varphi(r)}r^s \sum_{\psi} \overline{\psi}(-N)(L(s,\chi\psi) - M_s(\chi\psi)) \tag{3.79}
$$

valid for all values of s, where $M_s(\chi) = \sum_{n=1}^{N-1} \frac{\chi(n)}{n^s}$ is the weighted complete character sum.

Proof. Taking the Mellin transform of both sides of (3.76) and using (2.106), we obtain for $\sigma > 1$

$$
\Gamma(s) \sum_{0<n<\frac{N}{r}} \frac{\chi(n)}{(rn)^s} = \Gamma(s)r^{-s}L(s,\chi) - \Gamma(s)\frac{\overline{\chi}(r)}{\varphi(r)} \sum_{\psi} \overline{\psi}(-N)\Phi(\chi\psi, N, s),
$$

where

$$
\Phi(\chi, a, s) = \sum_{n=0}^{\infty} \frac{\chi(n)}{(n+a)^s} \tag{3.80}
$$

is the **Hurwitz-Lerch L-function** (cf. §5.2.1 and [Mor]), which is analytically continued over the whole plane. Hence the formula

$$\sum_{0<n<\frac{N}{r}} \frac{\chi(n)}{(rn)^s} = r^{-s}L(s,\chi) - \frac{\overline{\chi}(r)}{\varphi(r)} \sum_{\psi} \overline{\psi}(-N)\Phi(\chi\psi, N, s)$$

is valid for all s.

Or more simply, using the expression

$$\Phi(\chi\psi, N, s) = L(s, \chi\psi) - M_s(\chi\psi)$$

and appealing to the analyticity of the Dirichlet L-function furnished by the functional equation (1.8), we deduce (3.79) for all $s \in \mathbb{C}$.   □

In view of [Ber75-1, Example 2] with slight modifications (or [SUZ, p. 276]), $M_s(\chi)$ may be considered as known for $s = m \in \mathbb{N}$:

$$\begin{aligned} M_m(\chi) &= \frac{1}{m}\left(B_{m+1,\chi}(M) - B_{m+1,\chi}\right) \\ &= \frac{1}{m+1}\sum_{k=1}^{m}\binom{m}{k}B_{m+1-k,\chi}M^k. \end{aligned} \tag{3.81}$$

In view of (3.81), Theorem 3.6 entails the values at non-positive integers.

3.3.3   *Ramanujan's formula leading to the eta-transformation formula*

The following remarkable **formula of Ramanujan** is probably one of the most famous ones named after him, which is stated as Entry 21 (i), Chapter 14 of Ramanujan's Notebook II [Berndt8] (which is also stated as I, Entry 15 of Chapter 16 [Berndt7], cf. also IV, Entry 20 of [Berndt9]). The most extensive account of information surrounding Ramanujan's formula is [Be11] and [IO] regarding special values of the zeta-functions in question. We remark in passing that the intersection of references in these two excellent survey papers is a null set.

Let $\alpha > 0, \beta > 0$ satisfy the relation (cf. (1.2))

$$\alpha\beta = \pi^2 \tag{3.82}$$

and let n be any positive integer. Then

$$\alpha^{-n}\left\{\frac{1}{2}\zeta(2n+1)+\sum_{k=1}^{\infty}\frac{k^{-2n-1}}{e^{2\alpha k}-1}\right\} \tag{3.83}$$

$$= (-\beta)^{-n}\left\{\frac{1}{2}\zeta(2n+1)+\sum_{k=1}^{\infty}\frac{k^{-2n-1}}{e^{2\beta k}-1}\right\}$$

$$- 2^{2n}\sum_{j=0}^{n+1}(-1)^j\frac{B_{2j}}{(2j)!}\frac{B_{2n+2-2j}}{(2n+2-2j)!}\alpha^{n+1-j}\beta^j,$$

where B_m denotes the m-th Bernoulli number defined by

$$\frac{x}{e^x-1}=\sum_{m=0}^{\infty}\frac{B_m}{m!}x^m, \quad |x|<2\pi. \tag{3.84}$$

Formula (3.83) is true for all integers $n \neq 0$ by interpreting a vacuous sum as 0. It looks like at a first glance that the case $n \geq 0$ counts more, giving an analytic expression for zeta values at odd positive integers, but the negative case is equally important in that it expresses the automorphic property of the corresponding Eisenstein series; the case $n = 0$ is no less important as we shall see below.

By the procedure similar to the one given above for treating Lambert series, we may apply **Liouville's formula** (cf. Koshlyakov [KoshI]) to include the case $n = 0$.

$$\sum_{k=1}^{\infty}\frac{k^a}{e^{2\pi kx}-1}=\sum_{k=1}^{\infty}\sigma_a(k)e^{-2\pi kx} \tag{3.85}$$

where

$$\sigma_a(k)=\sum_{d|k}d^a \tag{3.86}$$

indicates the **sum-of-divisors function**, the sum of a-th powers of all divisors of k.

Incorporating (3.85) with $x = \alpha/\pi$, in (3.83), we may express Ramanujan's formula as its **rephrased Ramanujan formula**

$$\sum_{k=1}^{\infty}\sigma_{-2n-1}(k)e^{-2\pi kx}+(-1)^{n+1}x^{2n}\sum_{k=1}^{\infty}\sigma_{-2n-1}(k)e^{-\frac{2\pi k}{x}}=\mathrm{P}(x) \tag{3.87}$$

where

$$P(x) = \frac{(2\pi)^{2n+1}}{2x} \sum_{j=0}^{n+1} (-1)^j \frac{B_{2j}}{(2j)!} \frac{B_{2n+2-2j}}{(2n+2-2j)!} x^{2n+2-2j} \qquad (3.88)$$

$$+ \begin{cases} -\dfrac{1}{2}\zeta(2n+1)\left\{1 + (-1)^{n+1}x^{2n}\right\} & \text{if } n \geq 1, \\ \dfrac{1}{2}\log x & \text{if } n = 0, \end{cases}$$

valid for $\operatorname{Re} x > 0$.

Indeed, $P(x)$ is the residual function given as the sum of the residues:

$$P(x) = \sum_{\xi \in R} \operatorname*{Res}_{s=\xi} (2\pi)^{-s} \Gamma(s) \zeta(s) \zeta(s + 2n + 1) x^{-s}, \qquad (3.89)$$

where $R = \{-2n-1, -2n, -2n+1, -2n+3, \ldots, -3, -1, 0, 1\}$, and $s = 0$ is a double pole only when $n = 0$ (others are simple poles). Here (6.153) is to be used to express the residues in terms of Bernoulli numbers. The same remark applies to (3.83).

Ramanujan's formula (3.83), in the rephrased form (3.87), amounts to a modular relation and is equivalent to Bochner's formula (Theorem 3.4) (cf. also [KTY1, Example 1], where the reason why we should have the relation (3.82) is elucidated). It is also a special case of Theorem 3.9 with $a = 0$. Cf. Remark 3.2 above. And therefore we may include the case $n = 0$, in which case Liouville's formula (3.85) is nothing but the sum of the Taylor expansion of the logarithm function:

$$\sum_{k=1}^{\infty} \sigma_{-1}(k) q^k = \sum_{l,m=1}^{\infty} \frac{1}{l} q^{lm} = -\sum_{k=1}^{\infty} \log(1 - q^k) := F(\tau), \qquad (3.90)$$

say, where

$$q = e^{2\pi i \tau} \qquad (3.91)$$

indicates the local uniformization parameter and τ is in the upper-half plane $\mathfrak{H}$ defined by (3.100), i.e. $\operatorname{Im} \tau > 0$.

Dedekind introduced the **eta function** (1892) by

$$\eta(\tau) = q^{\frac{1}{24}} \prod_{k=1}^{\infty} (1 - q^k) = e^{\frac{\pi i \tau}{12}} \prod_{k=1}^{\infty} (1 - e^{2\pi i k \tau}), \quad \operatorname{Im} \tau > 0 \qquad (3.92)$$

and proved its transformation formula (3.94) (under the Spiegelung) and *a fortiori* (3.137). Note that this is the $\frac{1}{24}$-th root of the discriminant function

(3.103) below, whence it is an automorphic form of weight "1/2". The most famous **transformation formula** for $\eta(\tau)$ (under the Spiegelung (3.102)) reads

$$\log \eta \left(-\frac{1}{\tau} \right) = \log \eta(\tau) + \frac{1}{2} \log \frac{\tau}{i}, \tag{3.93}$$

or

$$\eta \left(-\frac{1}{\tau} \right) = \sqrt{\frac{\tau}{i}} \eta(\tau). \tag{3.94}$$

The most comprehensive account of the eta-transformation formula is [KKlam]. See the references therein. For the general transformation formula see (3.137) below.

Theorem 3.7 *Ramanujan would have proved the eta-transformation formula (3.94) via his formula (3.87).*

Proof. We put $x = \tau/i \, (\mathrm{Re}\, x > 0)$ in (3.87) and (3.88) with $n = 0$ to obtain (for $F(\tau)$ defined by (3.90))

$$F(\tau) = F \left(-\frac{1}{\tau} \right) + \frac{\pi}{12} \left(\frac{i}{\tau} - \frac{\tau}{i} \right) + \frac{1}{2} \log \frac{\tau}{i}, \tag{3.95}$$

which is the formula deduced by A. Weil [We1, p.401].

Hence we see that

$$F(\tau) = \frac{\pi i}{12} \tau - \log \eta(\tau). \tag{3.96}$$

Rewriting (3.95) in terms of η, on using (3.96), gives (3.93).   □

In (3.90) we summed the infinite sum $\sum_{l,m=1}^{\infty} \frac{1}{l} q^{lm}$ by grouping those l, m which give the same value n, say, of lm. If instead, we sum over m first, then we will get a geometric series, so that

$$F(\tau) = \sum_{l=1}^{\infty} \frac{1}{l(q^{-l} - 1)} = \frac{i}{2} \sum_{l=1}^{\infty} \frac{1}{l} (\cot \pi l \tau + i). \tag{3.97}$$

Considering $F(\tau) - F(-\frac{1}{\tau})$, we are led to

$$F(\tau) - F \left(-\frac{1}{\tau} \right) = \frac{i}{2} \sum_{l=1}^{\infty} \frac{1}{l} \left(\cot \pi l \tau + \cot \frac{\pi l}{\tau} \right). \tag{3.98}$$

Theorem 3.7 should be compared with Theorem 3.10 below. Formula (3.97) suggests a definition of the Dedekind sum based on cotangent values, and (3.98) led C. L. Siegel [Sie5] to apply the residue theorem to the function $z^{-1}f(\nu z)$, where $f(z) = \cot z \cot \frac{z}{\tau}$ and $\nu = (n + \frac{1}{2})\pi$ $(n = 0, 1, \ldots)$, giving a simple proof of (3.94). As Siegel's proof has its genesis in the modular relation, the calculation done in [Sie5] is to find the residual function (and the right-hand side of (3.98)).

This also assures the validity of Theorem 3.7, i.e. (3.93) was within easy reach of Ramanujan.

Remark 3.4 *Ogg [Ogg] continued the argument of Hecke and Weil [We1] to give the product expansion of the basic theta-function $\vartheta = \vartheta_3$. For a new treatment of Ramanujan's formula based on the Barnes multiple zeta-functions. cf. [KMaT].*

Chan [HH] considers the modular relation for the inverse of the eta-function, which is the generating function of the partition function.

3.3.4　*A brief account of modular forms*

It would be appropriate to give a brief review here of the theory of modular forms (or sometimes automorphic forms) in order to make clear the meaning and importance of the eta-transformation formula (3.94) or in general the transformation formula under the Spiegelung. There are many good books, but the most accessible ones for beginners would be Serre [Ser], Doi and Miyake [DM], Rankin [Ran] or Shimura [Shm1].

Let $\Gamma = \Gamma(1) = \mathrm{SL}(2, \mathbb{Z})$ denote the special linear group with integer coefficients:

$$\Gamma = \mathrm{SL}(2, \mathbb{Z}) = \left\{ \gamma = \begin{pmatrix} a & b \\ c & d \end{pmatrix} \middle| ad - bc = 1, \ a, b, c, d \in \mathbb{Z} \right\}. \tag{3.99}$$

Γ acts on the **upper half-plane**

$$\mathfrak{H} = \{ z \in \mathbb{C} \mid \mathrm{Im}\, z > 0 \} \tag{3.100}$$

by the linear fractional transformations

$$\gamma z = \frac{az + b}{cz + d}. \tag{3.101}$$

We note that Γ is generated by the translation $T\tau = \tau + 1$ and the **Spiegelung**

$$S\tau = -\frac{1}{\tau}, \tag{3.102}$$

whence it usually suffices to know the transformation formula under the Spiegelung.

Let κ be an even integer greater than or equal to 4. Let M_κ denote the space of modular forms $f(z)$ of weight κ with respect to the full modular group $SL_2(\mathbb{Z})$, where $f(z)$ is called a **modular form** with the above-mentioned description if it satisfies the following conditions:

M1. $f(z)$ is a holomorphic function on the upper half-plane $\mathfrak{H}$ and is holomorphic at infinity.

M2. $f(z) = (cz + d)^{-\kappa} f\left(\dfrac{az + b}{cz + d}\right)$ for all $\begin{pmatrix} a & b \\ c & d \end{pmatrix} \in SL_2(\mathbb{Z})$.

A modular form $f(z)$ is called a **cusp form** if it vanishes at infinity.

The **discriminant function** is a cusp form of weight 12, which is defined by

$$\Delta(z) = q \prod_{k=1}^{\infty} (1 - q^k)^{24} = \sum_{k=1}^{\infty} \tau(k) q^k, \tag{3.103}$$

which is 24-th power of the Dedekind eta-function (3.92) defined above. Here $\tau(k)$ indicates the celebrated **Ramanujan's function** (cf. [Ser, pp.97-98]).

The most important modular forms include **Eisenstein series** $G_\kappa(z)$ defined by

$$G_\kappa(z) = \sum_{m,n}' \frac{1}{(mz + n)^\kappa}, \quad (\kappa \geq 4) \tag{3.104}$$

where the summation runs over all pairs (m, n) of integers other than $(0, 0)$. Cf. Definition 4.1 below. It is well known that G_κ is a modular form of weight κ with Fourier expansion

$$G_\kappa(z) = 2\zeta(\kappa) + 2\frac{(2\pi i)^\kappa}{(\kappa - 1)!} \sum_{m=1}^{\infty} \sigma_{\kappa-1}(m) e^{2\pi i m z}, \tag{3.105}$$

and that M_κ has for its basis $\{G_4^a G_6^b\}$ with a, b non-negative integers satisfying $4a + 6b = \kappa$.

The discriminant function appears in the context of the most famous j-invariant

$$j = \frac{1728 g_2^3}{\Delta}, \tag{3.106}$$

where $g_2 = 60 G_2$.

Theorem 3.8 *Let $f(z) = \sum_{n=0}^{\infty} c(n)q^n$ be a modular form of weight κ. Then the zeta-function attached to f defined by*

$$\Phi_f(s) = \sum_{n=1}^{\infty} \frac{c(n)}{n^s}, \quad \sigma > \kappa \tag{3.107}$$

is continued to a meromorphic function and satisfies the functional equation

$$(2\pi)^{-s}\Gamma(s)\Phi_f(s) = (-1)^k(2\pi)^{-(\kappa-s)}\Gamma(\kappa - s)\Phi_f(\kappa - s). \tag{3.108}$$

Note that (3.114) and (3.141) below are special cases of (3.108).

3.3.5 *The Ramanujan-Guinand formula*

In this subsection we shall illustrate a remarkable principle that differentiation of Ramanujan's formula, or the modular relation, gives rise to automorphy of the corresponding automorphic form.

To this end we introduce the **Mellin inversion with shifted argument** $I_a(x)$ for a real and $x > 0$ by

$$I_a(x) = I_{a,n}(x) = \frac{1}{2\pi i} \int_{(\kappa)} (2\pi)^{-s}\Gamma(s)\varphi(s - a)x^{-s}\,\mathrm{d}s \tag{3.109}$$

where

$$\varphi(s) = \zeta(s)\zeta(s + 2n + 1), \tag{3.110}$$

with n a non-negative integer and $\kappa > \max\{1, 1 + a\}$. It is easily seen that

$$I_a(x) = \sum_{k=1}^{\infty} \sigma_{-2n-1}(k)k^a e^{-2\pi kx}. \tag{3.111}$$

For simplicity we put

$$I(x) = I_0(x) = \frac{1}{2\pi i} \int_{(\kappa)} (2\pi)^{-s}\Gamma(s)\varphi(s)x^{-s}\,\mathrm{d}s. \tag{3.112}$$

We note that this is nothing other than the Lambert series (3.85). In the sequel suppose a is a non-negative integer referring to the number of times of differentiation. Differentiating $I(x)$ a-times with respect to x, whereby we perform differentiation under integral sign, we have the additional factor

$$\prod_{j=0}^{a-1}(-s - j),$$

which is $(-1)^a \frac{\Gamma(s+a)}{\Gamma(s)}$, whence we deduce the remarkable formula

$$\frac{\mathrm{d}^a}{\mathrm{d}x^a} I(x) = (-2\pi)^a I_a(x), \tag{3.113}$$

i.e. a-times differentiation of the Lambert series (3.112) is effected by shifting the argument of $\varphi(s)$ by a in (3.112) and multiplying by $(-2\pi)^a$. In view of (3.113), the a-times differentiated form of Ramanujan's formula (3.87) amounts to a counterpart of the modular relation for $I_a(x)$.

We shall obtain it by incorporating (cf. (3.129)) the functional equation satisfied by $\varphi(s)$

$$(2\pi)^{-s}\Gamma(s)\varphi(s) = (-1)^n (2\pi)^{2n+s}\Gamma(-2n-s)\varphi(-2n-s). \tag{3.114}$$

Moving the line of integration to $\sigma = -\kappa_1$ ($\kappa_1 > 2n+1-a$), whereby noting that the horizontal integrals vanish in the limit as $|t| \to \infty$, we have

$$I_a(x) = \frac{1}{2\pi i} \int_{(-\kappa_1)} \Gamma(s)\varphi(s-a)(2\pi x)^{-s}\,\mathrm{d}s + \mathrm{P}_a(x), \tag{3.115}$$

where $\mathrm{P}(x) = \mathrm{P}_a(x)$ denotes the sum of residues of the integrand at its poles at $s = a-2n-1, a-2n, a-2n+1, a-2n+3, \ldots, 0, a+1$.

Substitute the formula

$$\varphi(s-a) = (-1)^n (2\pi)^{2n+2s-2a} \frac{\Gamma(a-2n-s)}{\Gamma(s-a)} \varphi(a-2n-s)$$

which follows from (3.114) in the integral, say, $J_a(x)$ on the right-hand side of (3.115). Then

$$J_a(x) \tag{3.116}$$
$$= \frac{(-1)^n}{2\pi i} \int_{(-\kappa_1)} \frac{\Gamma(s)\Gamma(a-2n-s)}{\Gamma(s-a)} \varphi(a-2n-s)(2\pi)^{2n+s-2a} x^{-s}\,\mathrm{d}s$$
$$= \frac{(-1)^n}{2\pi i} \int_{(a-2n+\kappa_1)} \frac{\Gamma(a-2n-s)\Gamma(s)}{\Gamma(-2n-s)} \varphi(s)(2\pi)^{-s-a} x^{s-a+2n}\,\mathrm{d}s.$$

We note that by the reciprocal relation for the gamma function, we have

$$\frac{\Gamma(a-2n-s)}{\Gamma(-2n-s)} = \frac{\Gamma(1+s+2n)}{\Gamma(1+s+2n-a)} \frac{\sin \pi s}{\sin \pi(s-a)} \tag{3.117}$$
$$= (-1)^a \frac{\Gamma(a+2n+1)}{\Gamma(s+2n+1-a)}.$$

Hence

$$J_a(x) \tag{3.118}$$
$$= \frac{(-1)^{n+a}}{2\pi i} \int_{(a-2n+\kappa_1)} \frac{\Gamma(s+2n+1)\Gamma(s)}{\Gamma(s+2n+1-a)} \varphi(s)(2\pi)^{-s-a} x^{s-a+2n}\, \mathrm{d}s.$$

Since the gamma factor can be computed as follows for $0 \le a \le 2n$,

$$\frac{\Gamma(s)\Gamma(s+2n+1)}{\Gamma(s+2n+1-a)} = \sum_{k=0}^{a} \binom{a}{k} \frac{(2n-k)!}{(2n-a)!} \Gamma(s+k), \tag{3.119}$$

we conclude from (3.115) and (3.118) that

$$I_a(x) = (-1)^{n+a}(2\pi)^{-a} x^{-a+2n} \sum_{k=0}^{a} \binom{a}{k} \frac{(2n-k)!}{(2n-a)!} \tag{3.120}$$
$$\times \frac{1\cdot}{2\pi i} \int_{(a-2n+\kappa_1)} \Gamma(s+k)\varphi(s) \left(\frac{2\pi}{x}\right)^{-s}\, \mathrm{d}s + \mathrm{P}_a(x).$$

Finally, we note that the integral on the right-hand side of (3.120) becomes

$$\frac{1}{2\pi i} \int_{(a-2n+\kappa_1+k)} \Gamma(s)\varphi(s-k) \left(\frac{2\pi}{x}\right)^{-s+k}\, \mathrm{d}s,$$

which is identical with

$$\left(\frac{2\pi}{x}\right)^{k} I_k\left(\frac{1}{x}\right).$$

Thus we have proved

Theorem 3.9 (Ramanujan-Guinand formula) *For the Mellin transform $I_a(x)$ with shifted argument as defined by (3.109), we have the modular relation for $n \ge 0$ and $0 \le a \le 2n$,*

$$I_a(x) = (-1)^{n+a}(2\pi)^{-a} x^{-a+2n} \sum_{k=0}^{a} \binom{a}{k} \frac{(2n-k)!}{(2n-a)!} \left(\frac{2\pi}{x}\right)^{k} I_k\left(\frac{1}{x}\right) \tag{3.121}$$
$$+ \mathrm{P}_a(x),$$

where $\mathrm{P}_a(x)$ is the residual function.

Particular cases are

$$\mathrm{P}_0(x) = \mathrm{P}(x),$$

given by (3.88), and

$$P_{2n}(x) = (2n)! \zeta(2n+2)(2\pi x)^{-2n-1} + \frac{1}{2}\zeta(-2n-1)(2\pi x) \qquad (3.122)$$

$$+ \frac{(-1)^n}{2}(2\pi)^{-2n}(2n)!\zeta(2n+1), \quad n \geq 1.$$

On the other hand, for $a = 2n + 1$, we have

$$I_{2n+1}(x) = (-1)^{n+1}(2\pi)^{-2n-1}x^{-2n-2}I_{2n+1}\left(\frac{1}{x}\right) + P_{2n+1}(x), \qquad (3.123)$$

where

$$P_{2n+1}(x) = (2n+1)!\zeta(2n+2)(2\pi x)^{-2n-2} - \frac{1}{2}\zeta(-2n-1) \ (n \geq 1). \ (3.124)$$

The meaning of this equality is discussed below.

Remark 3.5 *The case $a = 0$ is Ramanujan's formula (3.87), the case $a = 2n$ is Guinand's formula (cf. [KTY1, Theorem 3]), and the case $a = 2n+1$ gives, on writing $m = -n - 1$, the negative case ($m < 0$) of Ramanujan's formula (3.87).*

The special case of (3.122) with $n = 1$, i.e. once differentiated form of Ramanujan's formula, yields Terras' formula [Te2], [Te3]

$$\zeta(3) = \frac{2}{45}\pi^3 - 4\sum_{k=1}^{\infty} e^{-2\pi k}\sigma_{-3}(k)\left(2\pi^2k^2 + \pi k + \frac{1}{2}\right). \qquad (3.125)$$

We now state a corollary to Theorem 3.9, which makes it possible to perform numerical calculation of zeta values in a remarkably efficient way.

Corollary 3.2 ([KTY1, Corollary 1]) *For $n \geq 1$ we have*

$$\frac{1}{2}(2n)! \zeta(2n+1) + \frac{(2\pi)^{2n+1}B_{2n+2}}{2(2n+2)}\left(\frac{1}{2n+1} - (-1)^n\right) \qquad (3.126)$$

$$= (-1)^n(2\pi)^{2n}\sum_{r=1}^{2n+1}(r-1)!S(2n+1,r)A(1,r)$$

$$- \sum_{m=0}^{2n}\frac{(2n)!}{m!}(2\pi)^m\sum_{r=1}^{m+1}(r-1)!\,S(m+1,r)A(2n+1-m,r),$$

where

$$A(p,q) = \sum_{m=1}^{\infty}\frac{1}{m^p(e^{2\pi m} - 1)^q}$$

and $S(k, r)$ denote the **Stirling number** *of the second kind defined by*

$$x^k = \sum_{r=1}^{k} S(k, r) x(x-1) \cdots (x - r + 1).$$

Example 3.2 We have

$$\zeta(3) = \frac{2}{45}\pi^3 - 8\pi^2 \left\{ \sum_{k=1}^{\infty} \frac{1}{k} \left(1 + \frac{1}{2\pi k} + \frac{1}{(2\pi k)^2} \right) \frac{1}{e^{2\pi k}-1} \right. \tag{i}$$

$$\left. + \sum_{k=1}^{\infty} \frac{1}{k} \left(3 + \frac{1}{2\pi k} \right) \frac{1}{(e^{2\pi k} - 1)^2} + \sum_{k=1}^{\infty} \frac{2}{k} \frac{1}{(e^{2\pi k} - 1)^3} \right\}$$

$$= 1.20205690315959428539973816151144999076498629234049$$

$$88817922715553418382057863130901 8\ldots$$

correct up to 84 decimal places (by computing first 30 terms).

$$\zeta(5) = \frac{2^2 \pi^5}{3^3 \cdot 5 \cdot 7} - \frac{8\pi^3}{3} \sum_{k=1}^{\infty} \frac{1}{k^2} \left(1 + \frac{3}{2\pi k} + \frac{6}{(2\pi k)^2} + \frac{6}{(2\pi k)^3} \right) \frac{1}{e^{2\pi k} - 1} \tag{ii}$$

$$- \frac{8\pi^3}{3} \sum_{k=1}^{\infty} \frac{1}{k^2} \left(7 + \frac{9}{2\pi k} + \frac{6}{(2\pi k)^2} \right) \frac{1}{(e^{2\pi k} - 1)^2}$$

$$- 16\pi^3 \sum_{k=1}^{\infty} \frac{1}{k^2} \left(2 + \frac{1}{2\pi k} \right) \frac{1}{(e^{2\pi k} - 1)^3} - 16\pi^3 \sum_{k=1}^{\infty} \frac{1}{k^2} \frac{1}{(e^{2\pi k} - 1)^4}$$

$$= 1.03692775514336992633136548645703416805708091950191$$

$$2811974192677903803589786281484 5600\ldots.$$

Remark 3.6 *In the book [SC] Srivastava and Choi record that by their formula, they were able to compute the value of $\zeta(3)$ accurate up to 7 decimal places by taking only first 50 terms.*

It should be noted that their formula is a variant of Euler's classical formula and is easily derivable from [KKY1] or [KKY2] by differentiating and forming linear combinations of resulting formulas. Note that by their method one could not attain the acceleration of order $e^{2\pi}$,

$$e^{2\pi} = 535.5491655524764736503049326\ldots,$$

(cf. the Research Notices at the end of [KTY7]).

Remark 3.7 *The method of proof of Theorem 3.9, which was not perceived in [KTY1], the Mellin inversion with shifted argument is an additive* *version of the "pseudo modular relation principle" whose multiplicative analogues have been successfully applied in [KTY2]–[KTY5]. However, from our present view point, this is simply a processed modular relation with the processing gamma factor $\Gamma(s+a)$ (cf. (3.114) and (3.115)). In this special case, however, the main formula (3.113) is the manifestation of the statement of Razar [GoRa] that the differentiation of Lambert series essentially corresponds to the shift of the argument of the associated Dirichlet series (cf. Remark 3.1).*

Remark 3.8 *We have seen that the special case $n = 0, a = 0$ of Theorem 3.9, the Ramanujan-Guinand formula, is nothing other than the automorphic property of the Dedekind eta-function. This can be regarded as once differentiated form of Guinand's formula. We note that in the case of $n = 0$, the twice differentiated Guinand's formula, with a suitable modification, coincides with the automorphic property of the Eisenstein series E_2 (cf. e.g. Rankin [Ran, p.196]). It is related to F defined by (3.90) by way of*

$$E_2(\tau) + \frac{3}{\pi y} = 1 - 24 \sum_{k=1}^{\infty} \frac{kq^k}{1 - q^k}$$

$$= 1 - \frac{12}{\pi i} F'(\tau)$$

$$= \frac{12}{\pi i} \left(\log \eta(\tau) \right)'.$$

In order to grasp the meaning of Formula (3.123) in the light of the transformation formula for Eisenstein series, we are now in a position to interpret (3.123), once differentiated form of Guinand's formula (= $2n + 1$ times differentiated form of Ramanujan's formula = the negative case thereof), in terms of G_{2n+2}.

Stating (3.123) in explicit form

$$-2 \sum_{k=1}^{\infty} \sigma_{2n+1}(k)e^{-2\pi kx} + \frac{B_{2n+2}}{2n + 2}$$

$$= (-1)^{n+1} x^{-2n-2} \left\{ -2 \sum_{k=1}^{\infty} \sigma_{2n+1}(k)e^{-\frac{2\pi k}{x}} + \frac{B_{2n+2}}{2n + 2} \right\},$$

we see that it is nothing but

$$G_{2n+2}\left(-\frac{1}{ix} \right) = (ix)^{2n+2} G_{2n+2}(ix), \tag{3.127}$$

i.e. the automorphy of $G_{2n+2}(z)$.

The following theorem is to be compared with Theorem 3.7. It is remarkable that both are touched upon by Indian mathematicians.

Theorem 3.10 *The functional equation* (1.5) *for the Riemann zeta-function implies the functional equation* (3.114), *which in turn implies the transformation formula for the discriminant function:*

$$\Delta\left(-\frac{1}{\tau}\right) = \tau^{24}\Delta(\tau), \qquad (3.128)$$

whence follows (3.94) *up to the root of unity.*

Proof.   Proof is due to ([Cho]). According to Convention in §1.1, (3.114) should read

$$\Gamma(s)\varphi(s) = \Gamma(2n+1-s)\psi(2n+1-s), \qquad (3.129)$$

where n can be any integer. Then we may simply apply Theorem 3.2 to deduce (3.123) for $I_{a,n}(x)$ and the remaining thing is to compute the residual function $P(x)$ given by (3.89). We need the case $I_1(X) = I_{0,1}(x) = \sum_{k=1}^{\infty} \sigma_1(k)e^{-2\pi kx}$ in (3.111) and the modular relation reads

$$I_1(x) = -x^{-2}I_1\left(\frac{1}{x}\right) + \frac{1}{24}x^{-2} - \frac{1}{4\pi}x^{-1} + \frac{1}{24}$$

or

$$-\frac{1}{24} + I_1\left(\frac{1}{\tau}\right) = -\tau^2\left(-\frac{1}{24} + I_1(\tau)\right) - \frac{\tau}{4\pi}, \qquad (3.130)$$

or in the form of the Lambert series

$$\tau^2\left(1 - 24\sum_{k=1}^{\infty}\sigma_1(k)e^{-2\pi k\tau}\right) = 1 - 24\sum_{k=1}^{\infty}\sigma_1(k)e^{-2\pi k\frac{1}{\tau}} + \frac{6i\tau}{\pi}.$$

Since

$$\frac{1}{2\pi i}\frac{\Delta'}{\Delta}(\tau) = q\frac{\Delta'}{\Delta}(q) = 1 - 24I_1(\tau), \qquad (3.131)$$

(3.130) amounts to

$$\frac{\Delta'}{\Delta}(\tau) = \frac{1}{\tau^2}\frac{\Delta'}{\Delta}\left(-\frac{1}{\tau}\right) - \frac{12}{\tau},$$

whence integrating we deduce that

$$\log \Delta(\tau) = \log \Delta\left(-\frac{1}{\tau}\right) - 12 \log \tau + \log k,$$

k being a constant. Exponentiating, we find that

$$\frac{\Delta(\tau)}{\Delta\left(-\frac{1}{\tau}\right)} = \frac{k}{\tau^{12}}. \tag{3.132}$$

Putting $\tau = i$ gives $k = 1$, so that (3.130) amounts to (3.128), completing the proof. □

Exercise 3.4 Prove that the j-invariant defined by (3.106) is a modular form of weight 0, i.e.

$$j\left(-\frac{1}{\tau}\right) = j(\tau). \tag{3.133}$$

Solution Note that (3.106) reads

$$j(\tau) = \frac{1 + 240I_3(\tau)}{\Delta(\tau)}, \tag{3.134}$$

where

$$I_3(\tau) = I_{3,1}(\tau) = \sum_{k=1}^{\infty} \sigma_3(k) e^{-2\pi k \tau}.$$

By (3.123) with $a = 2$, we have

$$1 + 240I_3\left(\frac{1}{\tau}\right) = \tau^4(1 + 240I_3(\tau)). \tag{3.135}$$

Substituting (3.128) and (3.134) in (3.135) leads to (3.133).

3.3.6 *The reciprocity law for Dedekind sums*

For the history of Dedekind sums, cf. [Asa], [HiTa] and the beginning of §3.3.1. Since the 24-th power Δ of η is the more classically known fundamental discriminant function, which is a modular form of weight 12, it is natural that the η-function underwent much attention and has been a fundamental example of half-integral weight modular forms.

Another transformation formula is

$$\log \eta(z + b) = \log \eta(z) + \frac{\pi i b}{12}, \tag{3.136}$$

which is immediate from (3.92).

The general transformation formula for $\eta(z)$ for $c > 0$ reads

$$
\begin{aligned}
\log \eta(\gamma z) &= \log \eta \left(\frac{az + b}{cz + d} \right) \\
&= \log \eta(z) + \frac{1}{2} \log \frac{cz + d}{i} + \frac{\pi i}{12} \frac{a + d}{c} - \pi i s(c, d),
\end{aligned} \tag{3.137}
$$

where $s(h, k)$ is the celebrated **Dedekind sum** defined by

$$
s(h, k) = \sum_{m=1}^{k-1} \left(\left(\frac{m}{k} \right) \right) \left(\left(\frac{hm}{k} \right) \right) = \sum_{m \bmod k} \left(\left(\frac{m}{k} \right) \right) \left(\left(\frac{hm}{k} \right) \right), \tag{3.138}
$$

where $(h, k) = 1$, $k \in \mathbb{N}$ and $(((x))$ is the special notation used in Dedekind sums only and coincides with the sawtooth Fourier series $\psi(x)$ in (3.74).

Hence we have another expression

$$
s(h, k) = \sum_{m=1}^{k-1} \bar{B}_1 \left(\frac{m}{k} \right) \bar{B}_1 \left(\frac{hm}{k} \right) = \sum_{\substack{m \bmod k \\ m \not\equiv 0}} \bar{B}_1 \left(\frac{m}{k} \right) \bar{B}_1 \left(\frac{hm}{k} \right). \tag{3.139}
$$

The Dedekind sums obey the famous **reciprocity law**: For $h, k \in \mathbb{N}$, $(h, k) = 1$

$$
s(h, k) + s(k, h) = -\frac{1}{4} + \frac{1}{12} \left(\frac{h}{k} + \frac{1}{hk} + \frac{k}{h} \right). \tag{3.140}
$$

There exist many generalizations of (3.140) and proofs thereof for which we refer to [RG].

Theorem 3.11 *The functional equation (1.5) for the Riemann zeta-function implies the reciprocity relation (3.140) up to (3.136).*

We divide the proof of Theorem in two parts as stated in the following

Lemma 3.1 (i) *The functional equation (1.5) for the Riemann zeta-function $\zeta(s)$ implies*

$$
\Phi(s) = (2\pi)^{-s} \Gamma(s) \zeta(s) \zeta(s + 1) = \Phi(-s). \tag{3.141}
$$

(ii) *The functional equation (1.5) for the Riemann zeta-function implies the transformation formula (3.93).*

Proof of Lemma 3.1. Proof of (i) is elementary using the reciprocal formula and the duplication formula for the gamma function. Proof of (ii) can be found in [RG, pp.47-48], which amounts to the following. For any

$$g = \begin{pmatrix} a & b \\ c & d \end{pmatrix}$$

in Γ, we have

$$gS = \begin{pmatrix} a & b \\ c & d \end{pmatrix} \begin{pmatrix} 0 & 1 \\ -1 & 0 \end{pmatrix} = \begin{pmatrix} b & -a \\ d & -c \end{pmatrix}.$$

By comparing the two expressions for $\eta(\tau'')$, where $\tau'' = \frac{aS\tau + b}{cS\tau + d}$ on one hand and

$$\tau'' = \frac{b\tau - a}{d\tau - c}, \tag{3.142}$$

we obtain

$$s(d, c) - s(-c, d) = \frac{1}{12} \left(\frac{d}{c} + \frac{c}{d} \frac{ad - bc}{dc} \right) + \frac{1}{2\pi i} \log(-i), \tag{3.143}$$

on the other, whence by oddness $s(-h, k) = -s(h, k)$, we arrive at (3.140).

By Lemma 3.1, we complete the proof of Theorem 3.11.

3.4 Koshlyakov's method [KoshI]

This section will serve to get the reader familiar with the setting in number fields as well as fixing some notation, including the choice of the fundamental sequence (by incorporating the associated constants in them). We shall elucidate the trilogy of Koshlyalov and show that Koshlyakov's theory of Dedekind zeta-functions of quadratic fields [KoshI] reduces in the long run to Theorem 4.8: the $G_{1,1}^{1,1} \leftrightarrow G_{0,2}^{2,0}$ formula, or the Fourier-Bessel expansion (the K-Bessel series for the perturbed Dirichlet series in the context of [KTY7]). This trilogy hitherto has not been well-known compared with his other papers [Kosum] and [Kosvor] but is rich in contents and can be thought of as the compilation of his work. To expound it is beneficial in two respects. First, one may come to know this relatively unknown but important work, and secondly, to our benefit, we can exhibit three typical cases of the functional equation corresponding to the rational, imaginary and real quadratic field. The latter two corresponds to the zeta-functions of definite and indefinite quadratic forms, respectively. The zeta-functions of (positive) definite quadratic forms are the Epstein zeta-functions whose theory has been developed rather fully. We refer to [Te5], [Vista, Chapter 6] and [Vi2, Chapter 6] for the definite case, and refer to [Sie3] and [Sie4] for the latter. This view point may be found in [Coh] as well as in [Dav], which refers to [LanZT] for proofs.

3.4.1   *Dedekind zeta-function II*

We follow the notation in §3.2, which is a slight deviation of Koshlyakov's. We assume that the extension degree of the number field Ω satisfies $\varkappa = [\Omega : \mathbb{Q}] \leq 2$, whence Ω is either the rational or a quadratic field. Hence if Δ stands for the discriminant of Ω, then we may write $\Omega = \Omega(\sqrt{\Delta})$ under the convention that $\Delta = 1$ in the case of the prime field $\Omega = \mathbb{Q}$.

$$\begin{cases} \Delta = 1 & n = 1 \\ \Delta = \text{a square-free integer} & n = 2. \end{cases} \tag{3.144}$$

The residue ρ at $s = 1$ of the Dedekind zeta-function (cf. (3.38)) in this case may be expressed as

$$\rho = \frac{2^{r+1}\pi^{r_2} R h}{w\sqrt{|\Delta|}} = -\frac{2^{r+1}\pi^{r_2}\zeta_{\Omega}^{(r)}(s)}{\sqrt{|\Delta|}}, \tag{3.145}$$

where (r) means the r-th derivative. But since the rank of the unit group of Ω satisfies $r = r_1 + r_2 - 1 \leq 1$, the derivative appears only in the case of real quadratic fields.

(3.39), (3.40), (3.41) may be expressed concretely as

$$\chi(s) = \begin{cases} \Gamma\left(\dfrac{s}{2}\right)\pi^{-\frac{s}{2}}\zeta(s) & n = 1 \\[2ex] \Gamma(s)\left(\dfrac{2\pi}{\sqrt{|\Delta|}}\right)^{-s}\zeta_{\Omega}(s) & n = 2, \Delta < 0 \\[2ex] \Gamma^2\left(\dfrac{s}{2}\right)\left(\dfrac{\pi}{\sqrt{\Delta}}\right)^{-s}\zeta_{\Omega}(s) & n = 2, \Delta > 0 \end{cases} \tag{3.146}$$

$$= \chi(1 - s),$$

where

$$\varphi(s) = A^{-s}\zeta_\Omega(s) = \sum_{n=1}^{\infty} \frac{F_\Omega(n)}{(An)^s} \tag{3.147}$$

$$= \begin{cases} \pi^{-\frac{s}{2}}\zeta(s) = \displaystyle\sum_{n=1}^{\infty} \frac{1}{(\sqrt{\pi}n)^s} & n = 1 \\[2ex] \displaystyle\sum_{n=1}^{\infty} \frac{F_\Omega(n)}{\left(\frac{2\pi}{\sqrt{|\Delta|}}n\right)^s} & n = 2, \Delta < 0 \\[2ex] \displaystyle\sum_{n=1}^{\infty} \frac{F_\Omega(n)}{\left(\frac{\sqrt{\pi}}{\sqrt{|\Delta|}}n\right)^s} & n = 2, \Delta > 0. \end{cases}$$

3.5 Koshlyakov's functions

Koshlyakov introduced several functions starting from the K-function (kernel function), $K(x) = K_{r_1,r_2}(x)$ defined by (4.57), which plays the most prominent role in his theory; along with K, also introduced are X, Y, Z, L, M, N, see the following table ([KoshI, p.122]). We shall treat some (properties) of them.

Table 3.2   Koshlyakov's functions

ftn's	rat.	imag. quadr.	real quadr.
X	$2e^{-x^2}$	e^{-x}	$4K_0(2x)$
Y	e^{-x^2}	$\mathrm{si}(x)$	$J_0(2x)$
K	$\sqrt{\pi}e^{-x}$	$-2\mathrm{kei}_0\left(4\sqrt{\frac{x}{4}}\right)$	$4\mathrm{ker}_0\left(4\sqrt{x}\right)$
L	$\sqrt{\pi}\cos 2x$	$\frac{\pi}{2}J_0(2\sqrt{x})$	$2\left(K_0(4\sqrt{x}) - \frac{\pi}{2}Y_0(4\sqrt{x})\right)$
M	$\sqrt{\pi}\sin 2x$	$K_0(2\sqrt{x}) + \frac{\pi}{2}Y_0(2\sqrt{x})$	$\pi J_0(4\sqrt{x})$
N	$\frac{\sqrt{\pi}}{2}e^{-2x}$	$\frac{\sqrt{x}}{i}\left(\varepsilon K_1(2\bar{\varepsilon}\sqrt{x}) - \bar{\varepsilon}K_1(2\varepsilon\sqrt{x})\right)$	$\sqrt{x}\left(\varepsilon K_1(4\bar{\varepsilon}\sqrt{x}) + \bar{\varepsilon}K_1(4\varepsilon\sqrt{x})\right)$

3.5.1 *Koshlyakov's X-functions*

Exercise 3.5   Prove the formulas in Table 3.2 for the X-function defined by (3.43).

$$X_{1,0}(x) = 2e^{-x^2} \tag{3.148}$$

$$X_{0,1}(x) = e^{-x} \tag{3.149}$$

and

$$X_{2,0}(x) = 4K_0(2x). \tag{3.150}$$

Solution These follow from (2.93), (2.106), (2.109).

Exercise 3.6 Recall that R_j denote the residues at $j = 0, 1$ of the meromorphic function $\Gamma^{r_1}\left(\frac{s}{2}\right)\Gamma^{r_2}(s)A^{-s}\zeta_\Omega(s)z^{-s}$ ((3.45)). Prove in the case $\varkappa \leq 2$ that

$$R_0 = 2^{r_1}\zeta_\Omega^{(r)}(0), \quad R_1 = -2^{r_1}\zeta_\Omega^{(r)}(0)z^{-1}, \tag{3.151}$$

whence $P(x) = 2^{r_1}\zeta_\Omega^{(r)}(0) - 2^{r_1}\zeta_\Omega^{(r)}(0)z^{-1}$, which is the corrected form of [KoshI, p.119, l.11 from below].

Exercise 3.7 In the case $r_2 = 0\,(r_1 \geq 2)$, show that

$$R_0 = \mathrm{Res}\left(\Gamma\left(\frac{s}{2}\right)^{r_1} A^{-s}\zeta_\Omega(s)\,z^{-s}, s = 0\right) = 2^{r_1}\frac{\zeta_\Omega^{(r_1-1)}(0)}{(r_1-1)!}$$

and

$$R_1 = \mathrm{Res}\left(\Gamma\left(\frac{s}{2}\right)^{r_1} A^{-s}\zeta_\Omega(s)\,z^{-s}, s = 1\right) = -2^{r_1}\frac{1}{(r_1-1)!}\,\zeta_\Omega^{(r_1-1)}(0)\,z^{-1}.$$

Solution

$$\begin{aligned}
R_1 &= \mathrm{Res}\left(\Gamma\left(\frac{1-s}{2}\right)^{r_1} A^{-(1-s)}\,\zeta_\Omega(1-s)\,z^{-s}, s = 1\right)\\
&= \mathrm{Res}\left(\left(\frac{-2}{s-1}\right)^{r_1}\Gamma\left(\frac{3}{2}-\frac{s}{2}\right)^{r_1} A^{-(1-s)}\right.\\
&\qquad\left.\left(\cdots + \frac{\zeta_\Omega^{(r_1-1)}(0)}{(r_1-1)!}\,(1-s)^{r_1-1} + \cdots\right)z^{-s}, s = 1\right).
\end{aligned}$$

Note that in [BeI], $\mathrm{Res}\left(\Gamma(s)^{r_1} A^{-2s}\zeta_\Omega(s)\,z^{-s}, s = 0\right) = 2^{r_1-1}\frac{\zeta_\Omega^{(r_1-1)}(0)}{(r_1-1)!}$
and

$$\mathrm{Res}\left(\Gamma\left(\frac{s}{2}\right)^{r_1} A^{-s}\zeta_\Omega(s)\,z^{-s}, s = 1\right) = \Gamma\left(\frac{1}{2}\right)^{r_1} A^{-1}\lambda h z^{-1},$$

where λh is the residue of $\zeta_\Omega(s)$ at $s = 1$ given by (3.38). It may be also expressed in the case $\varkappa \leq 2$ as

$$\lambda h = -\frac{2^{r+1}\pi^{r_2}\zeta_\Omega^{(r)}(s)}{\sqrt{|\Delta|}}. \tag{3.152}$$

Remark 3.9 *In [KoshE], Koshlyakov mentions:*

"On [KoshI, p. 114], the erroneous value $\zeta_\Omega(0) = -1$ of the Dedekind zeta-function for the imaginary quadratic field is given. The true value is $\zeta_\Omega(0) = -\frac{h}{2}$ with the class number h in my previous paper $\cdots$."

However, the correction is valid except for the cases of the Gauss field $\Omega = \mathbb{Q}(i)$ $(w = 4)$ and the Eisenstein field $\Omega = \mathbb{Q}(\rho)$ $(w = 6)$, where $i = e^{\frac{2\pi i}{4}}$ and $\rho = e^{\frac{2\pi i}{6}}$ indicates the piervot'nyi primitive root of unity. In our case of $\varkappa \leq 2$, we may unify the values into the form

$$\zeta_\Omega(0) = -\frac{h}{w} \tag{3.153}$$

with understanding that $h = 1$, $w = 2$ in the rational field and $w = \infty$ in the real quadratic field. Of course, in the rational case, a more familiar expression for the values of the Riemann zeta-function at non-positive integral arguments is that of Euler (cf. [NTA, Proposition 5.1, p. 138]), $\zeta(0) = -\frac{B_0}{2} = -\frac{1}{2}$. Euler's formula is a consequence of the functional equation. Historically, Euler first prove the expression and conjectured the general form of the functional equation.

Using Exercises 3.5 and 3.6, we may state Corollary 3.1 in a more explicit form, which is [KoshI, (4.1)]. It combines three identities (3.155), (3.156), and (3.157) below.

Theorem 3.12 *In the case of the rational or a quadratic field, the unprocessed modular relation reads*

$$\sqrt{\rho} \left\{ 2^{r_1} \zeta_\Omega^{(r)}(0) - \sum_{k=1}^{\infty} F(k) X_{r_1, r_2}(Ak\rho) \right\}$$
$$= \frac{1}{\sqrt{\rho}} \left\{ 2^{r_1} \zeta_\Omega^r(0) - \sum_{k=1}^{\infty} F(k) X_{r_1, r_2}\left(Ak\frac{1}{\rho} \right) \right\}. \tag{3.154}$$

(3.154) leads, on incorporating (3.153), to Koshlyakov's formula [KoshI, (4.8)]:

$$\sqrt{\rho} \left\{ 1 + 2 \sum_{k=1}^{\infty} e^{-\pi k^2 \rho^2} \right\} = \frac{1}{\sqrt{\rho}} \left\{ 1 + 2 \sum_{k=1}^{\infty} e^{-\pi k^2 \frac{1}{\rho^2}} \right\} \tag{3.155}$$

or writing ρ^2 for z,

$$\sum_{k=1}^{\infty} e^{-\pi k^2 z} + \frac{1}{2} = z^{-\frac{1}{2}} \sum_{k=1}^{\infty} \left(e^{-\pi k^2 z^{-1}} + \frac{1}{2} \right),$$

the theta-transformation formula (1.72) (cf. §1.72).

In the imaginary quadratic case, (3.154) reads

$$-\zeta_\Omega(0) + \sum_{k=1}^{\infty} F(k) e^{-\frac{2\pi}{\sqrt{|\Delta|}} z} = z^{-1} \left(-\zeta_\Omega(0) + \sum_{k=1}^{\infty} F(k) e^{-\frac{2\pi}{\sqrt{|\Delta|}} z^{-1}} \right),$$

or

$$\sqrt{\rho} \left(h + w \sum_{k=1}^{\infty} F(k) e^{-\frac{2\pi}{\sqrt{|\Delta|}} \rho} \right)$$

$$= \frac{1}{\sqrt{\rho}} \left(h + w \sum_{k=1}^{\infty} F(k) e^{-\frac{2\pi}{\sqrt{|\Delta|}} \rho^{-1}} \right),$$

(3.156)

by (3.153). Formula (3.156) is the corrected form of Koshlyakov's [KoshI, (4.9)].

Finally, in the real quadratic case, (3.154) reads

$$\sqrt{\rho} \left\{ 4\zeta_\Omega'(0) - 4 \sum_{k=1}^{\infty} F(k) K_0 \left(\frac{2\pi k}{\sqrt{\Delta}} \rho \right) \right\}$$

$$= \frac{1}{\sqrt{\rho}} \left\{ 4\zeta_\Omega'(0) - 4 \sum_{k=1}^{\infty} F(k) K_0 \left(\frac{2\pi i}{\sqrt{\Delta}} \frac{1}{\rho} \right) \right\}.$$

(3.157)

Formula (3.157) is a special case of the $G_{0,2}^{2,0} \leftrightarrow G_{0,2}^{2,0}$ formula (§3.1.2). A more general case has been treated as [BeI, Example 3] and a more general form of (3.157) for general r_1 is given on [BeI, p.358].

Berndt treats on [BeI, p.358] $A^{-2s}\zeta_\Omega(2s)$ as the Dirichlet series $\varphi(s)$, so that $\chi(s) = \Gamma(s)^{r_1} \varphi(s) z^{-s}$, replaces z/A^2 by y, whence $z^{-\frac{1}{2}} = Ay^{-\frac{1}{2}}$, and so

$$\mathrm{Res}\left(\Gamma(s)^{r_1} A^{-2s} \zeta_\Omega(2s) \, z^{-s}, s = \frac{1}{2} \right) = \Gamma\left(\frac{1}{2} \right)^{r_1} \frac{1}{2} \lambda h y^{-\frac{1}{2}}.$$

Because of this replacement, the modular relation leads to a slightly different form ([BeI, p.358]). For this cf. [KTY7, §7].

3.5.2 *Koshlyakov's Y-function*

Complementary to the X-function is the Y-function which is the Mellin factor of the K-function in the sense that ([KoshI, (5.8)], Example 6.1):

$$\int_0^\infty X\left(\frac{a}{x}\right) Y(bx)\,dx = \frac{1}{b} K(ab), \quad a > 0, b > 0. \tag{3.158}$$

The Y-function is defined by ([KoshI, (5.4)])

$$Y(x) = Y_{r_1,r_2}(x)$$
$$= \frac{1}{2\pi i} \int_{(c)} \frac{\pi}{2\sin\left(\frac{\pi}{2}s\right) \Gamma^{r_1}\left(\frac{2-s}{2}\right) \Gamma^{r_2}(2-s)} x^{-s}\,ds \tag{3.159}$$
$$= \frac{1}{2\pi i} \int_{(c)} \frac{\Gamma(s)\Gamma(1-s)\cos\left(\frac{\pi}{2}s\right)}{\Gamma^{r_1}\left(\frac{2-s}{2}\right)\Gamma^{r_2}(2-s)} x^{-s}\,ds, \quad c > 0, x > 0,$$

or after some transformation,

$$Y(x) = Y_{r_1,r_2}(x) \tag{3.160}$$
$$= H^{1,1}_{2,r+3}\left(x \left|\begin{array}{c} (0,1),(\frac{1}{2},\frac{1}{2}) \\ (0,1),(\frac{1}{2},\frac{1}{2}),(0,\frac{1}{2}),\ldots,(0,\frac{1}{2}),(-1,1),\ldots,(-1,1) \end{array}\right.\right)$$
$$= \frac{1}{2} H^{1,1}_{1,r+2}\left(e^{\frac{\pi}{2}i}x \left|\begin{array}{c} (0,1) \\ (0,1),(0,\frac{1}{2}),\ldots,(0,\frac{1}{2}),(-1,1),\ldots,(-1,1) \end{array}\right.\right)$$
$$+ \frac{1}{2} H^{1,1}_{1,r+2}\left(e^{-\frac{\pi}{2}i}x \left|\begin{array}{c} (0,1) \\ (0,1),(0,\frac{1}{2}),\ldots,(0,\frac{1}{2}),(-1,1),\ldots,(-1,1) \end{array}\right.\right).$$

Though contradicting to our claim that the processing gamma factor is trivial, we may view Y as the case where the processing gamma factor ((7.24)) is $\Gamma(s\,|\,\Delta) = \frac{\Gamma(1-s)}{\Gamma^{r_1}\left(\frac{2-s}{2}\right)\Gamma^{r_2}(2-s)}$ and the functional equation (3.41) is the case of the imaginary quadratic field. This fact will show its effect in due course.

We distinguish three cases. In the case $r_1 = 1, r_2 = 0$,

$$Y_{1,0}(x) = \frac{1}{2\pi i} \int_{(c)} \Gamma\left(\frac{s}{2}\right) x^{-s} \frac{ds}{2} = G^{1,0}_{0,1}\left(x^2 \left|\begin{array}{c} - \\ 0 \end{array}\right.\right) = e^{-x^2}. \tag{3.161}$$

In the case $r_1 = 0, r_2 = 1$, by (3.160),

$$Y_{0,1}(x) = \frac{1}{2}\left(G^{1,1}_{1,2}\left(iz \left|\begin{array}{c} 0 \\ 0,-1 \end{array}\right.\right) + G^{1,1}_{1,2}\left(-iz \left|\begin{array}{c} 0 \\ 0,-1 \end{array}\right.\right)\right). \tag{3.162}$$

By (2.112),

$$G_{1,2}^{1,1}\left(z\,\middle|\,\begin{matrix}0\\s,-1\end{matrix}\right) = z^s\frac{\Gamma(1+s)}{\Gamma(2+s)}\,{}_1F_1\left(\begin{matrix}1+s\\2+s\end{matrix};-z\right).$$

By (2.122), this reduces further to

$$G_{1,2}^{1,1}\left(z\,\middle|\,\begin{matrix}0\\s,-1\end{matrix}\right) = z^s\frac{\Gamma(1+s)}{\Gamma(2+s)}(s+1)z^{-(s+1)}\gamma(s+1,z)$$

$$= z^{-1}\gamma(s+1,z). \tag{3.163}$$

In the special case $s = 0, r = 1$, (3.163) reduces to

$$G_{1,2}^{1,1}\left(z\,\middle|\,\begin{matrix}0\\0,-1\end{matrix}\right) = z^{-1}\gamma(1,z) = \frac{1-e^{-z}}{z}. \tag{3.164}$$

Hence

$$Y_{0,1}(x) = \frac{\sin(x)}{x} = \mathrm{si}(x), \tag{3.165}$$

the sinus cardinalis function which plays an important role in signal transmission (cf. [Spl]). Below we will give a modular relation version of the sampling theorem in Theorem 3.14.

Finally, in the case of

$$Y_{2,0}(x) = \frac{1}{2\pi i}\int_{(c)}\frac{\pi}{2\sin\left(\frac{\pi}{2}s\right)\Gamma\left(1-\frac{s}{2}\right)^2}\,x^{-s}\mathrm{d}s,$$

we rewrite the integrand as $\frac{\Gamma\left(\frac{s}{2}\right)}{\Gamma\left(1-\frac{s}{2}\right)}\frac{x^{-s}}{2}$ and then express $Y_{2,0}(x)$ as the G-function:

$$Y_{2,0}(x) = \frac{1}{2}H_{0,2}^{1,0}\left(x\,\middle|\,\begin{matrix}-\\(0,\frac{1}{2}),(0,\frac{1}{2})\end{matrix}\right) = G_{0,2}^{1,0}\left(x^2\,\middle|\,\begin{matrix}-\\0,0\end{matrix}\right) = J_0(2x)$$

by (2.110), $J_s(x)$ being the Bessel function defined by (2.24).

We recall that the X- and Y-functions play an important role in the derivation of the Riemann-Siegel integral formula [KTY7, §4.2], in that a recourse to the formula ([KoshI, (9.10)])

$$\int_1^\infty X(ax)Y(bx)\,x\,\mathrm{d}x = \frac{Z(a,b)}{a^2+b^2}, \quad a > 0, b > 0$$

was essential.

In what follows we shall deduce Theorem 3.14 as a special case of the main theorem of modular relations. One of the forms of the main theorem in Chapter 7 reads

Theorem 3.13 *If*

$$\Delta = \begin{pmatrix} \{(1-a_j, A_j)\}_{j=1}^n; & \{(a_j, A_j)\}_{j=n+1}^p \\ \{(b_j, B_j)\}_{j=1}^m & ;\{(1-b_j, B_j)\}_{j=m+1}^q \end{pmatrix} \in \Omega \qquad (3.166)$$

with its associated Gamma factor

$$\Gamma(s \,|\, \Delta) = \frac{\displaystyle\prod_{j=1}^m \Gamma(b_j + B_j s) \ \prod_{j=1}^n \Gamma(a_j - A_j s)}{\displaystyle\prod_{j=n+1}^p \Gamma(a_j + A_j s) \ \prod_{j=m+1}^q \Gamma(b_j - B_j s)}, \qquad (3.167)$$

then

$$\sum_{k=1}^\infty \frac{\alpha_k}{\lambda_k^s} \, H_{p,q+1}^{m+1,n}\left(z\lambda_k \ \middle|\ \begin{array}{l} \{(1-a_j, A_j)\}_{j=1}^n, \{(a_j, A_j)\}_{j=n+1}^p \\ (s,1), \{(b_j, B_j)\}_{j=1}^m, \{(1-b_j, B_j)\}_{j=m+1}^q \end{array} \right) \qquad (3.168)$$

$$= \sum_{k=1}^\infty \frac{\beta_k}{\mu_k^{r-s}} \, H_{q,p+1}^{n+1,m}\left(\frac{\mu_k}{z} \ \middle|\ \begin{array}{l} \{(1-b_j, B_j)\}_{j=1}^m, \{(b_j, B_j)\}_{j=m+1}^q \\ (r-s,1), \{(a_j, A_j)\}_{j=1}^n, \{(1-a_j, A_j)\}_{j=n+1}^p \end{array} \right)$$

$$+ \mathrm{P}(z)$$

is equivalent to the functional equation

$$\Gamma(s)\varphi(s) = \Gamma(r-s)\varphi(r-s), \qquad (3.169)$$

where $\mathrm{P}(z) = \sum_{k=1}^L \mathrm{Res}\left(\Gamma(w-s \,|\, \Delta)\, \chi(w)\, z^{s-w}, w = s_k \right)$ *indicates the residual function.*

We assume that $\varphi(s)$ has a simple pole at $s = 1$ with residue ρ_φ, so that $\chi(s) = \Gamma(s)\varphi(s)$ has simple poles at $s = 1$ and $s = 0$.

Formally we apply the case $m = 0, n = 1, p = 2, q = 2$ and $a_1 = 1, A_1 = 1, a_2 = \frac{1}{2}, A_2 = \frac{1}{2}$ $b_1 = -1, B_1 = 1, b_2 = \frac{1}{2}, B_2 = \frac{1}{2}$ of Theorem 3.13 with

$$\Gamma(w - s \,|\, \Delta) = \frac{\Gamma(1 + s - w))}{\Gamma(2 + s - w)\Gamma\left(\frac{1+s}{2} - \frac{1}{2}w\right)\Gamma\left(\frac{1-s}{2} + \frac{1}{2}w\right)}.$$

However, we appeal to (3.162) by putting $\Gamma(w - s \,|\, \Delta) = \frac{\Gamma(1-(w-s))}{\Gamma(2-(w-s))} \cos \frac{\pi}{2}(w-s)$ and expressing the cosine function as the sum of exponential functions. Then this amounts to the residual function is computed to be

$$P(z) = \frac{\Gamma(1+s)}{\Gamma(2+s)} \cos\frac{\pi s}{2} z^s \varphi(0) + \frac{\Gamma(s)}{\Gamma(1+s)} z^{s-1} \sin\frac{\pi s}{2} \rho_\varphi \qquad (3.170)$$

$$= \frac{1}{s+1} \cos\frac{\pi s}{2} z^s \varphi(0) + \frac{1}{s} z^{s-1} \sin\frac{\pi s}{2} \rho_\varphi.$$

The modular relation (3.168) amounts to the case $m = 0$, $n = 1$, $p = 1$, $q = 1$:

$$\sum_{k=1}^{\infty} \frac{\alpha_k}{\lambda_k^s} \left(G_{1,2}^{1,1}\left(iz\lambda_k \,\middle|\, \begin{matrix} 0 \\ s, -1 \end{matrix} \right) + G_{1,2}^{1,1}\left(-iz\lambda_k \,\middle|\, \begin{matrix} 0 \\ s, -1 \end{matrix} \right) \right) \qquad (3.171)$$

$$= \sum_{k=1}^{\infty} \frac{\beta_k}{\mu_k^{r-s}} \left(G_{1,2}^{2,0}\left(i\frac{\mu_k}{z} \,\middle|\, \begin{matrix} 2 \\ r-s, 1 \end{matrix} \right) + G_{1,2}^{2,0}\left(-i\frac{\mu_k}{z} \,\middle|\, \begin{matrix} 2 \\ r-s, 1 \end{matrix} \right) \right)$$

$$+ 2P(z).$$

The left-hand side function is (3.162), which is the sinus cardinalis function in (3.165).

On the right-hand side, by (2.117)

$$G_{1,2}^{2,0}\left(z \,\middle|\, \begin{matrix} 2 \\ r-s, 1 \end{matrix} \right) = e^{-z} z^{r-s} U(1, r-s, z).$$

By (2.123), this reduces further to

$$G_{1,2}^{2,0}\left(z \,\middle|\, \begin{matrix} 2 \\ r-s, 1 \end{matrix} \right) = e^{-z} z^{r-s} z^{1-(r-s)} e^z \Gamma(r-s-1, z)$$

$$= z\Gamma(r-s, z). \qquad (3.172)$$

In the special case $s = 0, r = 1$ of (3.171), since (3.172) reduces to

$$G_{1,2}^{2,0}\left(z \,\middle|\, \begin{matrix} 2 \\ 1, 1 \end{matrix} \right) = z\Gamma(1, z) = ze^{-z}, \qquad (3.173)$$

we conclude that

$$\frac{1}{2}\left(G_{1,2}^{2,0}\left(iz \,\middle|\, \begin{matrix} 2 \\ 1, 1 \end{matrix} \right) + G_{1,2}^{2,0}\left(-iz \,\middle|\, \begin{matrix} 2 \\ 1, 1 \end{matrix} \right) \right) \qquad (3.174)$$

$$= z \sin z.$$

(3.170) reduces for $s = 0$ to $\varphi(0) + \frac{\pi}{2} z^{-1} \rho_\varphi$. Thus we have proved

Theorem 3.14

$$\sum_{k=1}^{\infty} \alpha_k \operatorname{si}(z\lambda_k) = \frac{1}{z} \sum_{k=1}^{\infty} \beta_k \sin \frac{\mu_k}{z} + \varphi(0) + \frac{\pi}{2} z^{-1} \rho_\varphi. \tag{3.175}$$

This is interesting in the light of the sampling theorem, i.e. the left-hand side of (3.175) may be thought of as the Fourier series of the function $f(z)$ in question and the right-hand side is the sampling series as given in the sampling theorem.

We now consider the L-function which are relatives of K-functions (in §4.6) and are related through

$$L_{r_1,r_2}(x) = \frac{1}{2} \left(K_{r_1,r_2}(-ix) + K_{r_1,r_2}(ix) \right) \quad ([\text{KoshI}, (7.11)]). \tag{3.176}$$

It is defined by ([KoshI, (8.9)])

$$L_{r_1,r_2}(x) = \frac{1}{2\pi i} \int_{(c)} \frac{\pi}{2} G(1 - s) x^{-s} ds$$
$$= \frac{1}{2\pi i} \int_{(c)} \frac{\pi}{2} \frac{\Gamma^{r_1}\left(\frac{s}{2}\right) \Gamma^{r_2}(s)}{\Gamma^{r_1}\left(\frac{1-s}{2}\right) \Gamma^{r_2}(1 - s)} x^{-s} ds, \quad c > 0, \, x > 0. \tag{3.177}$$

Exercise 3.8 Prove the formulas for L-function in Table 3.2.

Solution The $L_{1,0}(x)$ is similar to $Y_{0,1}(x)$ and

$$L_{1,0}(x) = \frac{\pi}{2} \cdot 2G_{0,2}^{1,0}\left(x^2 \,\middle|\, \begin{matrix} - \\ 0, \frac{1}{2} \end{matrix} \right) = \pi x^{\frac{1}{2}} J_{\frac{1}{2}}(2x)$$

by (2.110), which further reduces e.g. by (2.27) to the required formula $L_{0,1}(x)$ is just the same as $Y_{2,0}(x)$.

Finally, $L_{2,0}(x)$ may be transformed by (2.93) into

$$L_{2,0}(x) = \frac{\pi}{2} H_{0,4}^{2,0}\left(x \,\middle|\, \begin{matrix} - \\ (0, \frac{1}{2}), (0, \frac{1}{2}), (\frac{1}{2}, \frac{1}{2}), (\frac{1}{2}, \frac{1}{2}) \end{matrix} \right) \tag{3.178}$$
$$= \frac{\pi}{2} 2G_{0,4}^{2,0}\left(x^2 \,\middle|\, \begin{matrix} - \\ 0, 0, \frac{1}{2}, \frac{1}{2} \end{matrix} \right)$$
$$= \pi \left(\frac{2}{\pi} K_0(4\sqrt{x}) - Y_0(4\sqrt{x}) \right)$$

by (9.21) in §9.1.2.

In [KoshE], Koshlyakov mentions:

"It should be noted that deduction of the functional equation for the Dedekind zeta-function from formulas in our general theory is a total vicious circle since the former are already used in deriving the latter."

Such deductions occur

p.129, ll. 1-12 from below

which turns out to be the same as

p.234, ll. 1-10 from below and p.235, ll. 1-16.

Chapter 4

Fourier-Bessel Expansion $H_{1,1}^{1,1} \leftrightarrow H_{0,2}^{2,0}$

The importance of the Hecke gamma transform in number theory cannot be overstated. Along with this, there has been applied equally effective method of the beta transform, or the binomial expansion, leading to the K-Bessel series. The purpose here is to elucidate the explicit and implicit use of the beta transform in the transformation of the perturbed Dirichlet series and reveal the hidden structure as the Fourier-Bessel expansion. As the first instance, we shall clarify Stark's method along with the Murty-Sinha theorem as a manifestation of the beta-transform. As the second and main part, we shall resurrect the long-forgotten important work of Koshlyakov by revealing that Koshlyakov's formula for the perturbed Dedekind zeta-functions for a real quadratic field is in fact Lipschitz summation formula and that for an imaginary quadratic field the K-Bessel function is intrinsic to the Hecke functional equation. Similarly, we shall elucidate Koshlyalov's K-series in terms of Kelvin functions.

4.1 Introduction

We begin by materializing the Murty-Sinha theorem, Theorem 4.1, when the original Dirichlet series are assumed to satisfy the functional equation, as their Fourier-Bessel expansion, Theorem 4.2.

Then we shall elucidate Stark's method for deriving the Dirichlet class number formula and the special values at negative integers of the Hurwitz zeta-function as a manifestation of the beta-transform. A remarkable feature is that there is no need to appeal to the functional equation. However, to derive Lerch's formula (4.15), we need to make a recourse to a formula of Ramanujan and Yoshimoto.

Then we will discuss why the work in [KoshI]-[KoshE] is dominated by the modified K-Bessel function. In particular, we shall elucidate [KoshII, pp.243-247] in this vein and show that Koshlyakov's formula for the perturbed Dedekind zeta-functions for a real quadratic field is in fact Lipschitz summation formula [Lip2] (cf. also [Ler3]). While in the imaginary quadratic case, Proposition 4.3 explains the situation where the K-Bessel function is intrinsic with the Hecke type functional equation (i.e. with the gamma factor $\Gamma(s)$).

In the end we shall elucidate the nature of Koshlyakov's σ-function in terms of the modified Kelvin functions.

4.2   Stark's method

4.2.1   *Murty-Sinha theorem*

One can find in the literature many works devoted to the "analytic continuation by Taylor expansion" of a class of zeta-functions. One direction is to appeal to the binomial expansion due to Wilton [Wil0], e.g. [Mik, (4.6), p.152], [MS], [Sta], etc. and another is to apply the Euler transformation [Gs], [Has], [So] with differencing, which has recently been elucidated in [XHW]. The paper of Murty and Sinha [MS] contains a general theorem which gives an analytic continuation for the perturbed Dirichlet series once an analytic continuation is known for the original Dirichlet series. We state the Murty-Sinha theorem and materialize it in view of the binomial expansion, or the beta transform.

Theorem 4.1   ([MS, Theorem 3.1]) *If the Dirichlet series*

$$\varphi(s) = \sum_{n=1}^{\infty} \frac{\alpha_n}{\lambda_n^s}$$

with abscissa σ_φ of absolute convergence admits a meromorphic continuation to the whole plane, then its perturbation

$$\varphi(s,a) = \sum_{n=0}^{\infty} \frac{\alpha_n}{(\lambda_n + a)^s}$$

also extends meromorphically to the whole plane and the poles of $\varphi(s,a)$ are contained in the positive integral translates of poles of $\varphi(s)$.

The proof of the Murty-Sinha theorem depends on the "binomial principle of analytic continuation" stated as

$$\varphi(s,a) = \sum_{r=0}^{\infty} \binom{-s}{r} \varphi(s+r)a^r \tag{4.1}$$

for $0 < |a| < 1$.

We may interpret (4.1) as the beta-transform (cf. [Vi2, p.34–35]) (which turns out to be an analytic expression for the binomial expansion),

$$\begin{aligned}
(1+x)^{-z} &= \frac{1}{2\pi i} \int_{(c)} \frac{\Gamma(z+s)\,\Gamma(-s)}{\Gamma(z)} x^s \mathrm{d}s \\
&= \frac{1}{\Gamma(z)} G^{1,1}_{1,1}\left(x^{-1} \,\middle|\, \begin{matrix} 1 \\ z \end{matrix} \right)
\end{aligned} \tag{4.2}$$

for $x > 0$, $-\xi < c < 0$, where (c) denotes the vertical Bromwich path $s = c + it$ with $-\infty < t < \infty$. Here $G^{1,1}_{1,1}$ is the Meijer G-function.

This beta-transform has been extensively used in the literature by many authors including Mellin, Barnes, Hardy, Berndt, Koshlyakov, Katsurada, Matsumoto et al., who referred to it as the Mellin-Barnes integral, which turns out to be the beta-function integral $H^{1,1}_{1,1}$, where H indicates the Fox H-function ((1.22)).

Let $\xi = \operatorname{Re} z > 1$, then the Lipschitz-Lerch transcendent is defined by

$$\phi(\lambda, a, z) = \sum_{n=0}^{\infty} \frac{e^{2\pi i n \lambda}}{(n+a)^z} \tag{4.3}$$

where $\lambda \in \mathbb{R}$ and $a \in \mathbb{C} - (\mathbb{Z} \cup \{0\})$, but we usually restrict to $\operatorname{Re} a > 0$. Cf. e.g. [SC, p.122]. The Hurwitz zeta-function is a special case of (4.3) with $\lambda \in \mathbb{Z}$:

$$\zeta(z,a) = \phi(\lambda, a, z) = \sum_{m=0}^{\infty} (m+a)^{-z}$$

for $a > 0$.

We illustrate Theorem 4.1 by

Example 4.1 We apply (4.2) to the Lipschitz-Lerch transcendent and obtain the integral representation

$$\phi(\lambda, a + x, s) = \frac{1}{2\pi i} \int_{(c)} \frac{\Gamma(-w)\Gamma(s+w)}{\Gamma(s)} \phi(\lambda, a, s+w) x^w dw. \qquad (4.4)$$

This is valid for $1 - \sigma < c < 0$.

Moving the line of integration to the right (thereby using Stirling's formula to make sure that the horizontal integrals vanish) up to $\mathrm{Re}\, w = a_K$. Here

$$K < a_K < K + 1, \quad K \in \mathbb{N}.$$

As a result we get

$$\phi(\lambda, a + x, s) = \frac{1}{2\pi i} \int_{(a_K)} \frac{\Gamma(-w)\Gamma(s+w)}{\Gamma(s)} \phi(\lambda, a, s+w)\, x^w dw$$

$$+ \frac{1}{\Gamma(s)} \sum_{k=0}^{K} \Gamma(s+k) \frac{(-1)^k}{k!} \phi(\lambda, a, s+k) x^k. \qquad (4.5)$$

The relation (4.5) holds true for $\sigma > -K$ and the integral is absolutely convergent. Thus it gives an analytic continuation for $\phi(\lambda, a + x, s)$ in that domain. The sum in (4.5) is

$$\sum_{k=0}^{K} \binom{-s}{k} \phi(\lambda, a, s+k) x^k.$$

This is the K-th partial sum of the Taylor series around a.

The Taylor expansion [SC, p.125] (for $|a| > 0$) due to Klusch [Kl1] is:

$$L(x, s, a + \xi) = \sum_{\nu=0}^{\infty} (-1)^\nu \frac{(s)_\nu}{\nu!} L(x, s + \nu, a) \xi^\nu \qquad (4.6)$$

in $|\xi| < |a|$, where $(s)_\nu = s(s+1) \cdots (s+\nu-1)$ denotes the Pochhammber symbol. Now (4.6) with $x \in \mathbb{Z}$ reduces to Wilton's formula [Wil0]

$$\zeta(s, a + \xi) = \sum_{\nu=0}^{\infty} (-1)^\nu \frac{(s)_\nu}{\nu!} \zeta(s + \nu, a)(\xi)^\nu. \qquad (4.7)$$

The relation (4.7) entails the generalized Ramaswami's formula [SC, pp.145-146]:

$$\zeta(s, 2a - 1) - 2^{1-s}\zeta(s, a) = \sum_{\nu=0}^{\infty} \frac{(s)_\nu}{\nu!} \frac{\zeta(s + \nu, a)}{2^{s+\nu}}. \tag{4.8}$$

This reduces to the Jensen-Ramaswami formula [Tit, (2.14.2)]

$$(1 - 2^{1-s})\zeta(s) = \sum_{\nu=0}^{\infty} \frac{(s)_\nu}{\nu!} \frac{\zeta(s + \nu)}{2^{s+\nu}}$$

$$= \sum_{\nu=0}^{\infty} (-1)^\nu \binom{-s}{\nu} \frac{\zeta(s + \nu)}{2^{s+\nu}}$$

in view of the well-known relations

$$\frac{(s)_\nu}{\nu!} = (-1)^\nu \frac{(-s)(-s - 1) \cdots (-s - \nu + 1)}{\nu!}$$

$$= (-1)^\nu \binom{-s}{\nu}.$$

The Taylor series being convergent for $|x| < |a|$, we see that (4.5) leads to a generalized Ramaswami's formula.

Theorem 4.2 *If the original Dirichlet series satisfy the functional equation, then the Murty-Sinha theorem amounts to the correspondence $H_{1,1}^{1,1}$ and its counterpart. Our Theorems 4.4, 4.5 and 4.6 illustrate this principle.*

4.2.2 *Stark's method*

We now move to consider the special case $\zeta(s, x) = \phi(1, x, s)$ of the Hurwitz zeta-function. While applying (4.2), we need to separate the term with $n = 0$ and apply it with $\frac{x}{n}$. Then we obtain, corresponding to (4.4),

$$\zeta(s, x) = \frac{1}{2\pi i} \int_{(c)} \frac{\Gamma(-w)\Gamma(s + w)}{\Gamma(s)} \zeta(s + w) x^w \mathrm{d}w \tag{4.9}$$

valid for $1 - c < \sigma$ with $c < 0$. In the same way, we obtain corresponding to (4.5),

$$\zeta(s, x) = x^{-s} + \frac{1}{2\pi i} \int_{(a_K)} \frac{\Gamma(-w)\Gamma(s + w)}{\Gamma(s)} \zeta(s + w) x^w dw$$

$$+ \sum_{k=0}^{K} \binom{-s}{k} \zeta(s + k) x^k \tag{4.10}$$

which is valid for $\sigma > -K$ and the integral is absolutely convergent and thus gives an analytic continuation for $\zeta(s, x)$ in that domain. If we move the line of integration far to the right as in the proof of Proposition 4.2, we obtain the Taylor expansion (4.7) with $a = 1$.

H. Stark [Sta] uses only first a few terms to deduce the Dirichlet class number formula and the special values of the relevant zeta and L-functions. The Dirichlet L-function has the representation in terms of the Hurwitz zeta-function

$$L(s, \chi) = \frac{1}{q^s} \sum_{a=1}^{q-1} \chi(a) \zeta\left(s, \frac{a}{q}\right) \tag{4.11}$$

where χ is a Dirichlet character mod q. Thus it suffices to give closed formulas for the latter. Moving further and comparing [Sta, (13)] and (4.5), we obtain

$$\sum_{n=1}^{\infty} \left((n + x)^{-s} - \sum_{j=0}^{K} \binom{-s}{k} n^{-n-j} x^j \right) \tag{4.12}$$

$$= \frac{1}{2\pi i} \int_{(a_K)} \frac{\Gamma(-w)\Gamma(s + w)}{\Gamma(s)} \zeta(s + w) x^w dw.$$

If one follows Stark's method more closely and write $K + 1$ for K in (4.5) and set $s = -K$, $0 \leq K \in \mathbb{Z}$ in (4.5) one can deduce

$$\zeta(-K, x) = x^K + \sum_{k=0}^{K+1} \binom{K}{k} \zeta(-K + k) x^k \tag{4.13}$$

with the understanding that the term with $k = K + 1$ is

$$\binom{-s}{K+1}\zeta(-s+K+1)\bigg|_{s=-K}x^{K+1}$$

$$= \frac{(-s)\cdots(-s-K)\zeta(-s+K+1)\big|_{s=-K}}{(K+1)!}x^{K+1} = \frac{x^{K+1}}{K+1}$$

on using the fact that $\operatorname{Res}\zeta(s) = 1$ and the term with $k = K$ is $\zeta(0)x^K$.

Proposition 4.1 *We have for any* $0 \leq K \in \mathbb{Z}$,

$$\zeta(-K, x) = \frac{x^{K+1}}{K+1} + \left(1 + \zeta(0)\right)x^K + \sum_{k=0}^{K-1}\binom{K}{k}\zeta(-K+k)\,x^k. \quad (4.14)$$

In particular, if we use $\zeta(0) = -\frac{1}{2}$, *then*

$$\zeta(0, x) = \frac{1}{2} - x = -B_1(x)$$

where $B_1(x)$ *is the first Bernoulli polynomial (cf. (3.72)).*

Proof. Putting $s = -K$ in (4.10) and noting that $\frac{1}{\Gamma(-K)} = 0$ for $0 \leq K \in \mathbb{Z}$, the proof follows. $\qquad\square$

If we use $\zeta(-n) = \frac{B_{n+1}}{n+1}$ for $0 \leq n \in \mathbb{Z}$, we have (from (4.14)),

$$\zeta(-K, x) = -\frac{B_{K+1}(x)}{K+1}.$$

Now we give a proof of Lerch's formula by using the well known facts that $\zeta'(0) = -\log\sqrt{2\pi}$ and

$$\zeta(s) = \frac{1}{s-1} - \gamma + O((s-1)).$$

Proposition 4.2 (Lerch's formula)

$$\zeta'(0, x) = \log\frac{\Gamma(x)}{\sqrt{2\pi}}. \quad (4.15)$$

Proof. Stark's method of differentiating the left-hand side of (4.12) and appealing to the Weierstrass product for the gamma function may in fact be simpler.

We use the integral representation (4.10). Putting $K = 1$ it becomes

$$\zeta(s, x) = x^{-s} + \frac{1}{2\pi i} \int_{(a_K)} \frac{\Gamma(-w)\Gamma(s+w)}{\Gamma(s)} \zeta(s+w)x^w dw$$

$$+ \zeta(s) - s\zeta(s+1)x. \tag{4.16}$$

Now differentiating and putting $s = 0$ in (4.16), we find as in Stark that those terms except for the integral, say I, becomes

$$\zeta'(0, x) = \log x + \zeta'(0) + \gamma x + \frac{d}{dt}I\Big|_{s=0}. \tag{4.17}$$

Clearly, only the term with $-\frac{\Gamma'(s)}{\Gamma^2(s)} = -\frac{s\Gamma'(s)-\Gamma(s+1)}{\Gamma^2(s+1)}$ counts, other terms vanish in view of $\frac{1}{\Gamma(0)} = 0$. Therefore we only need to evaluate

$$\frac{d}{dt}I\Big|_{s=0} = \frac{1}{2\pi i} \int_{(a_K)} \Gamma(-w)\Gamma(w)\zeta(w)x^w dw$$

$$= \sum_{n=1}^{\infty} \frac{1}{2\pi i} \int_{(a_K)} \frac{\pi}{w \sin \pi w} \left(\frac{x}{n}\right)^w dw$$

with $1 < a_K < 2$. Moving the line of integration far to the left, counting the residues and noting that the resulting integral tends to 0, we conclude that

$$\frac{d}{dt}I\Big|_{s=0} = \sum_{n=1}^{\infty} \sum_{k=2}^{\infty} \frac{1}{k}\left(-\frac{x}{n}\right)^k$$

$$= \sum_{k=2}^{\infty} \frac{\zeta(k)}{k}(-x)^k. \tag{4.18}$$

Now we recall the formula [Vista, Theorem 3.2, p.62] (also [SC, (2), p.159]) of Ramanujan and Yoshimoto with $\alpha = 1$, which is

$$\sum_{k=2}^{\infty} \frac{\zeta(k)}{k}(-x)^k = \log \Gamma(x+1) + \gamma x, \quad |x| < 1.$$

Thus (4.18) leads to

$$\frac{d}{dt}I\Big|_{s=0} = \log \Gamma(x+1) - \gamma x.$$

Now substituting this in (4.17) and appealing to the difference equation $\log \Gamma(x+1) = \log \Gamma(x) + \log x$ for $x > 0$, we conclude (4.15). $\qquad\square$

4.3 The main formula for modular relations

In this section, we state Theorem 4.3, the concrete version of our main theorem, Theorem 7.1 and give a number of applications. This will be applied in subsequent chapters.

We have two sets of Dirichlet series

$$\{\phi_h(s)\} \quad (1 \leq h \leq H) \quad \text{and} \quad \{\psi_i(s)\} \quad (1 \leq i \leq I)$$

that satisfy the generalized functional equation (4.19) in the following sense.

Let $\{\lambda_k^{(h)}\}_{k=1}^\infty$, $\{\mu_k^{(i)}\}_{k=1}^\infty$ denote sequences and $\{\alpha_k^{(h)}\}_{k=1}^\infty$, $\{\beta_k^{(i)}\}_{k=1}^\infty$ are complex sequences. Consider Dirichlet series

$$\varphi_h(s) = \sum_{k=1}^\infty \frac{\alpha_k^{(h)}}{\lambda_k^{(h)s}}$$

and

$$\psi_i(s) = \sum_{k=1}^\infty \frac{\beta_k^{(i)}}{\mu_k^{(i)s}},$$

with finite abscissa of absolute convergence σ_{φ_h} and σ_{ψ_i} respectively.

Let there exist a meromorphic function $\chi(s)$, which satisfies (for a real number r) the functional equation:

$$\chi(s) \tag{4.19}$$

$$= \begin{cases} \displaystyle\sum_{h=1}^{H} \frac{\displaystyle\prod_{j=1}^{M^{(h)}} \Gamma\left(d_j^{(h)} + D_j^{(h)} s\right) \prod_{j=1}^{N^{(h)}} \Gamma\left(c_j^{(h)} - C_j^{(h)} s\right)}{\displaystyle\prod_{j=N^{(h)}+1}^{P^{(h)}} \Gamma\left(c_j^{(h)} + C_j^{(h)} s\right) \prod_{j=M^{(h)}+1}^{Q^{(h)}} \Gamma\left(d_j^{(h)} - D_j^{(h)} s\right)} \varphi_h(s), \\ \qquad\qquad\qquad\qquad\qquad\qquad \mathrm{Re}(s) > \displaystyle\max_{1 \leq h \leq H}(\sigma_{\varphi_h}) \\[2em] \displaystyle\sum_{i=1}^{I} \frac{\displaystyle\prod_{j=1}^{\tilde{N}^{(i)}} \Gamma\left(e_j^{(i)} + E_j^{(i)}(r-s)\right) \prod_{j=1}^{\tilde{M}^{(i)}} \Gamma\left(f_j^{(i)} - F_j^{(i)}(r-s)\right)}{\displaystyle\prod_{j=\tilde{M}^{(i)}+1}^{\tilde{Q}^{(i)}} \Gamma\left(f_j^{(i)} + F_j^{(i)}(r-s)\right) \prod_{j=\tilde{N}^{(i)}+1}^{\tilde{P}^{(i)}} \Gamma\left(e_j^{(i)} - E_j^{(i)}(r-s)\right)} \psi_i(r-s), \\ \qquad\qquad\qquad\qquad\qquad\qquad \mathrm{Re}(s) < \displaystyle\min_{1 \leq i \leq I}(r - \sigma_{\psi_i}) \end{cases}$$

$$\left(C_j^{(h)}, D_j^{(h)}, E_j^{(i)}, F_j^{(i)} > 0\right).$$

We further assume that only finitely many of the poles s_k $(1 \leq k \leq L)$ of $\chi(s)$ are neither a pole of

$$\frac{\prod_{j=1}^{N^{(h)}} \Gamma\left(c_j^{(h)} - C_j^{(h)} s\right)}{\prod_{j=N^{(h)}+1}^{P^{(h)}} \Gamma\left(c_j^{(h)} + C_j^{(h)} s\right) \prod_{j=M^{(h)}+1}^{Q^{(h)}} \Gamma\left(d_j^{(h)} - D_j^{(h)} s\right)}$$

nor a pole of

$$\frac{\prod_{j=1}^{\tilde{M}^{(i)}} \Gamma\left(f_j^{(i)} - F_j^{(i)} r + F_j^{(i)} s\right)}{\prod_{j=\tilde{N}^{(i)}+1}^{\tilde{P}^{(i)}} \Gamma\left(e_j^{(i)} - E_j^{(i)} r + E_j^{(i)} s\right) \prod_{j=\tilde{M}^{(i)}+1}^{\tilde{Q}^{(i)}} \Gamma\left(f_j^{(i)} + F_j^{(i)} r - F_j^{(i)} s\right)}.$$

We suppose that for any real numbers u_1, u_2 $(u_1 < u_2)$

$$\lim_{|v| \to \infty} \Gamma(u + iv - s \,|\, \Delta)\, \chi(u + iv) = 0 \tag{4.20}$$

uniformly in $u_1 \leq u \leq u_2$.

We choose $L_1(s)$ so that the poles of

$$\frac{\prod_{j=1}^{n} \Gamma(a_j + A_j s - A_j w) \prod_{j=1}^{N^{(h)}} \Gamma\left(c_j^{(h)} - C_j^{(h)} w\right)}{\prod_{j=n+1}^{p} \Gamma(a_j - A_j s + A_j w) \prod_{j=N^{(h)}+1}^{P^{(h)}} \Gamma\left(c_j^{(h)} + C_j^{(h)} w\right)}$$

$$\times \frac{1}{\prod_{j=m+1}^{q} \Gamma(b_j + B_j s - B_j w) \prod_{j=M^{(h)}+1}^{Q^{(h)}} \Gamma\left(d_j^{(h)} - D_j^{(h)} w\right)}$$

lie on the right of $L_1(s)$, and those of

$$\frac{\prod_{j=1}^{m} \Gamma(b_j - B_j s + B_j w) \prod_{j=1}^{M^{(h)}} \Gamma\left(d_j^{(h)} + D_j^{(h)} w\right)}{\prod_{j=m+1}^{q} \Gamma(b_j + B_j s - B_j w) \prod_{j=M^{(h)}+1}^{Q^{(h)}} \Gamma\left(d_j^{(h)} - D_j^{(h)} w\right)}$$

$$\times \; \frac{1}{\displaystyle\prod_{j=n+1}^{p} \Gamma(a_j - A_j s + A_j w) \prod_{j=N^{(h)}+1}^{P^{(h)}} \Gamma\left(c_j^{(h)} + C_j^{(h)} w\right)}$$

lie on the left of $L_1(s)$, and choose $L_2(s)$ so that the poles of

$$\frac{\displaystyle\prod_{j=1}^{m} \Gamma(b_j - B_j s + B_j w) \prod_{j=1}^{\tilde{M}^{(i)}} \Gamma\left(f_j^{(i)} - F_j^{(i)} r + F_j^{(i)} w\right)}{\displaystyle\prod_{j=m+1}^{q} \Gamma(b_j + B_j s - B_j w) \prod_{j=\tilde{M}^{(i)}+1}^{\tilde{Q}^{(i)}} \Gamma\left(f_j^{(i)} + F_j^{(i)} r - F_j^{(i)} w\right)}$$

$$\times \; \frac{1}{\displaystyle\prod_{j=n+1}^{p} \Gamma(a_j - A_j s + A_j w) \prod_{j=\tilde{N}^{(i)}+1}^{\tilde{P}^{(i)}} \Gamma\left(e_j^{(i)} - E_j^{(i)} r + E_j^{(i)} w\right)}$$

lie on the left of $L_2(s)$, and those of

$$\frac{\displaystyle\prod_{j=1}^{n} \Gamma(a_j + A_j s - A_j w) \prod_{j=1}^{\tilde{N}^{(i)}} \Gamma\left(e_j^{(i)} + E_j^{(i)} r - E_j^{(i)} w\right)}{\displaystyle\prod_{j=n+1}^{p} \Gamma(a_j - A_j s + A_j w) \prod_{j=\tilde{N}^{(i)}+1}^{\tilde{P}^{(i)}} \Gamma\left(e_j^{(i)} - E_j^{(i)} r + E_j^{(i)} w\right)}$$

$$\times \; \frac{1}{\displaystyle\prod_{j=m+1}^{q} \Gamma(b_j + B_j s - B_j w) \prod_{j=\tilde{M}^{(i)}+1}^{\tilde{Q}^{(i)}} \Gamma\left(f_j^{(i)} + F_j^{(i)} r - F_j^{(i)} w\right)}$$

lie on the right of $L_2(s)$. Further, they squeeze a compact set $\mathcal{S}$ such that $s_k \in \mathcal{S}(1 \le k \le L)$.

Under these conditions the χ-function, **key-function**, $\mathrm{X}(s, z \,|\, \Delta)$ is defined by (3.2) above.

Then we have the following modular relation (cf. [Ts]) which is a traditional version of our main Theorem 7.1.

Theorem 4.3

$$X(s, z \mid \Delta) \tag{4.21}$$

$$
=
\begin{cases}
\displaystyle\sum_{h=1}^{H}\sum_{k=1}^{\infty} \frac{\alpha_k^{(h)}}{\lambda_k^{(h)s}} H_{p+P^{(h)},q+Q^{(h)}}^{m+M^{(h)},n+N^{(h)}}\left(z\lambda_k^{(h)} \left|\begin{array}{c} \end{array}\right.\right. \\
\qquad \{(1-a_j,A_j)\}_{j=1}^{n}, \{(1-c_j^{(h)}+C_j^{(h)}s,C_j^{(h)})\}_{j=1}^{N^{(h)}}, \\
\qquad \{(b_j,B_j)\}_{j=1}^{m}, \{(d_j^{(h)}+D_j^{(h)}s,D_j^{(h)})\}_{j=1}^{M^{(h)}}, \\
\qquad \{(a_j,A_j)\}_{j=n+1}^{p}, \{(c_j^{(h)}+C_j^{(h)}s,C_j^{(h)})\}_{j=N^{(h)}+1}^{P^{(h)}} \\
\qquad \{(1-b_j,B_j)\}_{j=m+1}^{q}, \{(1-d_j^{(h)}+D_j^{(h)}s,D_j^{(h)})\}_{j=M^{(h)}+1}^{Q^{(h)}} \Bigg) \\
\qquad\qquad \text{if } L_1(s) \text{ can be taken to the right of } \max_{1\le h\le H}(\sigma_{\varphi_h}) \\[2em]
\displaystyle\sum_{i=1}^{I}\sum_{k=1}^{\infty} \frac{\beta_k^{(i)}}{\mu_k^{(i)r-s}} H_{q+\tilde{Q}^{(i)},p+\tilde{P}^{(i)}}^{n+\tilde{N}^{(i)},m+\tilde{M}^{(i)}}\left(\frac{\mu_k^{(i)}}{z} \left|\begin{array}{c} \end{array}\right.\right. \\
\qquad \{(1-b_j,B_j)\}_{j=1}^{m}, \{(1-f_j^{(i)}+F_j^{(i)}(r-s),F_j^{(i)})\}_{j=1}^{\tilde{M}^{(i)}}, \\
\qquad \{(a_j,A_j)\}_{j=1}^{n}, \{(e_j^{(i)}+E_j^{(i)}(r-s),E_j^{(i)})\}_{j=1}^{\tilde{N}^{(i)}}, \\
\qquad \{(b_j,B_j)\}_{j=m+1}^{q}, \{(f_j^{(i)}+F_j^{(i)}(r-s),F_j^{(i)})\}_{j=\tilde{M}^{(i)}+1}^{\tilde{Q}^{(i)}} \\
\qquad \{(1-a_j,A_j)\}_{j=n+1}^{p}, \{(1-e_j^{(i)}+E_j^{(i)}(r-s),E_j^{(i)})\}_{j=\tilde{N}^{(i)}+1}^{\tilde{P}^{(i)}} \Bigg) \\
\qquad +\displaystyle\sum_{k=1}^{L} \text{Res}\Big(\Gamma(w-s\mid\Delta)\chi(w)\,z^{s-w}, w=s_k\Big) \\
\qquad\qquad \text{if } L_2(s) \text{ can be taken to the left of } \min_{1\le i\le I}(r-\sigma_{\psi_i}),
\end{cases}
$$

is equivalent to the functional equation (4.19).

We now apply the modular relation (4.21) in the following special case.

Table 4.1   Choice of parameters:

values	proc.gamma	LHSFE	RHSFE	values
$m+M^{(h)}=1$	$m=0$	$M^{(h)}=1$	$M^{(i)}=0$	$m+\tilde{M}^{(i)}=0$
$n+N^{(h)}=1$	$n=1$	$N^{(h)}=0$	$N^{(i)}=1$	$n+\tilde{N}^{(i)}=2$
$p+P^{(h)}=1$	$p=1$	$P^{(h)}=0$	$P^{(i)}=1$	$p+\tilde{P}^{(i)}=2$
$q+Q^{(h)}=1$	$q=0$	$Q^{(h)}=1$	$Q^{(i)}=0$	$q+\tilde{Q}^{(i)}=0$

Theorem 4.4   *The functional equation*

$$\Gamma(d_1+D_1 s)\varphi(s)=\Gamma(e_1+E_1(r-s))\psi(r-s), \tag{4.22}$$

with $\Gamma(s|\Delta) = \Gamma(a_1 - A_1 s)$, *is equivalent to*

$$X(s, z \,|\, \Delta) \tag{4.23}$$

$$= \begin{cases} \displaystyle\sum_{k=1}^{\infty} \frac{\alpha_k}{\lambda_k^s} H^{1,1}_{1,1}\left(z\lambda_k \,\middle|\, \begin{array}{l} (1 - a_1, A_1) \\ (d_1 + D_1 s, D_1) \end{array}\right) \\ \qquad\qquad\qquad if\ L_1(s)\ can\ be\ taken\ to\ the\ right\ of\ \displaystyle\max_{1 \le h \le H}(\sigma_{\varphi_h}) \\[2mm] \displaystyle\sum_{k=1}^{\infty} \frac{\beta_k}{\mu_k^{r-s}} H^{2,0}_{0,2}\left(\frac{\mu_k}{z} \,\middle|\, \begin{array}{l} - \\ (a_1, A_1), (e_1 + E_1(r - s), E_1) \end{array}\right) \\ \qquad + \displaystyle\sum_{k=1}^{L} \mathrm{Res}\Big(\Gamma(a_1 - A_1(w - s))\chi(w)\, z^{s-w}, w = s_k\Big) \\ \qquad\qquad\qquad if\ L_2(s)\ can\ be\ taken\ to\ the\ left\ of\ \displaystyle\min_{1 \le i \le I}(r - \sigma_{\psi_i}). \end{cases}$$

4.3.1   *Specification of Theorem 4.4*

We are ready to specify (4.23) with $D_1 = E_1 = A_1$. The right-hand side H-function of the modular relation (4.23) becomes

$$H^{1,1}_{1,1}\left(z\lambda_k \,\middle|\, \begin{array}{l} (1 - a_1, A_1) \\ (d_1 + A_1 s, A_1) \end{array}\right) = A_1^{-1} G^{1,1}_{1,1}\left((z\lambda_k)^{\frac{1}{A_1}} \,\middle|\, \begin{array}{l} 1 - a_1 \\ d_1 + A_1 s \end{array}\right)$$

$$= A_1^{-1}\Gamma(1 - a_1 + d_1 + A_1 s)\lambda_k^s (z\lambda_k)^{\frac{d_1}{A_1}}\left((z\lambda_k)^{\frac{1}{A_1}} + 1\right)^{-a_1 - d_1} \tag{4.24}$$

$$\times \frac{1}{\left(\lambda_k^{\frac{1}{A_1}} + (z^{-1})^{\frac{1}{A_1}}\right)^s}.$$

We write z for z^{-1} and for simplicity we choose $a_1 = d_1 = 0$ to reduce (4.24) into

$$H^{1,1}_{1,1}\left(z^{-1}\lambda_k \,\middle|\, \begin{array}{l} (1, A_1) \\ (A_1 s, A_1) \end{array}\right)$$

$$= A_1^{-1}\Gamma(1 + A_1 s)\lambda_k^s \frac{1}{\left(\lambda_k^{\frac{1}{A_1}} + z^{\frac{1}{A_1}}\right)^s}. \tag{4.25}$$

Similarly the series on right-hand side of (4.24) becomes for $z \to z^{-1}$ and $E_1 = A_1$:

$$H_{0,2}^{2,0}\left(\mu_k z \left|\begin{array}{c} - \\ (a_1, A_1), (e_1 + A_1(r-s), A_1) \end{array}\right.\right) \tag{4.26}$$

$$= A_1^{-1} G_{0,2}^{2,0}\left((\mu_k z)^{\frac{1}{A_1}} \left|\begin{array}{c} - \\ a_1, e_1 + A_1(r-s) \end{array}\right.\right)$$

$$= 2A_1^{-1}(\mu_k z)^{\frac{1}{2}(a_1 + e_1 + A_1(r-s))} K_{a_1 - e_1 - A_1(r-s)}\left(2(\mu_k z)^{\frac{1}{2A_1}}\right).$$

Further specifying $a_1 = d_1 = e_1 = 0$, we deduce the following corollary.

Corollary 4.1   *The functional equation*

$$\Gamma(A_1 s)\varphi(s) = \Gamma(A_1(r-s))\psi(r-s) \tag{4.27}$$

is equivalent to (with $\Gamma(w|\Delta)) = \Gamma(-A_1 w)$*)*

$$X(s, z^{-1}|\Delta) = \frac{1}{A_1}\Gamma(A_1 s)\sum_{k=1}^{\infty} \frac{\alpha_k}{\left(\lambda_k^{\frac{1}{A_1}} + z^{\frac{1}{A_1}}\right)^{A_1 s}}$$

$$= 2A_1^{-1} z^{\frac{1}{2}A_1(r-s)} \sum_{k=1}^{\infty} \frac{\beta_k}{\mu_k^{\frac{1}{2}A_1(r-s)}} K_{A_1(r-s)}\left(2(\mu_k z)^{\frac{1}{2A_1}}\right)$$

$$+ \sum_{k=1}^{L} \text{Res}\left(\Gamma(A_1(s-w))\Gamma(A_1 s)\varphi(s) z^{w-s}, w = s_k\right). \tag{4.28}$$

Example 4.2   A special case of Corollary 4.1 is the Dirichlet series satisfying the Riemann type functional equation with $A_1 = \frac{1}{2}$, i.e.,

$$\Gamma\left(\frac{1}{2}s\right)\varphi(s) = \Gamma\left(\frac{1}{2}(r-s)\right)\psi(r-s) \tag{4.29}$$

with a simple pole at $s = r > 0$ with residue ρ. Thus from (4.28) we have

$$2\Gamma\left(\frac{1}{2}s\right)\sum_{k=1}^{\infty} \frac{\alpha_k}{(\lambda_k^2 + z^2)^{\frac{1}{2}s}} = 4z^{\frac{1}{2}(r-s)}\sum_{k=1}^{\infty} \frac{\beta_k}{\mu_k^{\frac{1}{2}(r-s)}} K_{\frac{1}{2}(r-s)}(2\mu_k z)$$

$$+ \sum_{k=1}^{L} \text{Res}\left(\Gamma\left(\frac{1}{2}(s-w)\right)\Gamma\left(\frac{1}{2}w\right)\varphi(w) z^{w-s}, w = 0, r\right)$$

or

$$2\Gamma\left(\frac{1}{2}s\right)\sum_{k=1}^{\infty}\frac{\alpha_k}{(\lambda_k^2+z^2)^{\frac{1}{2}s}}$$

$$=4z^{\frac{1}{2}(r-s)}\sum_{k=1}^{\infty}\frac{\beta_k}{\mu_k^{\frac{1}{2}(r-s)}}K_{\frac{1}{2}(r-s)}(2\mu_k z)+2\Gamma\left(\frac{1}{2}s\right)\varphi(0)\,z^{-s}$$

$$+\rho\Gamma\left(\frac{1}{2}(s-r)\right)z^{r-s}. \tag{4.30}$$

In the case of the Riemann zeta-function, $\alpha_k = \beta_k = 1$, $\lambda_k = \sqrt{\pi}k$, $\mu_k = \sqrt{\pi}k$, $r = 1$ and $\rho = 1$, so that it satisfies (4.29) with $\varphi(s) = \psi(s) = \pi^{-\frac{s}{2}}\zeta(s)$. Hence expressing the left-hand side of (4.30) as the sum over all integers, we have

$$\Gamma\left(\frac{1}{2}s\right)\sum_{0\neq k\in\mathbb{Z}}\frac{1}{(\pi k^2+z^2)^{\frac{1}{2}s}} \tag{4.31}$$

$$=4z^{\frac{1}{4}(1-s)}\sum_{k=1}^{\infty}\frac{1}{k^{\frac{1}{2}(1-s)}}K_{\frac{1}{2}(1-s)}(2\sqrt{\pi}kz)+2\Gamma\left(\frac{1}{2}s\right)\zeta(0)\,z^{-s}.$$

That is,

$$\sum_{k\in\mathbb{Z}}\frac{1}{(\pi k^2+z^2)^{\frac{1}{2}s}}$$

$$=4\frac{z^{\frac{1}{4}(1-s)}}{\Gamma\left(\frac{1}{2}s\right)}\sum_{k=1}^{\infty}\frac{1}{k^{\frac{1}{2}(1-s)}}K_{\frac{1}{2}(1-s)}(2\sqrt{\pi}kz), \tag{4.32}$$

i.e. Watson's formula again. Cf. (1.28) and (1.25) above.

Example 4.3 Another more special case of Corollary 4.1 is the Dirichlet series satisfying the Hecke type functional equation with $A_1 = 1$, i.e.,

$$\Gamma(s)\,\varphi(s) = \Gamma(r-s)\,\psi(r-s) \tag{4.33}$$

with a simple pole at $s = r$ with residue ρ. In this case (4.28) gives

$$\Gamma(s)\sum_{k=1}^{\infty}\frac{\alpha_k}{(\lambda_k+z)^s} = 2z^{\frac{1}{2}(r-s)}\sum_{k=1}^{\infty}\frac{\beta_k}{\mu_k^{\frac{1}{2}(r-s)}}K_{r-s}\left(2\sqrt{\mu_k z}\right)$$

$$+\operatorname{Res}\left(\Gamma(s-w)\chi(w)\,z^{w-s},w=r\right),$$

which becomes

$$\Gamma(s) \sum_{k=1}^{\infty} \frac{\alpha_k}{(\lambda_k + z)^s} \tag{4.34}$$

$$= 2z^{\frac{1}{2}(r-s)} \sum_{k=1}^{\infty} \frac{\beta_k}{\mu_k^{\frac{1}{2}(r-s)}} K_{r-s}\left(2\sqrt{\mu_k}\right) + \rho\Gamma(s - r)\, z^{r-s}.$$

4.4 Dedekind zeta-function III

In this section we specify the above results to the Dedekind zeta-function of a number field Ω with degree $\varkappa = r_1 + 2r_2 \leq 2$ and discriminant Δ. For standard notation we refer to §3.4.1.

We recall the functional equation (3.1) as

$$\Gamma^{r_1}\left(\frac{1}{2}s\right)\Gamma^{r_2}(s)\varphi(s) = \Gamma^{r_1}\left(\frac{1}{2} - \frac{1}{2}s\right)\Gamma^{r_2}(1 - s)\varphi(1 - s) \tag{4.35}$$

with the coefficient A defined by (3.35).

Now Corollary 4.1 with $A_1 = 1$ gives

$$\Gamma(s) \sum_{k=1}^{\infty} \frac{\alpha(k)}{(Ak + z)^s}$$

$$= 2z^{\frac{1}{2}(1-s)} \sum_{k=1}^{\infty} \frac{\alpha(k)}{(Ak)^{\frac{1}{2}(1-s)}} K_{1-s}\left(2\sqrt{Akz}\right)$$

$$+ \Gamma(s)\varphi(0)\, z^{-s} + \rho\Gamma(s - 1)\, z^{1-s}. \tag{4.36}$$

Now if we replace $\frac{z}{A} = \omega$ with $\operatorname{Re}\omega > 0$, we get

$$A^{-s} \sum_{k=1}^{\infty} \frac{\alpha(k)}{(k + \omega)^s}$$

$$= \frac{2}{\Gamma(s)} \omega^{\frac{1}{2}(1-s)} \sum_{k=1}^{\infty} \frac{\alpha(k)}{k^{\frac{1}{2}(1-s)}} K_{1-s}\left(2A\sqrt{k\omega}\right) + A^{-s}\varphi(0)\,\omega^{-s}$$

$$+ A^{1-s}\rho\frac{\Gamma(s - 1)}{\Gamma(s)}\,\omega^{1-s},$$

which leads to

$$\sum_{k=1}^{\infty} \frac{\alpha(k)}{(k+\omega)^s} \tag{4.37}$$

$$= A^s \frac{2}{\Gamma(s)} \omega^{\frac{1}{2}(1-s)} \sum_{k=1}^{\infty} \frac{\alpha(k)}{k^{\frac{1}{2}(1-s)}} K_{1-s}\left(2A\sqrt{k\omega}\right) + \varphi(0)\,\omega^{-s} + A\rho\,\omega^{1-s}\frac{1}{s-1}.$$

Therefore in the case of an imaginary quadratic field one gets from (4.37) the following result.

Proposition 4.3 *The functional equation (4.33) with $r = 1$ is equivalent to the Fourier-Bessel series*

$$\sum_{k=1}^{\infty} \frac{\alpha(k)}{(k+\omega)^s}$$

$$= A^s \frac{2}{\Gamma(s)} \omega^{\frac{1}{2}(1-s)} \sum_{k=1}^{\infty} \frac{\alpha(k)}{k^{\frac{1}{2}(1-s)}} K_{1-s}\left(2A\sqrt{k\omega}\right)$$

$$+ \zeta_\Omega(0)\,\omega^{-s} + A\rho\,\omega^{1-s}\frac{1}{s-1}. \tag{4.38}$$

We shall now show that Proposition 4.3 gives the Lipschitz summation formula as in Koshlyakov [KoshII, p.245-246], where he takes a weighted sum of $\zeta_\Omega(s)$ (one reason may be to cancel the pole at $s = 1$). The essential reason being discussed in more detail in §4.5. For a character analogue of the Lipschitz summation formula, (cf. Berndt [Ber75-2]), as in [KoshII, p.243], we consider for $0 < \mathrm{Re}\,\omega < 1$ (in [KoshII, p.243], the limiting case $\mathrm{Re}\,\omega = 0$, i.e. $\omega = iw$, $0 < w < 1$ is considered)

$$\frac{1}{2} e^{\frac{\pi i s}{2}} \sum_{k=1}^{\infty} \frac{\alpha(k)}{(k+iw)^s} + \frac{1}{2} e^{-\frac{\pi i s}{2}} \sum_{k=1}^{\infty} \frac{\alpha(k)}{(k-iw)^s}$$

$$= \frac{1}{2} A^s \frac{2}{\Gamma(s)} \sum_{k=1}^{\infty} \frac{\alpha(k)}{k^{\frac{1}{2}(1-s)}} \tag{4.39}$$

$$\times \left(e^{\frac{\pi i s}{2}} (iw)^{\frac{1}{2}(1-s)} K_{1-s}\left(2A\sqrt{ikw}\right) + e^{\frac{-\pi i s}{2}} (-iw)^{\frac{1}{2}(1-s)} K_{1-s}\left(2A\sqrt{-ikw}\right) \right)$$

$$+ \zeta_\Omega(0)\,w^{-s}.$$

If one writes $\varepsilon = e^{\frac{\pi i}{4}}$, then the right-hand side of (4.39) becomes

$$= \frac{1}{2}A^s \frac{2w^{\frac{1}{2}(1-s)}}{\Gamma(s)} \sum_{k=1}^{\infty} \frac{\alpha(k)}{k^{\frac{1}{2}(1-s)}}$$

$$\left(\varepsilon^{s+1} K_{1-s}\left(2A\varepsilon\sqrt{kw}\right) + \bar{\varepsilon}^{s+1} K_{1-s}\left(2A\bar{\varepsilon}\sqrt{kw}\right) \right)$$

$$+ \frac{\zeta_\Omega(0)}{w^s}. \tag{4.40}$$

At this point, in conformity with Koshlyakov [KoshII, p.241] (cf. (3.40)), we introduce the perturbed Dedekind zeta-function

$$\zeta_\Omega(s,\omega) = -\frac{2\zeta_\Omega(0)}{w} + \sum_{k=1}^{\infty} \frac{\alpha(k)}{(k+\omega)^s}. \tag{4.41}$$

Then (4.39) and (4.40) give the Lipschitz summation formula

$$\sum_{n=-\infty}^{\infty}{}' \frac{\alpha(n)}{(w+in)^s} = e^{\frac{\pi i}{2}s}\zeta_\Omega(s,iw) + e^{-\frac{\pi i}{2}s}\zeta_\Omega(s,-iw) \tag{4.42}$$

$$= A^s \frac{w^{\frac{1}{2}(1-s)}}{\Gamma(s)} \sum_{k=1}^{\infty} \frac{\alpha(k)}{k^{\frac{1}{2}(1-s)}} \left(\varepsilon^{s+1} K_{1-s}\left(2A\varepsilon\sqrt{kw}\right) + \bar{\varepsilon}^{s+1} K_{1-s}\left(2A\bar{\varepsilon}\sqrt{kw}\right) \right)$$

$$- \frac{\zeta_\Omega(0)}{w^s}.$$

This is the corrected form of [KoshII, (23.15)] and here the prime on the summation sign means that the term corresponding to $n = 0$ is excluded.

The Dedekind zeta-function of a real quadratic field satisfies the functional equation (3.41) with $r_1 = 2, r_2 = 0$:

$$\Gamma^2\left(\frac{1}{2}s\right)\varphi(s) = \Gamma^2\left(\frac{1}{2} - \frac{1}{2}s\right)\varphi(1-s) \tag{4.43}$$

where $\varphi(s)$ is defined by (3.40).

In this form, there would be no proper modular relation with $H_{1,1}^{1,1}$ unless we combine the perturbed Dirichlet series, which we shall state in §4.5.

Here we view (4.43) to mean

$$\frac{\Gamma\left(\frac{1}{2}s\right)}{\Gamma\left(\frac{1-s}{2}s\right)}\varphi(s) = \frac{\Gamma\left(\frac{1-s}{2}s\right)}{\Gamma\left(\frac{1-(1-s)}{2}s\right)}\varphi(1-s). \tag{4.44}$$

As before we choose the values of the parameters as in the following:

Table 4.2   Choice of parameters

values	proc.gamma	LHSFE	RHSFE	values
$m + M^{(h)} = 1$	$m = 0$	$M^{(h)} = 1$	$M^{(i)} = 0$	$m + \tilde{M}^{(i)} = 0$
$n + N^{(h)} = 1$	$n = 1$	$N^{(h)} = 0$	$N^{(i)} = 1$	$n + \tilde{N}^{(i)} = 2$
$p + P^{(h)} = 1$	$p = 1$	$P^{(h)} = 0$	$P^{(i)} = 2$	$p + \tilde{P}^{(i)} = 3$
$q + Q^{(h)} = 1$	$q = 0$	$Q^{(h)} = 1$	$Q^{(i)} = 0$	$q + \tilde{Q}^{(i)} = 0$

In this set-up the modular relation reads

Theorem 4.5   *The functional equation*

$$\frac{\Gamma(d_1 + D_1 s)}{\Gamma(d_2 - D_2 s)}\varphi(s) = \frac{\Gamma(e_1 + E_1(r - s))}{\Gamma(e_2 - E_2(r - s))}\psi(r - s), \qquad (4.45)$$

with $\Gamma(s|\Delta) = \Gamma(a_1 - A_1 s)$, is equivalent to

$$X(s, z \,|\, \Delta) \qquad\qquad\qquad\qquad\qquad\qquad\qquad\qquad (4.46)$$

$$= \begin{cases} \displaystyle\sum_{k=1}^{\infty} \frac{\alpha_k}{\lambda_k^s} H_{1,1}^{1,1}\left(z\lambda_k \,\middle|\, \begin{matrix} (1 - a_1, A_1) \\ (d_1 + D_1 s, D_1) \end{matrix} \right) \\ \qquad\qquad\qquad \text{if } L_1(s) \text{ can be taken to the right of } \sigma_\varphi \\[2em] \displaystyle\sum_{k=1}^{\infty} \frac{\beta_k}{\mu_k^{r-s}} H_{0,3}^{2,0}\left(\frac{\mu_k}{z} \,\middle|\, \begin{matrix} - \\ (a_1, A_1), (e_1 + E_1(r - s), E_1), \end{matrix} \right. \\ \qquad\qquad\qquad\qquad\qquad\qquad \left. \begin{matrix} - \\ (1 - e_2 + E_2(r - s), E_2) \end{matrix} \right) \\ \quad + \displaystyle\sum_{k=1}^{L} \text{Res}\Big(\Gamma(a_1 - A_1(w - s))\chi(w)\, z^{s-w}, w = s_k \Big) \\ \qquad\qquad\qquad \text{if } L_2(s) \text{ can be taken to the left of } r - \sigma_\psi. \end{cases}$$

The right-hand side of the above contains the summand

$$H_{0,3}^{2,0}\left(\mu_k z \,\middle|\, \begin{matrix} - \\ (a_1, A_1), (e_1 + (1/2)A_1(r - s), (1/2)A_1), \end{matrix} \right.$$

$$\left. \begin{matrix} - \\ (1 - e_2 + (1/2)A_1(r - s), (1/2)A_1) \end{matrix} \right)$$

$$= H_{0,3}^{2,0}\left(\mu_k z \,\middle|\, \begin{matrix} - \\ (0,1), (\frac{1}{2}(1 - s), \frac{1}{2}), (\frac{1}{2} + \frac{1}{2}(1 - s), \frac{1}{2}) \end{matrix} \right). \qquad (4.47)$$

We may work with a slightly more general form of the right-hand side of the modular relation

$$H_{0,3}^{2,0}\left(z\left|{\begin{array}{c}-\\(s,\frac{1}{2}A_1),(a_1,A_1),(s+\frac{1}{2},\frac{1}{2}A_1)\end{array}}\right.\right) \tag{4.48}$$

$$= H_{1,4}^{3,0}\left(z\left|{\begin{array}{c}(s+\frac{1}{2},\frac{1}{2}A_1)\\(s,\frac{1}{2}A_1),(s+\frac{1}{2},\frac{1}{2}A_1),(a_1,A_1),(s+\frac{1}{2},\frac{1}{2}A_1)\end{array}}\right.\right)$$

$$= 2\sqrt{\pi}\,4^{-s}\,H_{1,3}^{2,0}\left(2^{A_1}z\left|{\begin{array}{c}(s+\frac{1}{2},\frac{1}{2}A_1)\\(2s,A_1),(a_1,A_1),(s+\frac{1}{2},\frac{1}{2}A_1)\end{array}}\right.\right)$$

$$= \frac{4^{-s}}{\sqrt{\pi}\,i}\left\{e^{(s+\frac{1}{2})\pi i}H_{0,2}^{2,0}\left(2^{A_1}\,e^{-\frac{\pi}{2}A_1 i}z\left|{\begin{array}{c}-\\(2s,A_1),(a_1,A_1)\end{array}}\right.\right)\right.$$

$$\left. - e^{-(s+\frac{1}{2})\pi i}H_{0,2}^{2,0}\left(2^{A_1}\,e^{\frac{\pi}{2}A_1 i}z\left|{\begin{array}{c}-\\(2s,A_1),(a_1,A_1)\end{array}}\right.\right)\right\}$$

$$= \frac{4^{-s}}{\sqrt{\pi}\,iA_1}\left\{e^{(s+\frac{1}{2})\pi i}G_{0,2}^{2,0}\left(2\,e^{-\frac{\pi}{2}i}z^{\frac{1}{A_1}}\left|{\begin{array}{c}-\\2s,a_1\end{array}}\right.\right)\right.$$

$$\left. - e^{-(s+\frac{1}{2})\pi i}G_{0,2}^{2,0}\left(2\,e^{\frac{\pi}{2}i}z^{\frac{1}{A_1}}\left|{\begin{array}{c}-\\2s,a_1\end{array}}\right.\right)\right\}$$

$$= \frac{2}{\sqrt{\pi}A_1}\left\{(2e^{-\frac{\pi}{2}i}z^{\frac{1}{A_1}})^{a_1}\left(\frac{e^{-\frac{\pi}{2}i}z^{\frac{1}{A_1}}}{2}\right)^s K_{2s-a_1}\left(2\sqrt{2}\,e^{-\frac{\pi i}{4}}z^{\frac{1}{2A_1}}\right)\right.$$

$$\left. + (2e^{\frac{\pi}{2}i}z^{\frac{1}{A_1}})^{a_1}\left(\frac{e^{\frac{\pi}{2}i}z^{\frac{1}{A_1}}}{2}\right)^s K_{2s-a_1}\left(2\sqrt{2}\,e^{\frac{\pi i}{4}}z^{\frac{1}{2A_1}}\right)\right\},$$

by (2.96), (2.99), (2.101) and (2.109). Equation (4.48) with $s \leftrightarrow \frac{1-s}{2}$ and $A_1 = 1$ takes the form:

$$H_{0,3}^{2,0}\left(z\left|{\begin{array}{c}-\\(\frac{1-s}{2},\frac{1}{2}),(a_1,1),(\frac{1-s}{2}+\frac{1}{2},\frac{1}{2})\end{array}}\right.\right)$$

$$= \frac{2}{\sqrt{\pi}}\left\{(2e^{-\frac{\pi}{2}i}z)^{a_1}\left(\frac{e^{-\frac{\pi}{2}i}z}{2}\right)^{\frac{1-s}{2}} K_{1-s-a_1}\left(2\,e^{-\frac{\pi i}{4}}\sqrt{2z}\right)\right.$$

$$\left. + (2e^{\frac{\pi}{2}i}z)^{a_1}\left(\frac{e^{\frac{\pi}{2}i}z}{2}\right)^{\frac{1-s}{2}} K_{1-s-a_1}\left(2\,e^{\frac{\pi i}{4}}\sqrt{2z}\right)\right\}. \tag{4.49}$$

Using $a_1 = 0$, the equation (4.49) becomes

$$H_{0,3}^{2,0}\left(z \left| \begin{array}{c} - \\ \left(\frac{1-s}{2}, \frac{1}{2}\right), (0,1), \left(\frac{1-s}{2} + \frac{1}{2}, \frac{1}{2}\right) \end{array} \right. \right) \tag{4.50}$$

$$= \frac{2}{\sqrt{\pi}} \left\{ \left(\frac{e^{-\frac{\pi}{2}i}z}{2} \right)^{\frac{1-s}{2}} K_{1-s}\left(2 e^{-\frac{\pi i}{4}} \sqrt{2z} \right) \right.$$

$$\left. + \left(\frac{e^{\frac{\pi}{2}i}z}{2} \right)^{\frac{1-s}{2}} K_{1-s}\left(2 e^{\frac{\pi i}{4}} \sqrt{2z} \right) \right\}.$$

This leads to an analogue of Koshlyakov's result, however with the basic sequences get squared (since we must choose $E_1 = 2$ in (4.45), we are forced to choose $A_1 = \frac{1}{2}$).

4.5 Elucidation of Koshlyakov's result in the real quadratic case

We elucidate the reason why the K-Bessel function appears for the real quadratic field Ω. Let us briefly recall his method. Koshlyakov considers the integral with the integrand being of the form

$$\Gamma(z)\varphi(z) = \Gamma(z)\frac{\Gamma^2\left(\frac{1-z}{2}\right)}{\Gamma^2\left(\frac{z}{2}\right)}\varphi(1-z) \tag{4.51}$$

which is in a form similar to that of (4.43). He applies the duplication formula to the first factor and cancels $\Gamma\left(\frac{z}{2}\right)$ in the denominator. Then he writes $\frac{1}{\Gamma\left(\frac{z}{2}\right)}$ as $\Gamma\left(1 - \frac{z}{2}\right)\frac{\sin\frac{\pi}{2}z}{\pi}$ and also expresses the product $\Gamma\left(\frac{z}{2} + \frac{1}{2}\right)\Gamma\left(\frac{1-z}{2}\right)$ as $\frac{\pi}{\sin\pi\frac{1-z}{2}}$. Then finally, he combines $\Gamma\left(\frac{1-z}{2}\right)$ and $\Gamma\left(1 - \frac{z}{2}\right)$ to give $2^{-(1-z)}\pi^{\frac{1}{2}}\Gamma(1-z)$. It follows that

$$\Gamma(z)\zeta_\Omega(z) = \tan\frac{\pi}{2}z\Gamma(1-z)\zeta_\Omega(1-z) \tag{4.52}$$

which is a reminiscent of the Wigert-Bellman divisor problem [Bel2]. Thus it suggests the following

Theorem 4.6 *The generating Dirichlet series for the Wigert-Bellman*

divisor problem

$$\zeta_\Omega(s,\omega)$$

$$= \frac{\omega^{-s}}{\Gamma(s)} \sum_{k=1}^\infty \frac{\alpha(k)}{k} \frac{1}{2\pi i} \int_{(c)} \tan \frac{\pi}{2} z \Gamma(1-z)\Gamma(s-z) \left(\frac{1}{k\omega}\right)^{-z} dz \qquad (4.53)$$

amounts to the Lipschitz summation formula

$$\frac{1}{2} \sum_{n=-\infty}^\infty \frac{\alpha(n)}{(w+in)^s} = \frac{1}{2} e^{\frac{\pi i}{2}s} \zeta_\Omega(s,iw) + \frac{1}{2} e^{-\frac{\pi i}{2}s} \zeta_\Omega(s,-iw) \qquad (4.54)$$

$$= A^s \frac{w^{\frac{1}{2}(1-s)}}{\Gamma(s)} \sum_{k=1}^\infty \frac{\alpha(k)}{k^{\frac{1}{2}(1-s)}} \left(\frac{\varepsilon^{s+1}}{i} K_{1-s}\left(2A\varepsilon\sqrt{kw}\right)\right.$$

$$\left. - \frac{\bar\varepsilon^{s+1}}{i} K_{1-s}\left(2A\bar\varepsilon\sqrt{kw}\right)\right).$$

This in fact is the corrected form of [KoshII, (23.16)]. Proving the above result amounts to showing that, in contrast to the case of imaginary quadratic fields in Proposition 4.3, the left-hand side cancels the denominator $\cos\frac{\pi}{2}z$, which is given in

$$e^{\frac{\pi i}{2}s} \zeta_\Omega(s,iw) + e^{-\frac{\pi i}{2}s} \zeta_\Omega(s,-iw). \qquad (4.55)$$

The integral expression for (4.55) contains the integrand

$$\frac{w^z}{\pi} \sin \frac{\pi}{2} z \Gamma(1-z) \zeta_\Omega(1-z)$$

and so applying the inverse Heaviside integral (2.109) one can deduce (4.54).

4.6   Koshlyakov's K-series

Let $G(s) = G_\Omega(s)$ denote the quotient of gamma factors in (3.41):

$$G(1-s) = \frac{\Gamma^{r_1}\left(\frac{s}{2}\right)\Gamma^{r_2}(s)}{\Gamma^{r_1}\left(\frac{1-s}{2}\right)\Gamma^{r_2}(1-s)} \qquad (4.56)$$

and consider the kernel, which we call **Koshlyakov's K-function**

$$K(x) = K_{r_1,r_2}(x) = \frac{1}{2\pi i} \int_{(c)} \frac{\pi}{2\cos\left(\frac{\pi}{2}s\right)} \frac{G(1-s)}{x^s} ds \qquad (4.57)$$

for $c > 0$ and $x > 0$ ($\mathrm{Re}(x) > 0$ being allowed).

Then Koshlyakov considers the basic series

$$\sigma(x) = \sigma_{r_1, r_2}(x) = \frac{A}{\pi} \sum_{n=1}^{\infty} F(n) K_{r_1, r_2}\left(A^2 x n\right) \tag{4.58}$$

$x > 0 \, (\mathrm{Re}(x) > 0)$. He transforms the series (4.58) by Cauchy's residue theorem and apply in the last stage a form of the beta transform (4.2) ([KoshI, (3.3)])

$$\frac{1}{2\pi i} \int_{(c)} \frac{\pi}{2 \sin\left(\frac{\pi}{2} w\right)} \frac{dw}{\alpha^w} = \frac{1}{1 + \alpha^s}, \quad \alpha > 0, \ 0 < c < 2. \tag{4.59}$$

He proves the following theorem.

Theorem 4.7 (Koshlyakov) *One has*

$$\sigma(x) = -\frac{1}{2}\rho - \frac{\zeta_\Omega(0)}{\pi} \frac{1}{x} + \frac{x}{\pi} \sum_{n=1}^{\infty} \frac{\alpha(n)}{n^2 + x^2} \tag{4.60}$$

for $\mathrm{Re}\, x > 0$, *where* ρ *is defined in* (3.38). *Thus one gets the closed forms for the K-function:*

$$K_{1,0}(x) = \sqrt{\pi}\, e^{-2x}. \tag{4.61}$$

$$K_{0,1}(x) = -2 \operatorname{kei}_0\left(4\sqrt{\frac{x}{4}}\right) = \frac{1}{i}\left(K_0\left(2\varepsilon\sqrt{x}\right) - K_0\left(2\bar{\varepsilon}\sqrt{x}\right)\right) \tag{4.62}$$

$$K_{2,0}(x) = 4 \operatorname{ker}_0(4\sqrt{x}) = 2\left(K_0\left(\varepsilon\sqrt{x}\right) + K_0\left(\bar{\varepsilon}\sqrt{x}\right)\right), \tag{4.63}$$

where kei_0 *and* ker_0 *are modified Kelvin functions ([ErdH, p.6], [PBM]) and*

$$\varepsilon = e^{\frac{\pi}{4}i}, \quad \bar{\varepsilon} = e^{-\frac{\pi}{4}i}. \tag{4.64}$$

The proof of (4.60) follows by taking the limit as $s \to 0$ in formulas (4.42). The main difference is that we express the resulting functions as Kelvin functions.

We turn to the elucidation of Koshlyakov's K-function (4.57). We distinguish three cases.

In the case of $\mathbb{Q}$ one has $r_1 = 1$ and (4.60) amounts to Example 4.2 and (4.61) corresponds to the degenerate case of the K-Bessel function.

$$K_{1,0}(x) = \frac{1}{2}\frac{1}{2\pi i}\int_{(c)} \Gamma\left(\frac{s}{2}\right)\Gamma\left(\frac{s}{2}+\frac{1}{2}\right) x^{-s}ds$$

$$= \frac{1}{2} H_{0,2}^{2,0}\left(x \,\middle|\, \begin{matrix} - \\ (0,\frac{1}{2}),(\frac{1}{2},\frac{1}{2}) \end{matrix}\right) \tag{4.65}$$

$$= G_{0,2}^{2,0}\left(x^2 \,\middle|\, \begin{matrix} - \\ 0,\frac{1}{2} \end{matrix}\right) = 2x^{\frac{1}{2}}K_{\frac{1}{2}}(2x),$$

where in the last step we applied (2.109), $K_s(x)$ being the K-Bessel function defined e.g. by (2.31) (cf. [Wat]). By (2.32), the right-hand side of (4.65) further reduces to $\sqrt{\pi}\,e^{-2x}$, and we have (4.61) ([KoshI, (2.4)]).

In the imaginary quadratic case, $r_1 = 0, r_2 = 1$, we need to locate Koshlyakov's formula [KoshI, (2.5)] (except for the middle member in (4.62)):

$$K_{0,1}(x) = \frac{1}{i}\left(K_0\left(2\bar{\varepsilon}\sqrt{x}\right) - K_0\left(2\varepsilon\sqrt{x}\right)\right). \tag{4.66}$$

Since

$$K_{0,1}(x) = \frac{1}{2} H_{1,3}^{2,1}\left(x \,\middle|\, \begin{matrix} (\frac{1}{2},\frac{1}{2}) \\ (\frac{1}{2},\frac{1}{2}),(0,1),(0,1) \end{matrix}\right). \tag{4.67}$$

Applying (2.96), we see that

$$K_{0,1}(x) = \frac{1}{4} H_{1,5}^{3,1}\left(\frac{x}{4} \,\middle|\, \begin{matrix} (\frac{1}{2},\frac{1}{2}) \\ (\frac{1}{2},\frac{1}{2}),(0,\frac{1}{2}),(\frac{1}{2},\frac{1}{2}),(0,\frac{1}{2}),(\frac{1}{2},\frac{1}{2}) \end{matrix}\right)$$

$$= \frac{1}{4} H_{0,4}^{3,0}\left(\frac{x}{4} \,\middle|\, \begin{matrix} - \\ (0,\frac{1}{2}),(\frac{1}{2},\frac{1}{2}),(\frac{1}{2},\frac{1}{2}),(0,\frac{1}{2}) \end{matrix}\right) \tag{4.68}$$

$$= \frac{1}{4} H_{1,5}^{4,0}\left(\frac{x}{4} \,\middle|\, \begin{matrix} (0,\frac{1}{2}) \\ (0,\frac{1}{2}),(\frac{1}{2},\frac{1}{2}),(0,\frac{1}{2}),(\frac{1}{2},\frac{1}{2}),(0,\frac{1}{2}) \end{matrix}\right).$$

Hence by (2.101)

$$K_{0,1}(x) = \frac{1}{8\pi i}\left\{ H^{4,0}_{0,4}\left(\frac{e^{-\frac{\pi}{2}i}x}{4} \,\middle|\, \begin{matrix} - \\ (0,\frac{1}{2}),(\frac{1}{2},\frac{1}{2}),(0,\frac{1}{2}),(\frac{1}{2},\frac{1}{2}) \end{matrix} \right) \right.$$

$$\left. - H^{4,0}_{0,4}\left(\frac{e^{\frac{\pi}{2}i}x}{4} \,\middle|\, \begin{matrix} - \\ (0,\frac{1}{2}),(\frac{1}{2},\frac{1}{2}),(0,\frac{1}{2}),(\frac{1}{2},\frac{1}{2}) \end{matrix} \right) \right\}$$

$$= \frac{1}{2i}\left\{ G^{2,0}_{0,2}\left(e^{-\frac{\pi}{2}i}x \,\middle|\, \begin{matrix} - \\ 0,0 \end{matrix} \right) - G^{2,0}_{0,2}\left(e^{\frac{\pi}{2}i}x \,\middle|\, \begin{matrix} - \\ 0,0 \end{matrix} \right) \right\}. \qquad (4.69)$$

In the last step we applied (2.96) and then (2.93). The equation (4.69) amounts to (4.66) by (4.72) above which is a special case of (2.109).

Exercise 4.1 Indicate the concrete procedure for the above computation.

Solution Applying (2.3) to the integrand, we deduce the second equality of (4.68). Multiplying the integrand by $\frac{\Gamma\left(\frac{w}{2}\right)}{\Gamma\left(\frac{w}{2}\right)}$, we see that

$$K_{0,1}(x) = \frac{1}{2\pi i}\int_{(c)} \frac{2^{2w-2}\left(\Gamma\left(\frac{w}{2}\right)\Gamma\left(\frac{w+1}{2}\right)\right)^2}{\Gamma\left(1-\frac{w}{2}\right)\Gamma\left(\frac{w}{2}\right)} x^{-w}\,dw \qquad (4.70)$$

$$= \frac{1}{2\pi i}\int_{(c)} \frac{\sin\left(\frac{\pi}{2}w\right)}{\pi} 2^{2w-2}\left(\Gamma\left(\frac{w}{2}\right)\Gamma\left(\frac{w+1}{2}\right)\right)^2 x^{-w}\,dw$$

by (2.3). Now by (2.4) and Euler's identity, the integrand is

$$\frac{e^{\frac{\pi i}{2}s} - e^{-\frac{\pi i}{2}s}}{2i}\Gamma(w)^2.$$

Hence

$$K_{0,1}(x) = \frac{1}{i}\left(\frac{1}{2\pi i}\int_{(c)} \Gamma(w)^2 2^{2w-1}(4e^{-\frac{\pi i}{4}}x)^{-w}\,dw \right. \qquad (4.71)$$

$$\left. - \frac{1}{2\pi i}\int_{(c)} \Gamma(w)^2 2^{2w-1}(4e^{\frac{\pi i}{4}}x)^{-w}\,dw \right)$$

which amounts to (4.66) by

$$G^{2,0}_{0,2}\left(\frac{u}{4} \,\middle|\, \begin{matrix} - \\ 0,0 \end{matrix} \right) = \frac{1}{2\pi i}\int_{(c)} \Gamma^2(s)\frac{u}{4}^{-s}\,ds = 2K_0\left(\sqrt{u}\right), \qquad (4.72)$$

$$\operatorname{Re}(c) > 0,$$

which is a special case of (2.109).

We observe that we are to transit from (4.68) to

$$K_{0,1}(x) = \frac{1}{2} G_{0,4}^{3,0}\left(\left(\frac{x}{4}\right)^2 \bigg|_{0,\frac{1}{2},\frac{1}{2},0}^{\quad -} \right). \tag{4.73}$$

We notice that the right-hand side is one of the modified Kelvin's functions:

$$\frac{1}{2} \cdot (-4) \operatorname{kei}_0\left(4\sqrt{\frac{x}{4}} \right)$$

[PBM, 13, p.675].

Modified Kelvin functions satisfy the relation

$$\ker_\nu(z) \pm i\operatorname{kei}_\nu(z) = e^{\mp \frac{i\nu}{2}\pi} K_\nu\left(ze^{\pm \frac{\pi i}{4}} \right), \tag{4.74}$$

whence

$$2i\operatorname{kei}_\nu(z) = e^{-\frac{\nu\pi}{2}i} K_\nu\left(ze^{\frac{\pi i}{4}} \right) - e^{\frac{\nu\pi}{2}i} K_\nu\left(ze^{-\frac{\pi i}{4}} \right). \tag{4.75}$$

Thus, in our case

$$K_{0,1}(x) = -2\operatorname{kei}_0\left(4\sqrt{\frac{x}{4}} \right) = \frac{1}{i}\left(K_0\left(2\varepsilon\sqrt{x}\right) - K_0\left(2\bar\varepsilon\sqrt{x}\right) \right). \tag{4.76}$$

Note that in deriving (4.76), we have given a proof of (4.75).

Finally, we deal with the real quadratic case, where

$$K_{2,0}(x) = \frac{1}{2\pi i} \int_{(c)} \frac{1}{2} \frac{\Gamma^2\left(\frac{w}{2}\right)\Gamma\left(\frac{1+w}{2}\right)}{\Gamma\left(\frac{1-w}{2}\right)} x^{-w} dw. \tag{4.77}$$

Formula (4.77) is already

$$K_{2,0}(x) = \frac{1}{2} H_{0,4}^{3,0}\left(x \bigg|_{(0,\frac{1}{2}),(0,\frac{1}{2}),(\frac{1}{2},\frac{1}{2}),(\frac{1}{2},\frac{1}{2})}^{\quad -} \right)$$

$$= G_{0,4}^{3,0}\left(x^2 \bigg|_{0,0,\frac{1}{2},\frac{1}{2}}^{\quad -} \right)$$

by (2.93), which is $4\ker_0(4\sqrt{x})$ ([PBM, 13, p.675]).

By (4.74), we have

$$2\ker_\nu(z) = e^{-\frac{\nu\pi i}{2}} K_\nu\left(ze^{\frac{\pi i}{4}} \right) + e^{\frac{\nu\pi i}{2}} K_\nu\left(ze^{-\frac{\pi i}{4}} \right), \tag{4.78}$$

whence it follows that

$$
\begin{aligned}
K_{2,0}(x) &= G_{0,4}^{3,0}\left(x^2 \,\Big|\, {\textstyle -\atop 0,0,\frac{1}{2},\frac{1}{2}}\right) \\
&= 4\ker_0(4\sqrt{x}) \\
&= 2\left(K_0(\varepsilon\sqrt{x}) + K_0(\bar{\varepsilon}\sqrt{x})\right),
\end{aligned}
\tag{4.79}
$$

i.e. [KoshI, (2.6)].

Exercise 4.2   As in Exercise 4.1, work out the above formal computation.

We transform the integrand into

$$
\begin{aligned}
\frac{1}{2}\frac{\left(\Gamma\left(\frac{w}{2}\right)\Gamma\left(\frac{w+1}{2}\right)\pi^{-\frac{1}{2}}2^{w-1}\right)^2 \pi 2^{2-2w}}{\Gamma\left(\frac{1-w}{2}\right)\Gamma\left(\frac{1+w}{2}\right)}\,x^{-w} \\
= 2\cos\left(\frac{\pi}{2}w\right)\Gamma^2(w)2^{-2w}x^{-w}
\end{aligned}
$$

by (2.3) and (2.4).

Then we use Euler's identity to obtain

$$
\begin{aligned}
K_{2,0}(x) &= 2\left(\frac{1}{2\pi i}\int_{(c)}2^{2w-1}\Gamma(w)^2\left(4^2 e^{\frac{\pi i}{2}}x\right)^{-w}\,\mathrm{d}w\right. \\
&\qquad\left. + \frac{1}{2\pi i}\int_{(c)}2^{2w-1}\Gamma(w)^2\left(4^2 e^{-\frac{\pi i}{2}}x\right)^{-w}\,\mathrm{d}w\right) \\
&= 2\left(K_0\left(4\varepsilon\sqrt{x}\right) + K_0\left(4\bar{\varepsilon}\sqrt{x}\right)\right)
\end{aligned}
$$

in view of (4.72), i.e. [KoshI, (2.6)] follows.

4.7   The Fourier-Bessel expansion $G_{1,1}^{1,1} \leftrightarrow G_{0,2}^{2,0}$

Apparently, the functional equation of Hecke's type has been most extensively studied and it has created such tremendous amount of results that it is natural to dwell on this case rather fully and more thoroughly than in other cases.

$$
\chi(s) = \begin{cases}
\Gamma(s)\,\varphi(s), & \mathrm{Re}(s) > \sigma_\varphi, \\
\Gamma(r-s)\,\psi(r-s), & \mathrm{Re}(s) < r - \sigma_\psi.
\end{cases}
\tag{4.80}
$$

The special case of (3.6) described in the title reads

$$\sum_{k=1}^{\infty} \frac{\beta_k}{\mu_k^{r-s}} \, G_{1,1}^{1,1}\left(\frac{\mu_k}{z} \, \middle| \, \begin{matrix} 1-a \\ r-s \end{matrix}\right) + \sum_{k=1}^{L} \mathrm{Res}\left(\Gamma(a-s+w)\,\chi(w)\,z^{s-w}, w=s_k\right)$$

$$= \sum_{k=1}^{\infty} \frac{\alpha_k}{\lambda_k^s} \, G_{0,2}^{2,0}\left(z\lambda_k \, \middle| \, \begin{matrix} - \\ s,a \end{matrix}\right). \tag{4.81}$$

Exercise 4.3   Check by (2.108) and (2.109) that (4.81) reduces to

$$2\sum_{k=1}^{\infty} \frac{\alpha_k}{\lambda_k^{\frac{s-a}{2}}} \, z^{\frac{s+a}{2}} K_{s-a}\left(2\sqrt{z\lambda_k}\right)$$

$$= \Gamma(a+r-s) \sum_{k=1}^{\infty} \beta_k \frac{z^a}{(\mu_k+z)^{a+r-s}}$$

$$+ \sum_{k=1}^{L} \mathrm{Res}\left(\Gamma(a-s+w)\,\chi(w)\,z^{s-w}, w=s_k\right).$$

(4.81) in the special case $a=0$ gives the **Fourier-Bessel expansion for the perturbed Dirichlet series** (or K-Bessel expansion):

Theorem 4.8   (Fourier-Bessel expansion) *The Fourier-Bessel expansion for the perturbed Dirichlet series*

$$\Gamma(r-s) \sum_{k=1}^{\infty} \beta_k \frac{1}{(\mu_k+z)^{r-s}}$$

$$= 2\sum_{k=1}^{\infty} \frac{\alpha_k}{\lambda_k^{\frac{s}{2}}} \, z^{\frac{s}{2}} K_s\left(2\sqrt{z\lambda_k}\right) \tag{4.82}$$

$$- \sum_{k=1}^{L} \mathrm{Res}\left(\Gamma(-s+w)\,\chi(w)\,z^{s-w}, w=s_k\right)$$

is equivalent to the functional equation.

Theorem 4.8 was first proved by Berndt [Be1] and is stated as [BK, Theorem 8.3, p.120]. In [BeIII] another proof is given and in [BeIV] a general form (with multiple gamma factor) is given. It is stated as [KTY7, (1.6)], one of the equivalent assertions to the functional equation ([KTY7, Theorem 1]) (for prototype of these, cf. Exercises 1.1 and 1.2). For its history, cf. §5.3 [KTY7]. As is stated there, the most essential ingredient in the Fourier-Bessel expansion is the beta transform or the beta integral for which, we refer to [Vi2, p.34]. There we referred to it as the Mellin-Barnes integral in [Vista, p.126] and its special case appeared also on [Vista, p.34, p.98 and p.196]. Since the Mellin-Barnes integrals refer to a much wider class of special functions ([ErdH, p.49]), the Fox H-functions, we shall speak about the **beta transform** (or beta integral) in the same spirit as the gamma transform ((1.42), cf. [Vi2, p.25]). In the present book, it is (2.108), which we restate in a specialized and concrete form

$$(1+x)^{-s} = \frac{1}{2\pi i} \int_{(c)} \frac{\Gamma(s-z)\,\Gamma(z)}{\Gamma(s)}\, x^{-z} \mathrm{d}z \qquad (4.83)$$

for $x > 0$, $0 < c < \sigma$. This has been used extensively in the context of the **perturbed Dirichlet series**

$$\varphi(s,a) = \sum a_n (\lambda_n + a)^{-s} \quad (a > 0, \sigma \gg 0). \qquad (4.84)$$

In addition to the papers [Be1], [BeIII], [BeIV], [BeVI], [CZS1], [CZS2], [Mat] and [PK], which were referred to in [KTY7], cf. also [HarMB] and [KoshI]-[KoshIII]). Also Katsurada and Matsumoto made an extensive use of the beta-transform in dealing with mean square of the Hurwitz type zeta-functions [KatM1]-[KatM3]. [KoshI, p. 116, p. 130], [KoshII, p.243] extensively uses a special case of (4.83) in the form ([Vi2, p.35])

$$\frac{1}{2\pi i} \int_{(c)} \frac{\pi}{2 \sin \frac{\pi}{2} z}\, x^{-z} \mathrm{d}z = \frac{1}{1+x^2}, \ x > 0, 0 < c < 2. \qquad (4.85)$$

The overall procedures consist of expressing the sum in the form of the integral (4.83) $G_{1,1}^{1,1}$, applying the functional equation, and then finally appealing to the inverse Heaviside integral $G_{0,2}^{2,0}$ or to some more general G-function according to the form of the processing gamma factor.

Exercise 4.4   Deduce the following from (4.82).

$$\sqrt{\pi} \sum_{k=1}^{\infty} \frac{\alpha_k}{\sqrt{\lambda_k}}\, e^{-2\sqrt{z\lambda_k}} = \Gamma\!\left(r - \tfrac{1}{2}\right) \sum_{k=1}^{\infty} \frac{\beta_k}{(\mu_k + z)^{r - \frac{1}{2}}}$$
$$+ \sum_{k=1}^{L} \mathrm{Res}\!\left(\Gamma\!\left(-\tfrac{1}{2} + w\right) \chi(w)\, z^{\frac{1}{2} - w}, w = s_k\right);$$

$$\sqrt{\pi}\, z^{-\frac{1}{2}} \sum_{k=1}^{\infty} \alpha_k e^{-2\sqrt{z\lambda_k}} = \Gamma\!\left(r + \tfrac{1}{2}\right) \sum_{k=1}^{\infty} \frac{\beta_k}{(\mu_k + z)^{r+\frac{1}{2}}} \tag{4.86}$$

$$+ \sum_{k=1}^{L} \operatorname{Res}\!\left(\Gamma\!\left(\tfrac{1}{2} + w\right) \chi(w)\, z^{-\frac{1}{2} - w}, w = s_k\right).$$

Exercise 4.5 Reviewing the proof of [Vi2, Proposition 2.1, p. 35], check that the beta transform (4.2) is an analytic expression for the binomial expansion.

To state the generalized Chowla-Selberg type formula (Theorem 4.9), we introduce new notation.

Let $Y = \left(\begin{array}{c|c} A & B \\ \hline {}^t B & C \end{array}\right)$ be a block decomposition with A an $n \times n$ matrix and B an $m \times m$ matrix. Set

$$D = C - {}^t B A^{-1} B.$$

In accordance with this decomposition, we decompose the vectors $\boldsymbol{g} = \begin{pmatrix} \boldsymbol{g}_1 \\ \boldsymbol{g}_2 \end{pmatrix}, \boldsymbol{h} = \begin{pmatrix} \boldsymbol{h}_1 \\ \boldsymbol{h}_2 \end{pmatrix}, \boldsymbol{g}_1, \boldsymbol{h}_1 \in \mathbb{Z}^n, \boldsymbol{g}_2, \boldsymbol{h}_2 \in \mathbb{Z}^m$.

Theorem 4.9 (generalized Chowla-Selberg type formula, cf. [Vista]) *Under the above notation, we have*

$$\Lambda(Y, \boldsymbol{g}, \boldsymbol{h}, s) \tag{4.87}$$

$$= \delta(\boldsymbol{g}_2)\, e^{-2\pi i \boldsymbol{g}_2 \cdot \boldsymbol{h}_2}\, \Lambda(A, \boldsymbol{g}_1, \boldsymbol{h}_1, s) + \delta(\boldsymbol{h}_1)\, \frac{1}{\sqrt{|A|}}\, \Lambda\!\left(D, \boldsymbol{g}_2, \boldsymbol{h}_2, s - \frac{n}{2}\right)$$

$$+ \frac{2 e^{-2\pi i \boldsymbol{g}_1 \cdot \boldsymbol{h}_1}}{\sqrt{|A|}} \sum_{\substack{\boldsymbol{a} \in \mathbb{Z}^n \\ \boldsymbol{a} + \boldsymbol{h}_1 \neq \boldsymbol{z}}} \sum_{\substack{\boldsymbol{b} \in \mathbb{Z}^m \\ \boldsymbol{b} + \boldsymbol{g}_2 \neq \boldsymbol{z}}} e^{2\pi i (-\boldsymbol{g}_1 \cdot \boldsymbol{a} + \boldsymbol{h}_2 \cdot \boldsymbol{b})} e^{-2\pi i A^{-1} B (\boldsymbol{b} + \boldsymbol{g}_2) \cdot (\boldsymbol{a} + \boldsymbol{h}_1)}$$

$$\times \sqrt{\frac{A^{-1}[\boldsymbol{a} + \boldsymbol{b}_1]}{D[\boldsymbol{b} + \boldsymbol{g}_2]}}^{\, s - \frac{n}{2}} K_{s - \frac{n}{2}}\!\left(2\sqrt{A^{-1}[\boldsymbol{a} + \boldsymbol{h}_1]\, D[\boldsymbol{b} + \boldsymbol{g}_2]}\, \pi\right).$$

The case $\boldsymbol{g} = \boldsymbol{h} = \boldsymbol{z}, n = m = 1$ is due to Chowla and Selberg [CS], [SC] (cf. also Bateman and Grosswald [BG]), the case $m = 1$ is due to Berndt [BeVI] and the general case with $\boldsymbol{g} = \boldsymbol{h} = \boldsymbol{z}$ is due to Terras [Te2, Example 4, p.208].

Corollary 4.2 *Let*

$$Z(s) = \sum_{m,n}{}' \frac{1}{(am^2 + 2bmn + cn^2)^s} \tag{4.88}$$

for $\sigma > 1$, where m, n run through integer values except for $(m,n) = (0,0)$ and $a, c > 0$ and the discriminant $\Delta = 4(b^2 - ac) < 0$. Then

$$\Gamma(s)Z(s) = 2\Gamma(s)\zeta(2s)a^{-s} + \frac{2^{2s}a^{s-1}\sqrt{\pi}}{|\Delta|^{s-\frac{1}{2}}}\Gamma\left(s - \frac{1}{2}\right)\zeta(2s-1) + Q(s), \tag{4.89}$$

where

$$Q(s) = \frac{4\pi^s 2^{s-\frac{1}{2}}}{\sqrt{a}|\Delta|^{\frac{s}{2}-\frac{1}{4}}} \sum_{n=1}^{\infty} n^{s-\frac{1}{2}}\sigma_{1-2s}(n)\cos n\pi\frac{b}{a}K_{s-\frac{1}{2}}\left(\frac{\pi\sqrt{|\Delta|}}{a}n\right). \tag{4.90}$$

Corollary 4.2 is often stated in slightly different form. Let

$$d = \frac{|\Delta|}{4} = ac - b^2, \ x = \frac{b}{a}, \ y = \frac{\sqrt{d}}{a} = \frac{\sqrt{|\Delta|}}{2a} > 0. \tag{4.91}$$

Then

$$\frac{2^{2s}a^{s-1}}{|\Delta|^{s-\frac{1}{2}}} = 2\frac{\left(\frac{y}{\sqrt{d}}\right)^{1-s}}{d^{s-\frac{1}{2}}}$$

and

Corollary 4.3 *(4.90) may be written as*

$$Q(s) = \frac{4\pi^s 2^{s-\frac{1}{2}}}{\sqrt{\frac{\sqrt{d}}{y}}4d^{\frac{s}{2}-\frac{1}{4}}} \sum_{n=1}^{\infty} n^{s-\frac{1}{2}}\sigma_{1-2s}(n)\cos 2\pi nx K_{s-\frac{1}{2}}(2\pi yn). \tag{4.92}$$

The special case of $s = 1$ of Corollary 4.2 gives rise to the Kronecker limit formula in the notation of Theorem 1.2.

$$\lim_{s\to 1}\left(Z(s) - \frac{\frac{2\pi}{\sqrt{|\Delta|}}}{s-1}\right)$$

$$= \frac{2\pi}{\sqrt{|\Delta|}}\left(2\gamma + \log\frac{a}{|\Delta|}\right) - \frac{4\pi}{\sqrt{|\Delta|}}\log\eta(\omega_1)\eta(\omega_2). \tag{4.93}$$

Definition 4.1 In conjunction with the Epstein zeta-function (4.88) the **Eisenstein series** is defined by

$$E(z;\alpha) = a^s Z(s) = \sum_{m,n}{}' \frac{1}{|m+n\alpha|^s}, \qquad (4.94)$$

for $\sigma > 1$, where m, n run through integer values except for $(m,n) = (0,0)$ $(a > 0)$ and $\alpha = x + iy$ with x, y in (4.91).

Theorem 4.10 (Kuzumaki) *We have the Fourier-Bessel expansion*

$$E(z;\alpha) = 2\zeta(2z) + \frac{2\sqrt{\pi}\Gamma(z-\frac{1}{2})}{\Gamma(z)}\zeta(2z-1)y^{-2z+1} \qquad (4.95)$$

$$+ \frac{4}{\Gamma(z)} \sum_{m,n=1}^{\infty} (ny)^{-2z}\frac{\cos(2\pi mnx)}{m}I_{m,n}(z,\alpha),$$

where $\alpha = x + iy$ $(y > 0)$ and

$$I_{m,n}(z,\alpha) = \frac{1}{\sqrt{\pi}}G_{0,2}^{2,0}\left((\pi mny)^2 \,\middle|\, \begin{matrix} - \\ z, \; \frac{1}{2} \end{matrix}\right). \qquad (4.96)$$

We have several well-known expressions for $G_{0,2}^{2,0}$ and so

$$I_{m,n}(z,\alpha) = e^{-2\pi mny}(2\pi mny)^{2z}U(z, 2z; 4\pi mny). \qquad (4.97)$$

Under (4.100), (4.95) reduces to the Chowla-Selberg integral formula in Corollary 4.2 above

$$I_{m,n}(z,\alpha) = 2\pi^z(mny)^{z+\frac{1}{2}}K_{z-\frac{1}{2}}(2\pi mny), \qquad (4.98)$$

while (4.102) leads to the Whittaker function series expression

$$I_{m,n}(z,\alpha) = (\pi mny)^z W_{0,z-\frac{1}{2}}(4\pi mny). \qquad (4.99)$$

The expressions depend on the G-function hierarchy

$$G_{0,2}^{2,0}\left(x^2 \left|\begin{array}{c} - \\ z, \ \frac{1}{2} \end{array}\right.\right) = 2x^{z+\frac{1}{2}}K_{z-\frac{1}{2}}(2x) \tag{4.100}$$

$$= \sqrt{\pi}x^z W_{0,z-\frac{1}{2}}(4x)$$

$$= 2\sqrt{\pi}x^{z+\frac{1}{2}}e^{2x}G_{1,2}^{2,0}\left(4x \left|\begin{array}{c} \frac{1}{2} \\ z-\frac{1}{2}, \ -z+\frac{1}{2} \end{array}\right.\right) \tag{4.101}$$

$$= \sqrt{\pi}e^{-2x}(2x)^{2z}U(z,2z;4x) \tag{4.102}$$

$$= \frac{\sqrt{\pi}e^{-2x}x^z}{\Gamma(z)\Gamma(1-z)}G_{1,2}^{2,1}\left(4x \left|\begin{array}{c} 1 \\ z, \ 1-z \end{array}\right.\right). \tag{4.103}$$

Our task is now to ascend the hierarchy from (4.100) to (4.103) to prove Theorem 4.11. Before turning to the proof, we give the Katsurada's beta-transform (cf. (4.2)). By the difference equation we have for any x and $K \in \mathbb{N}$

$$\Gamma(x) = \frac{\Gamma(K+x)}{(K-1+x)(K-2+x)\cdots(1+x)x} \tag{4.104}$$

$$= \frac{(-1)^K\Gamma(K+x)\Gamma(-K+1-x)}{\Gamma(-x+1)}$$

$$= \frac{(-1)^K\Gamma(K+x)}{(K-1)!}B(-K+1-x, \ K),$$

where $B(x,y)$ is the beta function which satisfies

$$B(x,y) = \frac{\Gamma(x)\Gamma(y)}{\Gamma(x+y)} = \int_0^1 t^{x-1}(1-t)^{y-1}\,dt.$$

Hence

$$\Gamma(x) = \frac{(-1)^K\Gamma(K+x)}{(K-1)!}\int_0^1 t^{-K-x}(1-t)^{K-1}\,dt, \tag{4.105}$$

which we refer to as the **Katsurada transform** ([Kat5]).

Theorem 4.11

$$E(z;\alpha) = 2\zeta(2z) + \frac{2\sqrt{\pi}\,\Gamma(z-\frac{1}{2})}{\Gamma(z)}\zeta(2z-1)y^{-2z+1} \tag{4.106}$$

$$+ \frac{4}{\Gamma(z)}\sum_{m,n=1}^{\infty}(ny)^{-2z}\frac{\cos(2\pi mnx)}{m}e^{-2\pi mny}$$

$$\times \sum_{k=0}^{K-1}\frac{(-1)^k}{2^{2k}k!}(z)_k(1-z)_k(\pi mny)^{-k+z}$$

$$+ \frac{4}{\Gamma(z)}\sum_{m,n=1}^{\infty}(ny)^{-2z}\frac{\cos(2\pi mnx)}{m}\frac{(-1)^K}{(K-1)!}e^{-2\pi mny}\frac{(2\pi mny)^{2z}}{\Gamma(z)\Gamma(1-z)}$$

$$\times \int_0^1 G_{1,2}^{2,1}\left(\frac{4\pi mny}{t}\;\middle|\;\begin{matrix}1-z-K\\0,\,1-2z\end{matrix}\right)t^{-K-z}(1-t)^{K-1}\;dt.$$

Proof of Theorem 4.11 To make a clear comparison with Katsurada's reasoning ([Kat5]), we use (4.97) in what follows (we may equally use other expressions in Theorem 4.10). We assume that $0 < \operatorname{Re} z = \xi < 1$ and the line of integration (c_1') satisfies

$$-\xi < c' < \min\{0, 1-2\xi\}.$$

Under this condition, we have from (2.111)

$$U(z,2z;4\pi mny) = \frac{1}{2\pi i}\int_{(c')}\frac{\Gamma(s+z)\Gamma(-s)\Gamma(1-2z-s)}{\Gamma(z)\Gamma(1-z)}(4\pi mny)^s\;ds.$$

Moving the line of integration to the left up to $\operatorname{Re} s = c_K$ we encounter the residues at $s = -k - z, k = 0, 1, \ldots, K-1$, where $-\xi - K < c_K < -\xi - K + 1$, $K \in \mathbb{N}$. The sum of residues is

$$\sum_{k=0}^{K-1}\frac{(-1)^k\Gamma(k+z)\Gamma(1+k-z)}{k!\Gamma(z)\Gamma(1-z)}(4\pi mny)^{-k-z}.$$

Hence

$$U(z,2z;4\pi mny) \tag{4.107}$$

$$= \sum_{k=0}^{K-1}\frac{(-1)^k}{k!}(z)_k(1-z)_k(4\pi mny)^{-k-z}$$

$$+ \frac{1}{2\pi i}\int_{(c_K)}\frac{\Gamma(s+z)\Gamma(-s)\Gamma(1-2z-s)}{\Gamma(z)\Gamma(1-z)}(4\pi mny)^s\;ds.$$

The restriction on ξ can be relaxed to $-K < \xi < K+1$.

Using (4.105) for $\Gamma(s+z)$ in (4.107), we have

$$U(z, 2z; 4\pi mny)$$

$$= \sum_{k=0}^{K-1} \frac{(-1)^k}{k!} (z)_k (1-z)_k (4\pi mny)^{-k-z}$$

$$+ \frac{1}{2\pi i} \int_{(c_K)} \frac{\Gamma(s+z+K)\Gamma(-s)\Gamma(1-2z-s)}{\Gamma(z)\Gamma(1-z)}$$

$$\times (4\pi mny)^s \frac{(-1)^K}{(K-1)!} \int_0^1 t^{-K-s-z}(1-t)^{K-1} \, dt \; ds$$

$$= \sum_{k=0}^{K-1} \frac{(-1)^k}{k!} (z)_k (1-z)_k (4\pi mny)^{-k-z}$$

$$+ \frac{(-1)^K}{(K-1)!} \frac{1}{\Gamma(z)\Gamma(1-z)} \int_0^1 t^{-K-z}(1-t)^{K-1} I(t) \, dt,$$

where

$$I(t) = \frac{1}{2\pi i} \int_{(c_K)} \Gamma(s+z+K)\Gamma(-s)\Gamma(1-2z-s) \left(\frac{4\pi mny}{t} \right)^s \, ds.$$

Since in the integral $I(t)$ all the poles of $\Gamma(s+z+K)$ are to the left side and the all poles of $\Gamma(-s)$ and $\Gamma(1-2z-s)$ are to the right side of the line c_K, it is the G-function:

$$I(t) = G_{1,2}^{2,1} \left(\frac{4\pi mny}{t} \middle| \begin{array}{c} 1-z-K \\ 0, \; 1-2z \end{array} \right).$$

Hence

$$U(z, 2z; 4\pi mny) \tag{4.108}$$

$$= \sum_{k=0}^{K-1} \frac{(-1)^k}{k!} (z)_k (1-z)_k (4\pi mny)^{-k-z} + \frac{(-1)^K}{(K-1)!} \frac{1}{\Gamma(z)\Gamma(1-z)}$$

$$\times \int_0^1 t^{-K-z}(1-t)^{K-1} G_{1,2}^{2,1} \left(\frac{4\pi mny}{t} \middle| \begin{array}{c} 1-z-K \\ 0, \; 1-2z \end{array} \right) \, dt,$$

which leads to

$$I_{m,n}(z, \alpha) \tag{4.109}$$

$$= e^{-2\pi mny} \sum_{k=0}^{K-1} \frac{(-1)^k}{2^{2k}k!} (z)_k (1-z)_k (\pi mny)^{-k+z}$$

$$+ \frac{(-1)^K}{(K-1)!} e^{-2\pi mny} \frac{(2\pi mny)^{2z}}{\Gamma(z)\Gamma(1-z)}$$

$$\times \int_0^1 G_{1,2}^{2,1}\left(\frac{4\pi mny}{t} \,\middle|\, \begin{matrix} 1-z-K \\ 0,\ 1-2z \end{matrix}\right) t^{-K-z}(1-t)^{K-1}\ dt,$$

which leads, in view of (4.97), to (4.106), completing the proof.

4.8 Bochner-Chandrasekharan and Narasimhan formula

[ChN2] (along with [ChN3] for the multiple gamma case) seems to be the most fundamental work on equivalent formulations of the Hecke type functional equation, centering around (the summability of) the Riesz sums, whose treatment was excluded in [KTY7]. We make a closer analysis of the Riesz sums, incorporating the work of Anna Walfisz in §9.1.2, and in the following example we elucidate the formula in [ChN2, Lemma 6] which was proved in a more general form in Bochner and Chandrasekharan [BCh1]. Its prototype is due to Hardy [HarMB], [HarRam]. For the completion of Hardy's argument [HarMB], see [CZS2].

Example 4.4 (Bochenr-Chandrasekharan and Narasimhan [ChN2]) In this example we use the residual function for the sum of residues.

Table 4.3   Relation among variables

old	λ_k	μ_k	s	z
new	$2\pi\lambda_k$	$2\pi\mu_k$	$2\sqrt{2\pi}z$	$\frac{s^2}{8\pi}$

With slight changes (table of variables) in (4.82), we have

$$2\sum_{k=1}^{\infty} \frac{\alpha_k}{(2\pi\lambda_k)^{\frac{\nu}{2}}} \left(\frac{s^2}{8\pi}\right)^{\frac{\nu}{2}} K_\nu\left(s\sqrt{\lambda_k}\right)$$

$$= \Gamma(r - \nu) \sum_{k=1}^{\infty} \beta_k \frac{(8\pi)^{r-\nu}}{(s^2 + 16\pi^2 \mu_k)^{r-\nu}}$$

$$+ \frac{1}{2\pi i} \int_C \Gamma(z - \nu) \chi(z) \left(\frac{s^2}{8\pi} \right)^{\nu-z} dz, \tag{4.110}$$

where C is a curve encircling all of s_k, $1 \le k \le L$.

Exercise 4.6 Deduce the following from Exercise 4.3

$$\frac{2}{(4\pi)^\nu} \sum_{k=1}^{\infty} \frac{\alpha_k}{\lambda_k^\nu} \left(s\sqrt{\lambda_k} \right)^{\nu-\rho} K_{\nu-\rho} \left(s\sqrt{\lambda_k} \right)$$

$$= 2^\rho \Gamma(r - \nu + \rho) \sum_{k=1}^{\infty} \beta_k \frac{(8\pi)^{r-\nu}}{(s^2 + 16\pi^2 \mu_k)^{r-\nu+\rho}} \tag{4.111}$$

$$+ \frac{1}{2\pi i} \int_C (8\pi)^{z-\nu} \chi(z) 2^\rho \Gamma(z - \nu + \rho + 1) s^{2\nu-2z-2\rho-2} dz.$$

Also prove it by applying the operator $\left(-\frac{1}{s} \frac{d}{ds} \right)^\rho$ ($0 \le \rho \in \mathbb{Z}$), with the formula (2.34) and (2.6) in mind.

Now we choose $\nu = -\frac{1}{2}, \rho = 0$ in (4.111) to deduce that the left-hand side becomes, on appealing to (2.33)

$$2\sqrt{2\pi} \sum_{k=1}^{\infty} \alpha_k \frac{1}{s} e^{-s\sqrt{\lambda_k}},$$

while the right-hand side

$$\Gamma\left(r + \frac{1}{2} \right) (8\pi)^{r+\frac{1}{2}} \sum_{k=1}^{\infty} \frac{\beta_k}{(s^2 + 16\pi^2 \mu_k)^{r+\frac{1}{2}}}$$

$$+ \frac{1}{2\pi i} \int_C \chi(z) (8\pi)^{z+\frac{1}{2}} \Gamma\left(z + \frac{1}{2} \right) s^{-2z-1} dz.$$

Hence

$$\sum_{k=1}^{\infty} \alpha_k \frac{1}{s} e^{-s\sqrt{\lambda_k}}$$

$$= 2^{3r} \Gamma\left(r + \frac{1}{2} \right) \pi^{r-\frac{1}{2}} \sum_{k=1}^{\infty} \frac{\beta_k}{(s^2 + 16\pi^2 \mu_k)^{r+\frac{1}{2}}} \tag{4.112}$$

$$+ \frac{1}{2\pi i} \int_C \chi(z) (2\pi)^z \pi^{-\frac{1}{2}} 2^{2z} \Gamma\left(z + \frac{1}{2} \right) s^{-2z-1} dz.$$

We note that the residual function can be written as

$$\frac{1}{2\pi i} \int_C \chi(z)(2\pi)^z \frac{\Gamma(2z+1)}{\Gamma(z+1)} s^{-2z-1} \mathrm{d}z$$

by (2.3).

Then we apply the operator $\left(-\frac{1}{s}\frac{d}{ds}\right)^\rho$ again to both sides of (4.111) to deduce that

Corollary 4.4   ([ChN2, Lemma 6])

$$\left(-\frac{1}{s}\frac{d}{ds}\right)^\rho \sum_{k=1}^\infty \alpha_k \frac{1}{s} e^{-s\sqrt{\lambda_k}}$$

$$= 2^{3r+\rho} \Gamma\left(r + \rho + \frac{1}{2}\right) \pi^{r-\frac{1}{2}} \sum_{k=1}^\infty \frac{\beta_k}{(s^2 + 16\pi^2\mu_k)^{r+\rho+\frac{1}{2}}} \tag{4.113}$$

$$+ \frac{1}{2\pi i} \int_C \frac{\chi(z)(2\pi)^z \Gamma(2z+2s+1)2^{-\rho}}{\Gamma(z+\rho+1)} s^{-2z-2\rho-1} \mathrm{d}z$$

by (2.7).

From the shape of (4.113), it is clear that [ChN2] started from the special case of $\nu = -1/2$ of (4.110), which is (4.112), and applied the operator in question.

It is plain, however, from the argument that (4.111) is a proper formulation of (4.113), and the operator originates from the Bessel function theory (2.34), a standpoint gained only from a more general framework of ours.

Chapter 5

The Ewald Expansion or the Incomplete Gamma Series

In this chapter, we state a special case of the main formula whose specifications will give almost all existing Ewald expansions in the G-function hierarchy. Here we mean by an Ewald expansion those formulas given by $H_{1,2}^{2,0} \leftrightarrow H_{1,2}^{1,1}$ or its special cases, especially the incomplete gamma expansion, which is the genesis of the name. We shall treat the case of a single gamma factor which includes the Riemann type and the Hecke type. For the Riemann type cf. (4.29) while for the Hecke type, cf. (4.33). We shall state, however, a unified form starting from the functional equation (4.27).

In the case of the Riemann type, in view of (2.116), we state the confluent hypergeometric function series and specify them as Ewald expansions and for the Hecek type we state the formula in terms of G-functions and specify it in several variants, elucidating the Atkinson-Berndt theorem.

5.1 Ewald expansion for zeta-functions with a single gamma factor

In this section, we consider the general modular relation associated to Dirichlet series that satisfy the functional equation corresponding to that of the Riemann zeta-function and refer to the Siegel-Hecke approach to the Hamburger theorem. It would be appropriate to state some introductory remarks here on the characterization of the Riemann zeta- and allied functions from the functional equation of the Riemann type in reminiscence of H. Hamburger who first made the characterization of the Riemann zeta-function. The authors of [BochCh], [ChM1] and [ChM2] are concerned with bounding the number of linearly independent solutions to the functional equation (5.3), thus leading to the uniqueness of the solution. [ChM2] proves some theorems that necessitates the reflection r to be 1. Use is made of the gaps between the basic sequences $\{\lambda_k\}, \{\mu_k\}$ in the light of [Lev]. In all these investigations, a special case of (5.17) plays an essential role, which in turn was suggested by Siegel's simplest proof [Sieg1] of Hamburger's theorem. Here one sees the importance of exhausting those relations that are equivalent to the functional equation. They may have further use in the future. On the other hand, [Boch1957] goes back to a more general situation of multiple gamma factors but with $r = 0$. On p.30 [Boch1957], in the last passage, Bochner remarks that the method is modeled on [AZW1], which in itself is based on older models and that in the case of an imaginary quadratic field, the relation in question was first given by A. Z. Walfisz [AZW1]. These older models are those of G. H. Hardy [HarCP], [HarDP]. Walfisz developed his method further in [AZW2], [AZW3], [SzW] and [SzW] mainly in order to give explicit formulas for the Riesz sums and omega results (i.e. estimates from below for the resulting error term) for the Piltz divisor problem in an algebraic number field, i.e. he studied the coefficients of the power of the Dedekind zeta-function.

Given two Dirichlet series

$$\varphi(s) = \sum_{k=1}^{\infty} \frac{\alpha_k}{\lambda_k^s}, \tag{5.1}$$

$$\psi(s) = \sum_{k=1}^{\infty} \frac{\beta_k}{\mu_k^s} \tag{5.2}$$

with finite abscissas of absolute convergence σ_φ and σ_ψ, respectively, we assume that they satisfy the functional equation with a simple gamma factor:

$$\chi(s) = \begin{cases} \Gamma(Cs)\varphi(s), & \mathrm{Re}(s) > \sigma_\varphi \\ \Gamma(C(r-s))\psi(r-s), & \mathrm{Re}(s) < r - \sigma_\psi, \end{cases} \tag{5.3}$$

or

$$\chi(s) = \begin{cases} \Gamma\left(s \,\middle|\, \begin{matrix} - \\ (0,C) \end{matrix} ; \begin{matrix} - \\ - \end{matrix} \right)\varphi(s), & \mathrm{Re}(s) > \sigma_\varphi \\ \Gamma\left(r - s \,\middle|\, \begin{matrix} - \\ (0,C) \end{matrix} ; \begin{matrix} - \\ - \end{matrix} \right)\psi(1-s), & \mathrm{Re}(s) < r - \sigma_\psi. \end{cases} \tag{5.4}$$

By Theorem 4.3, we have the following modular relation:

Theorem 5.1

$$\sum_{k=1}^{\infty} \frac{\alpha_k}{\lambda_k^s} H\!\left(z\lambda_k \,\middle|\, \left(\begin{array}{cc} - & ;\ - \\ (Cs,C)\,; & - \end{array}\right) \oplus \Delta\right)$$
$$= \sum_{k=1}^{\infty} \frac{\beta_k}{\mu_k^{r-s}} H\!\left(\frac{\mu_k}{z} \,\middle|\, \left(\begin{array}{cc} - & ;\ - \\ (C(r-s),C)\,; & - \end{array}\right) \oplus \Delta^*\right)$$
$$+ \sum_{k=1}^{L} \mathrm{Res}\!\left(\Gamma(w-s\,|\,\Delta)\,\chi(w)\,z^{s-w},\, w=s_k\right).$$

In the traditional notation, if

$$\Delta = \left(\begin{array}{cc} \{(1-a_j,A_j)\}_{j=1}^{n}; & \{(a_j,A_j)\}_{j=n+1}^{p} \\ \{(b_j,B_j)\}_{j=1}^{m} & ;\{(1-b_j,B_j)\}_{j=m+1}^{q} \end{array}\right) \in \Omega \qquad (5.5)$$

with its associated Gamma factor

$$\Gamma(s\,|\,\Delta) = \frac{\displaystyle\prod_{j=1}^{m}\Gamma(b_j+B_js)\ \prod_{j=1}^{n}\Gamma(a_j-A_js)}{\displaystyle\prod_{j=n+1}^{p}\Gamma(a_j+A_js)\ \prod_{j=m+1}^{q}\Gamma(b_j-B_js)}, \qquad (5.6)$$

we have

$$\sum_{k=1}^{\infty}\frac{\alpha_k}{\lambda_k^s} H_{p,q+1}^{m+1,n}\!\left(z\lambda_k \,\middle|\, \begin{array}{c} \{(1-a_j,A_j)\}_{j=1}^{n},\{(a_j,A_j)\}_{j=n+1}^{p} \\ (Cs,C),\{(b_j,B_j)\}_{j=1}^{m},\{(1-b_j,B_j)\}_{j=m+1}^{q} \end{array}\right) \qquad (5.7)$$
$$= \sum_{k=1}^{\infty}\frac{\beta_k}{\mu_k^{r-s}} H_{q,p+1}^{n+1,m}\!\left(\frac{\mu_k}{z} \,\middle|\, \begin{array}{c} \{(1-b_j,B_j)\}_{j=1}^{m},\{(b_j,B_j)\}_{j=m+1}^{q} \\ (C(r-s),C),\{(a_j,A_j)\}_{j=1}^{n},\{(1-a_j,A_j)\}_{j=n+1}^{p} \end{array}\right)$$
$$+ \sum_{k=1}^{L}\mathrm{Res}\!\left(\Gamma(w-s\,|\,\Delta)\,\chi(w)\,z^{s-w},\,w=s_k\right).$$

In what follows we shall specify the formula in the following theorem, which is a special case of (5.7), to deduce Ewald expansions.

Theorem 5.2 [Ts] *For zeta-functions that satisfy the functional equation* (5.3), *we have a general Ewald expansion* $H_{1,2}^{2,0} \leftrightarrow H_{1,2}^{1,1}$ *which is equivalent* *to (5.3):*

$$\sum_{k=1}^{\infty} \frac{\alpha_k}{\lambda_k^s} H_{1,2}^{2,0}\left(z\lambda_k \left| \begin{array}{c} (a, A) \\ (Cs, C), (b, B) \end{array}\right.\right)$$

$$= \sum_{k=1}^{\infty} \frac{\beta_k}{\mu_k^{r-s}} H_{1,2}^{1,1}\left(\frac{\mu_k}{z} \left| \begin{array}{c} (1 - b, B) \\ (C(r - s), C), (1 - a, A) \end{array}\right.\right) \qquad (5.8)$$

$$+ \sum_{k=1}^{L} \text{Res}\left(\frac{\Gamma(b - Bs + Bw)}{\Gamma(a - As + Aw)} \chi(w) z^{s-w}, w = s_k\right)$$

and its special case

$$\frac{1}{C} \sum_{k=1}^{\infty} \frac{\alpha_k}{\lambda_k^s} G_{1,2}^{2,0}\left((z\lambda_k)^{\frac{1}{C}} \left| \begin{array}{c} a \\ Cs, b \end{array}\right.\right)$$

$$= \frac{1}{C} \sum_{k=1}^{\infty} \frac{\beta_k}{\mu_k^{r-s}} G_{1,2}^{1,1}\left(\left(\frac{\mu_k}{z}\right)^{\frac{1}{C}} \left| \begin{array}{c} 1 - b \\ C(r - s), 1 - a \end{array}\right.\right) \qquad (5.9)$$

$$+ \sum_{k=1}^{L} \text{Res}\left(\frac{\Gamma(b - Cs + Cw)}{\Gamma(a - Cs + Cw)} \chi(w) z^{s-w}, w = s_k\right)$$

(5.9) is the special case of (5.8) with $A = B = C$ (cf. (2.93)). Theorem 5.2 is due to the second author, which has a lot of applications.

5.1.1 *Confluent hypergeometric series imply incomplete gamma series, Ewald expansions*

Theorem 5.3 *Assumptions being as above, we have the confluent hypergeometric expansion*

$$\frac{1}{C} z^s \sum_{k=1}^{\infty} \alpha_k e^{-(z\lambda_k)^{1/C}} U\left(a - b, Cs - b + 1, (z\lambda_k)^{1/C}\right)$$

$$= \frac{1}{C} (z^{-1})^{r-s} \frac{\Gamma(b + C(r - s))}{\Gamma(a + C(r - s))} \sum_{k=1}^{\infty} \beta_k \, {}_1F_1\left(\begin{array}{c} b + C(r - s) \\ a + C(r - s) \end{array}; -\left(\frac{\mu_k}{z}\right)^{1/C}\right)$$

$$+ \sum_{k=1}^{L} \text{Res}\left(\frac{\Gamma(b - Cs + Cw)}{\Gamma(a - Cs + Cw)} \chi(w) z^{s-w}, w = s_k\right),$$

$$(5.10)$$

entailing the incomplete gamma series, or the Ewald expansion

$$\sum_{k=1}^{\infty} \frac{\alpha_k}{\lambda_k^s} \Gamma\left(Cs, (z\lambda_k)^{1/C}\right)$$

$$= \sum_{k=1}^{\infty} \frac{\beta_k}{\mu_k^{r-s}} \left(\Gamma(C(r-s)) - \Gamma\left(C(r-s), \left(\frac{\mu_k}{z}\right)^{1/C}\right)\right) \quad (5.11)$$

$$+ \sum_{k=1}^{L} \operatorname{Res}\left(\frac{1}{-s+w} \chi(w) z^{s-w}, w = s_k\right).$$

Proof. (5.10) follows from (5.9) in view of (2.117) and (2.111).

(5.11) is a specification of (5.10) with $a = 1, b = 0$ in view of (2.120) and (2.113). □

Corollary 5.1

(i) The Riemann type functional equation (5.3) with $C = \frac{1}{2}$ entails

$$\sum_{k=1}^{\infty} \frac{\alpha_k}{\lambda_k^s} \Gamma\left(\frac{s}{2}, z^2\lambda_k^2\right) = \sum_{k=1}^{\infty} \frac{\beta_k}{\mu_k^{r-s}} \left(\Gamma\left(\frac{r-s}{2}\right) - \Gamma\left(\frac{r-s}{2}, \frac{\mu_k^2}{z^2}\right)\right)$$

$$+ \sum_{k=1}^{L} \operatorname{Res}\left(\frac{1}{-s+w} \chi(w) z^{s-w}, w = s_k\right). \quad (5.12)$$

(ii) The Hecke type functional equation (5.3) entails

$$\sum_{k=1}^{\infty} \frac{\alpha_k}{\lambda_k^s} \Gamma(s, z\lambda_k) = \sum_{k=1}^{\infty} \frac{\beta_k}{\mu_k^{r-s}} \left(\Gamma(r-s) - \Gamma\left(r-s, \frac{\mu_k}{z}\right)\right)$$

$$+ \sum_{k=1}^{L} \operatorname{Res}\left(\frac{1}{-s+w} \chi(w) z^{s-w}, w = s_k\right). \quad (5.13)$$

Formula (5.13) is stated as [KTY7, Theorem 1, (1.6)] which is a basis for the Riemann-Siegel integral formula developed in [KTY7, §4]; for historical remarks, cf. also [KTY7, §5.4]. The incomplete gamma series have a long and rich history, dating back to Riemann. Terras renewed this theory, cf. [Te5] (also [Te1], [Te2], [Te3], [Te4]), after the works of Russian mathematicians Kuz'min ([Kuz'1], [Kuz'2]), Linnik ([Lin1], [Lin2]) and Lavrik ([Lav1], [Lav2]). [GV] uses Lavrik's method and introduces the incomplete multiple gamma function, There is a whole book ([CZ]) devoted to the theory of the incomplete gamma function. Some of the results stated there could be interpreted within the G-function hierarchy.

Prudnikov [PBM, 7.11.4.10, p. 584].

$$\Psi(n,b;z) = \frac{1}{(n-1)(2-b)_{n-1}}\left(\frac{d}{dz}\right)^{n-1}\left(z^{n-b}e^z\Gamma(b-1,z)\right)$$

is noteworthy in relation to the Hamburger theorem due to Bochner (and Chandrasekharan) [Boch1957] ([BCh1]).

5.1.2   *Bochner-Chandrasekharan formula as* $H_{1,2}^{2,0} \leftrightarrow H_{1,2}^{1,1}$

Along with (5.9), there are many other specifications of (5.7). We state one which leads to the formula of Bochner and Chandrasekharan et al.

Theorem 5.4   *Assumptions being as above, we have*

$$\frac{1}{B}\sum_{k=1}^{\infty}\frac{\alpha_k}{\lambda_k^s}(z\lambda_k)^{b/B}e^{-(z\lambda_k)^{1/B}}$$
$$= \frac{1}{C}(z^{-1})^{r-s}\sum_{k=1}^{\infty}\beta_k\,{}_1\Psi_1\left(\begin{matrix}(b+B(r-s),\frac{B}{C})\\(Cr,1)\end{matrix};-\left(\frac{\mu_k}{z}\right)^{1/C}\right) \qquad (5.14)$$
$$+ \sum_{k=1}^{L}\mathrm{Res}\left(\frac{\Gamma(b-Bs+Bw)}{\Gamma(Cw)}\chi(w)\,z^{s-w},w=s_k\right).$$

Proof.   Choosing $a = Cs, A = C$ and apply the reduction-augmentation formula (2.96), we obtain an intermediate formula

$$\sum_{k=1}^{\infty}\frac{\alpha_k}{\lambda_k^s}\,H_{0,1}^{1,0}\left(z\lambda_k\,\middle|\,\begin{matrix}-\\(b,B)\end{matrix}\right)$$
$$= \sum_{k=1}^{\infty}\frac{\beta_k}{\mu_k^{r-s}}\,H_{1,2}^{1,1}\left(\frac{\mu_k}{z}\,\middle|\,\begin{matrix}(1-b,B)\\(C(r-s),C),(1-Cs,\overset{\bullet}{C})\end{matrix}\right) \qquad (5.15)$$
$$+ \sum_{k=1}^{L}\mathrm{Res}\left(\frac{\Gamma(b-Bs+Bw)}{\Gamma(Cw)}\chi(w)\,z^{s-w},w=s_k\right),$$

which is of interest in its own right.   □

Berndt [BeIII, Theorem 10.1] has generalized the classical result of Szegö's [SzLP] pertaining to Laguerre polynomials, which is incorporated as [BK, Theorem 8.8], and which we state it as Corollary 5.2), (iii). We first introduce the Laguerre polynomials.

Definition 5.1 For $0 \le n \in \mathbb{Z}$ let

$$L_n^\alpha(x) = \frac{1}{n!} e^x x^{-\alpha} \frac{d^n}{d^n x}(e^x x^{n+\alpha}) \tag{5.16}$$

be the **Laguerre polynomial**.

Corollary 5.2

(i) *The Riemann type functional equation (5.3) with $C = \frac{1}{2}$ entails*

$$2\sum_{k=1}^\infty \frac{\alpha_k}{\lambda_k^s} e^{-z^2 \lambda_k^2} = 2\frac{\Gamma\left(\frac{r-s}{2}\right)}{\Gamma\left(\frac{r}{2}\right)} z^{s-r} \sum_{k=1}^\infty \beta_k \, {}_1F_1\left(\frac{r-s}{2} ; -\frac{\mu_k^2}{z^2}\right) \tag{5.17}$$

$$+ \sum_{k=1}^L \mathrm{Res}\left(\frac{\Gamma\left(-\frac{s}{2} + \frac{w}{2}\right)}{\Gamma(\frac{w}{2})} \chi(w) z^{s-w}, w = s_k\right).$$

(ii) *The Hecke type functional equation (5.3) with $C = 1$ on entails*

$$\sum_{k=1}^\infty \frac{\alpha_k}{\lambda_k^s} e^{-z\lambda_k} = \frac{\Gamma(r-s)}{\Gamma(r)} z^{s-r} \sum_{k=1}^\infty \beta_k \, {}_1F_1\left(\begin{matrix} r-s \\ r \end{matrix} ; -\frac{\mu_k}{z}\right)$$

$$+ \sum_{k=1}^L \mathrm{Res}\left(\frac{\Gamma(-s+w)}{\Gamma(w)} \chi(w) z^{s-w}, w = s_k\right). \tag{5.18}$$

(iii) *The Hecke type functional equation (5.3) with $C = 1$ and $2\pi\lambda_k$ in place of λ_k and so entails*

$$\sum_{k=1}^\infty \alpha_k (2\pi\lambda_k)^n e^{-2\pi z\lambda_k} = n! \, z^{-r-n} \sum_{k=1}^\infty \beta_k \, L_n^{(r-1)}\left(\frac{2\pi\mu_k}{z}\right) \tag{5.19}$$

$$+ \sum_{k=1}^L \mathrm{Res}\left(\frac{\Gamma(n+w)}{\Gamma(w)} \chi(w) z^{-n-w}, w = s_k\right).$$

Proof. (i) Indeed, we have

$$2\sum_{k=1}^\infty \frac{\alpha_k}{\lambda_k^s} e^{-(z\lambda_k)^2} = 2\left(z^{-1}\right)^{r-s} \sum_{k=1}^\infty \beta_k \, {}_1\Psi_1\left(\begin{matrix} (\frac{r-s}{2}, 1) \\ (\frac{r}{2}, 1) \end{matrix} ; -\left(\frac{\mu_k}{z}\right)^2\right)$$

$$+ \sum_{k=1}^L \mathrm{Res}\left(\frac{\Gamma\left(-\frac{s}{2} + \frac{w}{2}\right)}{\Gamma(\frac{w}{2})} \chi(w) z^{s-w}, w = s_k\right). \tag{5.20}$$

(ii) This is a special case of (2.51) with $B = C = 1$, $b = 0$ combined with (2.51).

(iii) This is the special case of Theorem 5.4 with $B = C = 1$, $b = 0$, Choosing $s = -n$ and appealing to the well-known relation

$$L_n^\alpha(z) = \frac{(-1)^n}{n!} \Psi(-n, \alpha + 1; z) \tag{5.21}$$

$$= \frac{(\alpha + 1)_n}{n!} \Phi(-n, \alpha + 1; z)$$

$$= {}_1F_1\left(\begin{array}{c} -n \\ \alpha + 1 \end{array}; z\right)$$

completes the proof in view of (5.21) and

$$\Gamma(-n - \alpha) = \frac{(-1)^n}{(\alpha + 1)_n} \tag{5.22}$$

for integer $n \geq 0$ by (2.41). □

The special case of (5.18) with $s = 0$ is a formula due to Bochner [Boch1957] (which is first stated in [BCh1]) and is restated in Berndt [BeIV].

Exercise 5.1   Prove that the special case $C = 1$ of (5.9)

$$\sum_{k=1}^\infty \frac{\alpha_k}{\lambda_k^s} G_{1,2}^{2,0}\left(z\lambda_k \left|\begin{array}{c} b \\ s, a \end{array}\right.\right)$$

$$= \sum_{k=1}^\infty \frac{\beta_k}{\mu_k^{r-s}} G_{1,2}^{1,1}\left(\frac{\mu_k}{z} \left|\begin{array}{c} 1 - a \\ r - s, 1 - b \end{array}\right.\right) \tag{5.23}$$

$$+ \sum_{k=1}^L \mathrm{Res}\left(\frac{\Gamma(a - s + w)}{\Gamma(b - s + w)} \chi(w) z^{s-w}, w = s_k\right)$$

implies

$$
\sum_{k=1}^{\infty} \alpha_k \left\{ \frac{\Gamma(a-s)}{\Gamma(a)} z^s \, {}_1F_1\left(\begin{matrix} 1-a \\ 1-a+s \end{matrix} ; -z\lambda_k \right) \right.
$$

$$
\left. + \frac{\Gamma(s-a)}{\Gamma(s)} \frac{z^a}{\lambda_k^{s-a}} \, {}_1F_1\left(\begin{matrix} 1-s \\ 1-s+a \end{matrix} ; -z\lambda_k \right) \right\}
$$

$$
= \frac{\Gamma(a+r-s)}{\Gamma(a+r)} z^{s-r} \sum_{k=1}^{\infty} \beta_k \, {}_1F_1\left(\begin{matrix} a+r-s \\ a+r \end{matrix} ; -\frac{\mu_k}{z} \right) \tag{5.24}
$$

$$
+ \sum_{k=1}^{L} \mathrm{Res}\left(\frac{\Gamma(a-s+w)}{\Gamma(a+w)} \chi(w) z^{s-w}, w = s_k \right)
$$

and that (5.24) further reduces to

$$
\sum_{k=1}^{\infty} \frac{\alpha_k}{\lambda_k^s} G_{1,2}^{2,0}\left(z\lambda_k \left| \begin{matrix} 1 \\ s,0 \end{matrix} \right. \right)
$$

$$
= \sum_{k=1}^{\infty} \frac{\beta_k}{\mu_k^{r-s}} G_{1,2}^{1,1}\left(\frac{\mu_k}{z} \left| \begin{matrix} 1 \\ r-s,0 \end{matrix} \right. \right) \tag{5.25}
$$

$$
+ \sum_{k=1}^{L} \mathrm{Res}\left(\frac{\Gamma(-s+w)}{\Gamma(1-s+w)} \chi(w) z^{s-w}, w = s_k \right).
$$

Solution We apply the formula in Exercise 2.11 and (2.111) to write (5.23) as

$$
\sum_{k=1}^{\infty} \alpha_k \left\{ \frac{\Gamma(a-s)}{\Gamma(b-s)} z^s \, {}_1F_1\left(\begin{matrix} 1-b+s \\ 1-a+s \end{matrix} ; -z\lambda_k \right) \right.
$$

$$
\left. + \frac{\Gamma(s-a)}{\Gamma(b-a)} \frac{z^a}{\lambda_k^{s-a}} \, {}_1F_1\left(\begin{matrix} 1-b+a \\ 1-s+a \end{matrix} ; -z\lambda_k \right) \right\}
$$

$$
= \frac{\Gamma(a+r-s)}{\Gamma(b+r-s)} z^{s-r} \sum_{k=1}^{\infty} \beta_k \, {}_1F_1\left(\begin{matrix} a+r-s \\ b+r-s \end{matrix} ; -\frac{\mu_k}{z} \right)
$$

$$
+ \sum_{k=1}^{L} \mathrm{Res}\left(\frac{\Gamma(a-s+w)}{\Gamma(b-s+w)} \chi(w) z^{s-w}, w = s_k \right),
$$

which leads to (5.24), on putting $b = a + s$.

(5.25) is the special case of (5.24) with $a = 0, b = 1$.

5.2   Atkinson-Berndt Abel mean

Example 5.1 (Atkinson-Berndt Abel mean)   In [BeIV], Berndt proves the following result, which is an extension of Atkison's [Atk2], both of which were elucidated briefly in [CZS3].

Suppose for simplicity that χ has at most simple poles at s_k with residue ρ_k, $1 \leq k \leq L$, respectively. For every $\delta > 0$ ($\mathrm{Re}\,\delta > 0$) suppose that the weighted Lambert series

$$\tilde{\varphi}(s, \delta) = \sum_{k=0}^{\infty} \frac{\alpha_k}{\lambda_k^s} e^{-\lambda_k \delta}$$

converges for $\sigma > \sigma_0$, where $\sigma_0 \leq \sigma_\varphi$. Then for $\sigma > \sigma_0, s \neq s_k$,

$$\lim_{\delta \to 0} \left\{ \tilde{\varphi}(s, \delta) - \sum_{k=1}^{L} \frac{\rho_k \Gamma(s_k - s)}{\Gamma(s_k)} \delta^{s-s_k} \right\} = \varphi(s). \qquad (5.26)$$

Here the limit $\delta \to 0$ means $\delta \to +0$ inside ($\mathrm{Re}\,\delta > 0$), which therefore suggests the name of Abel mean.

In this section, we shall use Theorem 5.4 to show that the Atkinson-Berndt Abel mean is nothing but another proof of the functional equation.

Corollary 5.3   *The Atkinson-Berndt Abel mean (5.26) for $\sigma < \min\{r, r - \sigma_\psi\}$ leads to the counterpart of the functional equation (4.80):*

$$\lim_{\delta \to 0} \left\{ \tilde{\varphi}(s, \delta) - \sum_{k=1}^{L} \frac{\rho_k \Gamma(s_k - s)}{\Gamma(s_k)} \delta^{s-s_k} \right\} = \frac{\Gamma(r - s)}{\Gamma(s)} \psi(r - s). \qquad (5.27)$$

Proof.   Indeed, for $\sigma > \sigma_\varphi$ (and $r/2$), we may take the limit in two different ways. Theorem 5.4 gives

$$\tilde{\varphi}(s, \delta) = \delta^{s-r} \frac{\Gamma(r - s)}{\Gamma(r)} \sum_{k=1}^{\infty} b_k {}_1F_1\left(\begin{matrix} r - s \\ r \end{matrix} ; -\frac{\mu_k}{\delta} \right)$$
$$+ \sum_{k=1}^{L} \frac{\Gamma(s_k - s)}{\Gamma(s_k)} \rho_k \delta^{s-s_k}. \qquad (5.28)$$

Now we apply ([ErdH, (3),p.276])

$$\Phi(a, c; x) = \frac{\Gamma(c)}{\Gamma(c - a)} (-x)^{-a} \left\{ 1 + O\big(|x|^{-1}\big) \right\}, \ \mathrm{Re}\,x \to -\infty \qquad (5.29)$$

to (5.28) to deduce (5.27) for $\sigma < \min\{r, r - \sigma_\psi\}$. $\square$

Exercise 5.2 By computing

$$\lim_{\delta \to 0} \left\{ \Gamma(s)\tilde{\varphi}(s, \delta) - \int_0^1 x^{s-1} P(x + \delta) \, \mathrm{d}x \right\} \qquad (5.30)$$

in two different ways, deduce (5.26). Appeal to Exercise 2.9 for necessary formulas.

Solution Let

$$\tilde{\varphi}(0, \delta) = \sum_{k=1}^{\infty} a_k e^{-\lambda_k \delta}, \quad \tilde{\psi}(0, \delta) = \sum_{k=1}^{\infty} b_k e^{-\mu_k \delta}$$

denote the Lambert series (cf. Part I, Chapter 2, Section Lambert series above, [KTY7] for details). Berndt considers the Mellin transform

$$\int_0^{\infty} x^{s-1} \tilde{\varphi}(0, x + \delta) \, \mathrm{d}x$$

of $\tilde{\varphi}(0, x + \delta)$ and, following Riemann's idea, splits the integral into two parts $(0, 1)$ and $(1, \infty)$, and in the latter integral, he uses the modular relation (after effecting the change of variable $x + \delta \leftrightarrow (x + \delta)^{-1}$)

$$\tilde{\varphi}(0, x) = x^{-r}\tilde{\psi}(0, x^{-1}) + \mathrm{P}(x),$$

where

$$\mathrm{P}(x) = \sum_{k=1}^{L} \rho_k x^{-s_k},$$

whence it follows that

$$\Gamma(s)\tilde{\varphi}(s, \delta) - \int_0^1 x^{s-1} \mathrm{P}(x + \delta) \, \mathrm{d}x$$

$$= \int_0^1 x^{s-1}(x + \delta)^{-r}\tilde{\psi}\left(0, \frac{1}{x + \delta}\right) \mathrm{d}x + \int_1^{\infty} x^{s-1}\tilde{\varphi}(x + \delta) \, \mathrm{d}x. \qquad (5.31)$$

We note that this is also valid for $\delta = 0$, reading for $\sigma > \sigma_\varphi$

$$\int_0^1 x^{s-1-r}\tilde{\psi}\left(0, \frac{1}{x}\right) \mathrm{d}x + \int_1^{\infty} x^{s-1}\tilde{\varphi}(0, x) \, \mathrm{d}x = \Gamma(s)\tilde{\varphi}(s) - \sum_{k=1}^{L} \frac{\rho_k}{s - s_k} \qquad (5.32)$$

since $\int_0^1 x^{s-1} \mathrm{P}(x) \, \mathrm{d}x = \sum_{k=1}^L \frac{\rho_k}{s-s_k}$. By analytic continuation, (5.32) is valid for all $s \neq s_k$. By (2.71), we have

$$\Gamma(s)\tilde{\varphi}(s,\delta) - \int_0^1 x^{s-1} \mathrm{P}(x+\delta) \, \mathrm{d}x \tag{5.33}$$

$$= \Gamma(s)\tilde{\varphi}(s,\delta) - \sum_{k=1}^L \frac{\rho_k}{s-s_k} {}_2F_1\left(\begin{matrix} s_k, s_k - s \\ s_k - s + 1 \end{matrix}; -\delta\right)$$

$$- s^{-1} \sum_{k=1}^L \rho_k \frac{\Gamma(s+1)\Gamma(s_k-s)}{\Gamma(s_k)} \delta^{s-s_k}.$$

Equating (5.32) and (5.33), taking the limit $\delta \to 0$, whereby using (2.81), we conclude that

$$\lim_{\delta \to 0} \left\{ \Gamma(s)\tilde{\varphi}(s,\delta) - \sum_{k=1}^L \frac{\rho_k}{s-s_k} - s^{-1} \sum_{k=1}^L \rho_k \frac{\Gamma(s+1)\Gamma(s_k-s)}{\Gamma(s_k)} \delta^{s-s_k} \right\}$$

$$= \Gamma(s)\varphi(s) - \sum_{k=1}^L \frac{\rho_k}{s-s_k},$$

where we used (2.72). whence on eliminating the common term,

$$\lim_{\delta \to 0} \left\{ \Gamma(s)\tilde{\varphi}(s,\delta) - \Gamma(s) \sum_{k=1}^L \rho_k \frac{\Gamma(s_k-s)}{\Gamma(s_k)} \delta^{s-s_k} \right\} = \Gamma(s)\varphi(s), \tag{5.34}$$

i.e. (5.26).

Remark 5.1   *On account of* (2.73), $B_z(\alpha, \beta)$ *indicates the incomplete beta function, Formula* (2.81) *looks like a modular relation, and we might hope this may entail a proof of the modular relation.*

5.2.1   *Landau's exposition*

We now mention Landau's exposition [LanE] of Euler's prototype work [Eu] preceding Riemann by about 110 years. We use the function in Corollary 1.3, which has appeared in various notation in the literature, often as $\beta(s)$ or $\eta(s)$. We temporarily label it by $\phi(s)$.

$$\phi(s) := (1 - 2^{1-s})\zeta(s) = \sum_{n=1}^\infty \frac{(-1)^{n-1}}{n^s}.$$

Then the functional equation for the Riemann zeta-function may be stated as

$$\frac{\phi(1-s)}{\phi(s)} = \frac{-\Gamma(s)\,(2^s - 1)}{(2^{s-1} - 1)\,\pi^s} \cos\frac{\pi s}{2}. \tag{5.35}$$

Euler [Eu] essentially conjectured (and verified for $s = \frac{2k+1}{2}$) the Abel mean formula

$$\lim_{x \to 1-} \frac{\text{Li}_{1-s}(-x)}{\text{Li}_s(-x)} = -\pi^{-s}\Gamma(s)\frac{(1 - 2^s)}{(1 - 2^{s-1})} \cos\frac{\pi n}{2}, \tag{5.36}$$

for those values of s different from the zeros of the denominators.

We shall now use a sledgehammer to hit a nail down, essentially following [LanE]. The starting point is

$$\lim_{x \to 1-} \left(x^w \Phi(x, s, w) - \Gamma(1 - s)\left(\log\frac{1}{x}\right)^{s-1} \right) = \zeta(s, w), \tag{5.37}$$

with $\Phi(x, s, w)$ denoting the **Hurwitz-Lerch zeta-function**, which is a special case of the generalized Hurwitz-Lerch zeta-function defined below:

$$\Phi(x, s, w) = \Phi_1(-\log x, s, w). \tag{5.38}$$

Cf. (3.80) for the L-function.

Following Mellin [Mel] and [KTT1], we introduce the **generalized Hurwitz-Lerch zeta-function** by

$$\Phi_A(z, s, w) = \sum_{n=0}^{\infty} \frac{e^{-z(n+w)^A}}{(n + w)^s}. \tag{5.39}$$

We recall the expansion due to Johnson in the formulation of [KTT1].

Proposition 5.1 (Johnson [Joh]) *The generalized Hurwitz-Lerch zeta-function admits the expansion*

$$\Phi_A(z, s, a) = \sum_{r=0}^{\infty} \zeta(s - Ar, a)\frac{(-z^A)^r}{r\,!} + \frac{\Gamma\left(\frac{1-s}{A}\right)}{A}\,z^{s-1}, \tag{5.40}$$

the series on the right being convergent at the origin.

The prototype of (5.40) is due to Mellin [Mel, (77), p.40], which reads

$$\Phi_A(z,s,w) = \frac{1}{A}\Gamma\left(\frac{1-s}{A}\right)z^{\frac{s-1}{A}} + \sum_{\nu=0}^{k}\zeta(s-A\nu,w)\frac{(-z)^{\nu}}{\nu!} \qquad (5.41)$$
$$+ \frac{1}{2\pi i}\int_{(c)}\Gamma(\omega)\zeta(A\omega+s,w)z^{-\omega}d\omega.$$

Formula (5.37) is a special case of (5.40) or (5.41). A very special case of (5.37) is

$$\lim_{x\to 1-}\left(\text{Li}_s(x) - \Gamma(1-s)\left(\log\frac{1}{x}\right)^{s-1}\right) = \zeta(s) \qquad (5.42)$$

valid for those values of s which are not natural numbers.

5.2.2   *Screened Coulomb potential*

Now we confine ourselves to the Epstein zeta-function to elucidate the crystal symmetry. Since this topic has been thoroughly expounded in [Vista, Chapter 6] and [Vi2, Chapter 6], we refer to them for basic notions and use the notation given therein.

Here, appealing to [CZS3], we clarify the meaning of screened Coulomb potential considered by Borweins [BB2], resp. by Chaba-Pathria [CP1]-[CP4], Hall [Hal], Hautot [Hau] in 2- and 3-dimensional lattice structures.

Let $Y = (y_{ij})$ be a positive definite $n \times n$ real symmetric matrix. For $g, h \in \mathbb{R}^n$ define the general Epstein zeta-function (of Hurwitz-Lerch type) by

$$Z(Y,g,h,s) = \sum_{\substack{a\in\mathbb{Z}^n \\ a+g\neq z}} \frac{e^{2\pi i h\cdot a}}{Y[a+g]^s}, \quad \sigma > \frac{n}{2}, \qquad (5.43)$$

and incorporate the completion

$$\Lambda(Y,g,h,s) = \pi^{-s}\Gamma(s)Z(Y,g,h,s). \qquad (5.44)$$

Theorem 5.5   *The (completed) Epstein zeta-function $\Lambda(Y,g,h,s)$ satisfies the functional equation of Hecke type*

$$\Lambda(Y,g,h,s) = \frac{1}{\sqrt{|Y|}}e^{-2\pi i g\cdot h}\Lambda\left(Y^{-1},h,-g,\frac{n}{2}-s\right). \qquad (5.45)$$

Cf. Example 5.3 below.

Proof of the functional equations by induction on the size n of the coefficient matrix Y.

For positive definite $(n + m) \times (n + m)$ symmetric real matrix Y, let

$$Y = \begin{pmatrix} A & B \\ {}^tB & C \end{pmatrix}, g = \begin{pmatrix} g_1 \\ g_2 \end{pmatrix}, h = \begin{pmatrix} h_1 \\ h_2 \end{pmatrix}, \tag{5.46}$$

and

$$D = C - {}^tBA^{-1}B, \tag{5.47}$$

where, A is a positive definite $n \times n$ symmetric real matrix and C is a positive definite $m \times m$ symmetric real matrix, then

$$\Lambda(Y, g, h, s)$$
$$= \delta(g_2) \, e^{-2\pi i g_2 \bullet h_2} \Lambda(A, g_1, h_1, s) + \delta(h_1) \, \frac{1}{\sqrt{|A|}} \, \Lambda(D, g_2, h_2, s - \tfrac{n}{2})$$
$$+ \frac{2 \, e^{-2\pi i g_1 \bullet h_1}}{\sqrt{|A|}} \, H_{n,m}(Y, g, h, s) \tag{5.48}$$

where

$$H_{n,m}(Y, g, h, s)$$
$$= \sum_{\substack{a \in \mathbb{Z}^n \\ a+h_1 \neq 0}} \sum_{\substack{b \in \mathbb{Z}^m \\ b+g_2 \neq 0}} e^{2\pi i(-g_1 \bullet a + h_2 \bullet b)} e^{-2\pi i A^{-1} B(b+g_2) \bullet (a+h_1)}$$
$$\times \sqrt{\frac{A^{-1}[a + h_1]}{D[b + g_2]}}^{s - \frac{n}{2}} K_{s - \frac{n}{2}} \left(2\sqrt{A^{-1}[a + h_1] D[b + g_2]} \, \pi \right). \tag{5.49}$$

By the assumption of induction

$$\Lambda(A, g_1, h_1, s) = \frac{e^{-2\pi i g_1 \bullet h_1}}{\sqrt{|A|}} \, \Lambda\big(A^{-1}, h_1, -g_1, \tfrac{n}{2} - s\big) \tag{5.50}$$

and

$$\Lambda\big(D, g_2, h_2, s - \tfrac{n}{2}\big) = \frac{e^{-2\pi i g_2 \bullet h_2}}{\sqrt{|D|}} \, \Lambda\big(D^{-1}, h_2, -g_2, \tfrac{n+m}{2} - s\big) \tag{5.51}$$

we have

$$\Lambda(Y, g, h, s)$$
$$= \delta(g_2) \frac{e^{-2\pi i g \bullet h}}{\sqrt{|A|}} \Lambda\left(A^{-1}, h_1, -g_1, \tfrac{n}{2} - s\right)$$
$$+ \delta(h_1) \frac{e^{-2\pi i g_2 \bullet h_2}}{\sqrt{|Y|}} \Lambda\left(D^{-1}, h_2, -g_2, \tfrac{n+m}{2} - s\right) \tag{5.52}$$
$$+ \frac{2\, e^{-2\pi i g_1 \bullet h_1}}{\sqrt{|A|}} H_{n,m}(Y, g, h, s).$$

On the other hand, if we set

$$Y^{-1} = \begin{pmatrix} F & {}^tG \\ G & H \end{pmatrix}, \quad U = \begin{pmatrix} O & I_n \\ I_m & O \end{pmatrix}, \tag{5.53}$$

then, we have

$${}^tU Y^{-1} U = \begin{pmatrix} H & G \\ {}^tG & F \end{pmatrix}, \tag{5.54}$$

and

$$H = D^{-1}, F - {}^tG H^{-1} G = A^{-1}, H^{-1} G = -{}^tB A^{-1}. \tag{5.55}$$

Therefore, we have

$$\frac{e^{-2\pi i g \bullet h}}{\sqrt{|Y|}} \Lambda\left(Y^{-1}, h, -g, \tfrac{n+m}{2} - s\right)$$
$$= \frac{e^{-2\pi i g \bullet h}}{\sqrt{|Y|}} \Lambda\left({}^tU Y^{-1} U, U^{-1} h, -{}^tU g, \tfrac{n+m}{2} - s\right)$$
$$= \delta(h_1) \frac{e^{-2\pi i g_2 \bullet h_2}}{\sqrt{|Y|}} \Lambda\left(D^{-1}, h_2, -g_2, \tfrac{n+m}{2} - s\right) \tag{5.56}$$
$$+ \delta(g_2) \frac{e^{-2\pi i g \bullet h}}{\sqrt{|A|}} \Lambda\left(A^{-1}, h_1, -g_1, \tfrac{n}{2} - s\right)$$
$$+ \frac{2\, e^{-2\pi i g_1 \bullet h_1}}{\sqrt{|A|}} H_{m,n}\left({}^tU Y^{-1} U, U^{-1} h, -{}^tU g, \tfrac{n+m}{2} - s\right).$$

Since

$$
\begin{aligned}
H_{m,n}&\bigl({}^tUY^{-1}U, U^{-1}h, -{}^tUg, \tfrac{n+m}{2} - s\bigr) \\
&= \sum_{\substack{b\in\mathbb{Z}^m \\ b-g_2\neq 0}} \sum_{\substack{a\in\mathbb{Z}^n \\ a+h_1\neq 0}} e^{2\pi i(-h_2\bullet b - g_1\bullet a)} e^{2\pi i\, {}^t B A^{-1}(a+h_1)\bullet(b-g_2)} \\
&\quad\times \sqrt{\frac{D[b-g_2]}{A^{-1}[a+h_1]}}^{\;\frac{n}{2}-s} K_{\frac{n}{2}-s}\Bigl(2\sqrt{D[b-g_2]A^{-1}[a+h_1]}\,\pi\Bigr) \\
&= H_{n,m}(Y, g, h, s),
\end{aligned}
\tag{5.57}
$$

we obtain

$$
\Lambda(Y, g, h, s) = \frac{e^{-2\pi i g\bullet h}}{\sqrt{|Y|}}\,\Lambda\bigl(Y^{-1}, h, -g, \tfrac{n+m}{2} - s\bigr).
\tag{5.58}
$$

By **screened Coulomb potentials**, we meant the Dirichlet series

$$
S(\epsilon; a) := \sum_{\substack{r\in\mathbb{Z}^3 \\ r\neq o}} \frac{\exp(2\pi i\epsilon\cdot\boldsymbol{r} - ar^2)}{r^2}
\tag{5.59}
$$

or

$$
T(\epsilon; a) := \sum_{\substack{r\in\mathbb{Z}^3 \\ r+\epsilon\neq o}} \frac{\exp(-a|\boldsymbol{r}+\boldsymbol{\epsilon}|^2)}{|\boldsymbol{r}+\boldsymbol{\epsilon}|^2},
\tag{5.60}
$$

where $\boldsymbol{r}$ signifies the position vector (x, y, z), r its length $\sqrt{x^2 + y^2 + z^2}$, $\epsilon \in \mathbb{R}^3$ giving rise to the additive character and $a > 0$ (eventually $\operatorname{Re} a > 0$ is allowed). "Screened" means that they decay exponentially as $a \to \infty$ (cf. Chaba-Pathria [CP2], (48) and (54), respectively). Other ones are defined in Corollary 5.4 below.

Screened Coulomb potential was first studied by Hautot [Hau] and then extensively by Chaba-Pathria [CP1]-[CP4]. The 2-dimensional case was taken up by Borweins. We shall interpret all of these results as special cases (Abel mean) of our general result, which in turn is a slight generalization (i.e. the poles being taken into account) of Terras [Te5, Exercise 17 pp.81–82] and Ewald [Ew].

Theorem 5.6 *For $c > 0$ (or $\operatorname{Re} c > 0$) we have*

$$\Lambda(Y, g, h, s) = c^s \Lambda(cY, g, h, s) \tag{5.61}$$

$$= \frac{1}{\sqrt{|Y|}} e^{-2\pi i g \cdot h} \sum_{\substack{a \in \mathbb{Z}^n \\ a+h \neq z}} e^{-2\pi i g \cdot a} \frac{\Gamma\left(\frac{n}{2} - s, \frac{\pi}{c} Y^{-1}[a+h]\right)}{\pi^{\frac{n}{2}-s} Y^{-1}[a+h]^{\frac{n}{2}-s}} + \delta(h) \frac{1}{\sqrt{|Y|}} \frac{c^{s-\frac{n}{2}}}{s - \frac{n}{2}}$$

$$+ \sum_{\substack{a \in \mathbb{Z}^n \\ a+g \neq z}} e^{2\pi i h \cdot a} \frac{\Gamma(s, \pi c Y[a+g])}{\pi^s Y[a+g]^s} - \delta(g) e^{-2\pi i g \cdot h} \frac{c^s}{s},$$

where $\delta(\cdot)$ is the delta symbol and $\Gamma(s, z)$ signifies the incomplete gamma function.

Proof. Recall that the multi-dimensional theta function defined by

$$\Theta(Y, g, h, t) = \sum_{a \in \mathbb{Z}^n} e^{2\pi i h \bullet a} e^{-\pi Y[a+g] t}, \quad t > 0 \tag{5.62}$$

satisfies the transformation formula

$$\Theta(Y, g, h, t) = \frac{e^{-2\pi i g \bullet h}}{\sqrt{|Y|} t^{\frac{n}{2}}} \Theta\left(Y^{-1}, h, -g, t^{-1}\right). \tag{5.63}$$

By the gamma transform, we have

$$\Lambda(Y, g, h, s) = \int_0^\infty \left(\Theta(Y, g, h, t) - \delta(g) e^{-2\pi i g \bullet h}\right) t^{s-1} \mathrm{d}t. \tag{5.64}$$

We follow the method of Riemann of dividing the interval into two $(0, 1)$ and $(1, \infty)$ and making the change of variable in the integral over $(0, 1)$ to get

$$\int_0^1 = \int_1^\infty \left(\Theta\left(Y, g, h, u^{-1}\right) - \delta(g) e^{-2\pi i g \bullet h}\right) u^{-s-1} \mathrm{d}u. \tag{5.65}$$

Now using (5.63) and rewriting slightly, we get

$$\int_0^1 \tag{5.66}$$

$$= \frac{e^{-2\pi i g \bullet h}}{\sqrt{|Y|}} \int_1^\infty \left(\Theta\left(Y^{-1}, h, -g, u\right) - \delta(h) e^{2\pi i g \bullet h}\right) u^{\frac{n}{2}-s-1} \mathrm{d}u$$

$$+ \delta(h) \frac{1}{\sqrt{|Y|}} \int_1^\infty u^{\frac{n}{2}-s-1} \mathrm{d}u - \delta(g) e^{-2\pi i g \bullet h} \int_1^\infty u^{-s-1} \mathrm{d}u.$$

$$\Lambda(Y, g, h, s) \tag{5.67}$$

$$= \int_1^\infty \left(\Theta(Y, g, h, t) - \delta(g)\, e^{-2\pi i g \bullet h} \right) t^{s-1} \mathrm{d}t + \delta(h)\, \frac{1}{\sqrt{|Y|}} \frac{1}{s - \frac{n}{2}}$$

$$\frac{e^{-2\pi i g \bullet h}}{\sqrt{|Y|}} \int_1^\infty \left(\Theta(Y^{-1}, h, -g, u) - \delta(h)\, e^{2\pi i g \bullet g} \right) u^{\frac{n}{2} - s - 1} \mathrm{d}u$$

$$- \delta(g)\, e^{-2\pi i g \bullet h} \frac{1}{s}.$$

This proves (5.61) with $c = 1$ in view of the definition of the incomplete gamma function. (5.61) is a restatement in terms of $c^s \Lambda(cY, g, h, s)$. □

Exercise 5.3 Theorem 5.6 entails those results corresponding to Example 5.1 and Theorem 5.5.

Solution Indeed, for $0 < \sigma < \frac{n}{2}$, we let $c \to 0+$ to obtain

$$\lim_{c \to 0+} \left\{ \sum_{\substack{a \in \mathbb{Z}^n \\ a+g \neq z}} e^{2\pi i h \cdot a} \frac{\Gamma(s, \pi c Y[a+g])}{\pi^s Y[a+g]^s} + \delta(h) \frac{1}{\sqrt{|Y|}} \frac{c^{s-\frac{n}{2}}}{s - \frac{n}{2}} \right\}$$

$$= \Lambda(Y, g, h, s),$$

which is a counterpart of Example 5.1.

On the other hand, for $\sigma < 0$, let $c \to \infty$ to deduce (5.45), as stated by Terras [Te5, p.82].

Corollary 5.4 *For $a > 0$, $n = 3$ and $\epsilon \notin \mathbb{Z}^3$,*

(i)
$$S(\epsilon; a) = a + A(\epsilon) - \pi \sum_{r \in \mathbb{Z}^3} \frac{\operatorname{erfc}\left(\epsilon \frac{\pi}{\sqrt{a}} |r + \epsilon| \right)}{|r + \epsilon|},$$

(ii)
$$V(\epsilon; a) := \sum_{r \in \mathbb{Z}^3} \frac{\exp(-a|r + \epsilon|)}{|r + \epsilon|}$$

$$= \frac{4\pi}{a^2} + \frac{A(\epsilon)}{\pi} - \frac{a^2}{4\pi^3} \sum_{\substack{r \in \mathbb{Z}^3 \\ r \neq z}} \frac{\exp(2\pi i \epsilon \cdot r)}{r^2 \left(r^2 + \frac{a^2}{4\pi^2} \right)},$$

where $A(\epsilon) = S(\epsilon; 0) = \sum_{\substack{r \in \mathbb{Z}^3 \\ r \neq z}} \frac{\exp(2\pi i \epsilon \cdot r)}{r^2}$.

(iii) $$T(\epsilon; a) = \frac{2\pi^{3/2}}{\sqrt{a}} + B(\epsilon) - \pi \sum_{\substack{r \in \mathbb{Z}^3 \\ r \neq z}} \frac{\exp(2\pi i \epsilon \cdot r)\mathrm{erfc}\left(\frac{\pi}{\sqrt{a}}r\right)}{r},$$

(iv) $$U(\epsilon; a) := \sum_{\substack{r \in \mathbb{Z}^3 \\ r \neq z}} \frac{\exp(2\pi i \epsilon \cdot r - ar)}{r}$$

$$= a + \frac{B(\epsilon)}{\pi} - \frac{a^2}{4\pi^3} \sum_{r \in \mathbb{Z}^3} \frac{1}{|r + \epsilon|^2 \left(|r + \epsilon|^2 + \frac{a^2}{4\pi^2}\right)},$$

where $B(\epsilon) = \lim_{a \to 0+} \left(T(\epsilon; a) - 2\pi^{3/2}a^{-1/2}\right)$, *and where* $\mathrm{erfc}(z) =$ *is the error function defined by* (2.13).

Proof.   Formulas (i) and (iii) are direct specializations of the formula in Theorem 5.6 with (2.13) taken into account, while formulas in (ii) and (iv) are Laplace transforms of (i) and (iii), respectively, when the formulas (2.15) or more general (2.14) incorporated:

We choose $Y = I$ (identity matrix), once and for all, so that $Y[a] = I[a] = |a|^2$, and $c = \frac{a}{\pi}$ (we write r for a).

To prove (i) we choose $s = 1, g = z, h = \epsilon$. Then

$$\pi^{-1} \sum_{\substack{r \in \mathbb{Z}^3 \\ r \neq z}} \frac{e^{2\pi i \epsilon \cdot r}}{r^2} = \sum_{r \in \mathbb{Z}^3} \frac{\mathrm{erfc}(\frac{\pi}{\sqrt{a}}|r + e|)}{|r + e|} - \frac{a}{\pi} + \pi^{-1} \sum_{\substack{r \in \mathbb{Z}^3 \\ r \neq z}} \frac{e^{2\pi i \epsilon \cdot r} e^{-ar^2}}{r^2},$$

which is Chaba-Pathria [CP2], (48), p.1419.

Taking the Laplace transform $\mathcal{L}[\cdot](p)$ of (i) with (2.14) and (2.16) in mind, we conclude that

$$\frac{4\pi^2}{a^2}\Lambda(I, z, \epsilon, 1) = \frac{4\pi^2}{a^2} \sum_{\substack{r \in \mathbb{Z}^3 \\ r + \epsilon \neq z}} \frac{\exp(-a|r + e|)}{|r + e|}$$

$$+ \frac{1}{\pi} \sum_{\substack{r \in \mathbb{Z}^3 \\ r \neq z}} \frac{\exp(-2\pi i \epsilon \cdot r)}{r^2(r^2 + \frac{a^2}{4\pi^2})} - \frac{1}{\pi}\left(\frac{4\pi^2}{a^2}\right)^2,$$

where we put $p = \frac{a^2}{4\pi^2}$. This proves (ii), i.e. [CP2, (62)].

It may be worth while recording the general Laplace transform of the formula in Theorem 5.6.

$$\frac{4\pi^2}{a^2}\Lambda(Y,\boldsymbol{g},\boldsymbol{h},s) \tag{5.68}$$

$$= \frac{1}{\sqrt{|Y|}}e^{-2\pi i\boldsymbol{g}\cdot\boldsymbol{h}} \sum_{\substack{\boldsymbol{r}\in\mathbb{Z}^n \\ \boldsymbol{r}+\boldsymbol{h}\neq\boldsymbol{z}}} \frac{2(\pi^2 Y^{-1}[\boldsymbol{r}+\boldsymbol{h}])^{\frac{n/2-s}{2}}}{\pi^{n/2-s}(Y^{-1}[\boldsymbol{r}+\boldsymbol{h}])^{n/2-s}} \left(\frac{a}{2\pi}\right)^{\frac{n}{2}-s-2}$$

$$\times K_{\frac{1}{2}-s}\left(2\sqrt{\pi^2 Y^{-1}[\boldsymbol{r}+\boldsymbol{h}](\frac{a^2}{4\pi^2})}\right)$$

$$+ \delta(\boldsymbol{h})\frac{1}{\sqrt{|Y|}}\frac{\pi^{\frac{n}{2}-s}}{s-\frac{n}{2}}\left(\frac{a^2}{4\pi^2}\right)^{\frac{n}{2}-s-1}\Gamma\left(s-\frac{n}{2}+1\right)$$

$$+ \Gamma(s)\frac{4\pi^2}{a^2}\sum_{\substack{\boldsymbol{r}\in\mathbb{Z}^n \\ \boldsymbol{r}+\boldsymbol{g}\neq\boldsymbol{z}}} e^{-2\pi i\boldsymbol{h}\cdot\boldsymbol{r}}\frac{1}{\pi^s(Y[\boldsymbol{r}+\boldsymbol{g}])^s}\left(1-\frac{1}{(1+\frac{a^2}{4\pi^2}(Y[\boldsymbol{r}+\boldsymbol{g}])^{-1})^s}\right)$$

$$- \delta(\boldsymbol{g})e^{-2\pi i\boldsymbol{g}\cdot\boldsymbol{h}}\frac{\pi^{-s}}{s}\Gamma(s+1)p^{-s-1}.$$

Formula (iv) is a special case of (5.68) with $\boldsymbol{g}=\boldsymbol{\epsilon}$, $\boldsymbol{h}=\boldsymbol{z}$, $s\to 1$ ($c=a/\pi$, $n=3,\dots$). We note that $B(\boldsymbol{\epsilon})=\pi U(\boldsymbol{\epsilon};0)$.
The same specialization yields that

$$Z(I,\boldsymbol{\epsilon},\boldsymbol{z},1)=\pi\sum_{\substack{\boldsymbol{r}\in\mathbb{Z}^3 \\ \boldsymbol{r}\neq\boldsymbol{z}}}\frac{e^{-2\pi i\boldsymbol{\epsilon}\cdot\boldsymbol{r}}e^{-\pi a^{-1/2}r}}{r} - 2\frac{\pi^{\frac{3}{2}}}{a^{\frac{1}{2}}} + \sum_{\substack{\boldsymbol{r}\in\mathbb{Z}^3 \\ \boldsymbol{r}+\boldsymbol{\epsilon}\neq\boldsymbol{z}}}\frac{e^{-a|\boldsymbol{r}+\boldsymbol{e}|^2}}{|\boldsymbol{r}+\boldsymbol{e}|^2}, \tag{5.69}$$

which amounts to (iii). Of course, the Laplace transform of (5.69) leads to (iv). This completes the proof. $\qquad\square$

Some comments on the limiting cases as $a\to 0$ (or $\boldsymbol{\epsilon}\to\boldsymbol{z}$) of above formulas are in order.

Remark 5.2 (i) *Chaba-Pathria [CP2] state that the limiting case as $\boldsymbol{\epsilon}\to\boldsymbol{z}$ of (i)*

$$S(0;a)=\sum_{\substack{\boldsymbol{r}\in\mathbb{Z}^3 \\ \boldsymbol{r}\neq\boldsymbol{z}}}\frac{e^{-ar^2}}{r^2}=\frac{2\pi^{3/2}}{a^{1/2}}+C_3+a-\pi\sum_{\substack{\boldsymbol{r}\in\mathbb{Z}^3 \\ \boldsymbol{r}\neq\boldsymbol{z}}}\frac{\mathrm{erfc}(\pi a^{-1/2}r)}{r^2} \tag{5.70}$$

appeared already in [CP1], but (5.70) does not follow from (i), and to deduce *it, we need to appeal to our Theorem 5.6 with $g = h = z$, $s \to 1$:*

$$Z(I, z, z, 1) = \lim_{s \to 1} Z(I, z, z, s) \tag{5.71}$$

$$= \sum_{\substack{r \in \mathbb{Z}^3 \\ r \neq z}} \frac{e^{-ar^2}}{r^2} - 2\frac{\pi^{3/2}}{a^{1/2}} - a + \pi \sum_{\substack{r \in \mathbb{Z}^3 \\ r \neq z}} \frac{\operatorname{erfc}(\pi a^{-1/2} r)}{r^2}.$$

Chaba-Pathria [CP2, (50)] introduce the constant (introduced in [CP1])

$$C_3 = \lim_{a \to 0+} \left(S(0, a) - 2\pi^{3/2} a^{-1/2} \right),$$

but cannot state its source. We have an advantage over them to the effect that C_3 is exactly the special value at $s = 1$ of the (analytic continuation of the) Epstein zeta-function $Z(I, z, z, s) = \sum_{\substack{r \in \mathbb{Z}^3 \\ r \neq z}} \frac{1}{r^{2s}}$:

$$C_3 = Z(I, z, z, 1). \tag{5.72}$$

(ii) In the case of $B(\epsilon)$ we have the same advantage of assigning the definite value $Z(I, \epsilon, z, 1)$ to it:

$$B(\epsilon) = Z(I, \epsilon, z, 1). \tag{5.73}$$

Chaba-Pathria [CP2] give Formula (55) connecting $B(\epsilon)$ to $\pi \sum'_r \frac{e^{-2\pi i \epsilon \cdot r}}{r}$, but it is not clear how they deduce this. Only through (5.73), we get the following, on resorting to the functional equation (5.45)

$$B(\epsilon) = Z(I, \epsilon, z, 1) = \sqrt{\pi} Z\left(I, z, -\epsilon, \frac{1}{2} \right). \tag{5.74}$$

We now turn to the case $n = 2$ and recover Borweins' result [BB2].

Corollary 5.5 *For $\operatorname{Re} a > 0$, we have*

(i)
$$Z\left(I, \epsilon, z, \frac{1}{2} \right) = \sum_{\substack{r \in \mathbb{Z}^2 \\ r \neq z}} e^{-2\pi i \epsilon \cdot r} \frac{\operatorname{erfc}\left(\frac{\pi}{\sqrt{a}} r \right)}{r}$$

$$+ \sum_{\substack{r \in \mathbb{Z}^2 \\ r + e \neq z}} \frac{\operatorname{erfc}(\sqrt{a}|r + e|)}{|r + e|} - 2\sqrt{\frac{\pi}{a}},$$

(i)′
$$Z\left(I, \epsilon, z, \frac{1}{2}\right) = \sum_{\substack{r \in \mathbb{Z}^2 \\ r \neq z}} e^{-2\pi i \epsilon \cdot r} \frac{\operatorname{erfc}\left(\sqrt{a}r\right)}{r}$$

$$+ \sum_{\substack{r \in \mathbb{Z}^2 \\ r+\epsilon \neq z}} \frac{\operatorname{erfc}\left(\frac{\pi}{\sqrt{a}}|r+e|\right)}{|r+e|} - 2\sqrt{\frac{a}{\pi}},$$

(ii)
$$\sum_{\substack{r \in \mathbb{Z}^2 \\ r \neq z}} e^{-2\pi i \epsilon \cdot r} \frac{e^{ar}}{r} = Z\left(I, \epsilon, z, \frac{1}{2}\right) + a$$

$$- \sum_{\substack{r \in \mathbb{Z}^2 \\ r+\epsilon \neq z}} \left(\frac{1}{r_\epsilon} - \frac{1}{\sqrt{r_\epsilon^2 + \frac{a^2}{4\pi^2}}}\right),$$

where $r_\epsilon = |r+e|$. *We have further*

(iii)
$$\sum_{\substack{r \in \mathbb{Z}^2 \\ r \neq z}} \frac{e^{-ar}}{r} = \frac{2\pi}{a} + Z\left(I, z, z, \frac{1}{2}\right) + a$$

$$- \sum_{\substack{r \in \mathbb{Z}^2 \\ r \neq z}} \left(r^{-1} - \left(r^2 + \frac{a^2}{4\pi^2}\right)^{-1/2}\right),$$

(iii)′
$$\lim_{a \to 0+} \left(\sum_{\substack{r \in \mathbb{Z}^2 \\ r \neq z}} \frac{e^{-ar}}{r} - \frac{2\pi}{a}\right) = 4\zeta\left(\frac{1}{2}\right)\beta\left(\frac{1}{2}\right).$$

Proof. Formula (i) is a specialization of our formula in Theorem 5.6 with $n = 2$, $Y = I$, $g = \epsilon$, $h = z$, $s \to \frac{1}{2}$, while (i)′ is a restatement of (i) with $\frac{\pi^2}{a}$ in place of a. Formula (i) is Chaba-Pathria [CP2, (33)] and Formula (i)′ is their (39).

Formula (ii) is Chaba-Pathria [CP2, (41)] and Borweins' [BB2, (8)]. Both Chaba-Pathria and Borweins deduce (ii) using Fetter et al's technique. However, as is pointed out by Chaba-Pathria, we may prove (ii) by taking the Laplace transform $\mathcal{L}[\cdot](p)$ and put $p = \frac{a^2}{4\pi^2}$ in the same way as we proved Corollary 5.4.

To prove Formula (iii), we use a specialization of our formula in Theorem

5.6 with $n = 2$, $Y = I$, $g = h = z$, $s \to \frac{1}{2}$:

$$Z\left(I, z, z, \frac{1}{2}\right) = \sum_{\substack{r \in \mathbb{Z}^2 \\ r \neq z}} \frac{\mathrm{erfc}\left(\frac{\pi}{\sqrt{a}}r\right)}{r} + \sum_{\substack{r \in \mathbb{Z}^2 \\ r \neq z}} \frac{\mathrm{erfc}(\sqrt{a}r)}{r} \qquad (5.75)$$

$$- 2\sqrt{\frac{\pi}{a}} - 2\sqrt{\frac{a}{\pi}}.$$

Taking the Laplace transforms of (5.75), we obtain (iii). Formula (iii) is Chaba-Pathria [CP2, (42)] and Borweins' [BB2, (9)].

Formula (iii)$'$ follows from (iii) and is denoted by D ([CP2, (36)]), which should be defined as (iii)$'$. It is given as Borweins' [BB2, (7)].

Corollary 5.6   *For the special choice* $\epsilon = \frac{1}{2} = \left(\frac{1}{2}, \frac{1}{2}\right)$, *we write* $E_\epsilon(a) = E_-(a)$ *and let* $\rho = \frac{1}{2}\sqrt{(2m+1)^2 + (2n+1)^2}$ *instead of* r. *Then we have*

$$\lim_{x \to 0+} E_-(x + i\theta) = i\theta + \gamma\left(\frac{1}{2}\right) - \sum_{m,n}\left(\frac{1}{\rho} - \frac{1}{\sqrt{\rho^2 - \frac{\theta^2}{4\pi^2}}}\right). \qquad (5.76)$$

This is Borweins' formula (11), from which their Abel mean formulas for

$$S_\pm(\theta) = \sum_{n=1}^{\infty} \frac{(\pm 1)^n r(n)}{\sqrt{n}} \sin(\sqrt{n}\theta),$$

$$C_\pm(\theta) = \sum_{n=1}^{\infty} \frac{(\pm 1)^n r(n)}{\sqrt{n}} \cos(\sqrt{n}\theta)$$

can be read off immediately, where $r(n)$ stands for the number of representations of n as the sum of two squares.

More general questions concerning $S_\epsilon(\theta) = \sum'_r \frac{e^{-2\pi i \epsilon \cdot r}}{r} e^{2\pi i r \theta}$ and generalizations could be considered similarly.

Chapter 6

The Riesz Sums

In this chapter we shall elucidate various identities which are either disguised forms of Riesz sums as modular relations or integrated forms thereof.

6.1 Various modular relations

6.1.1 *Riesz sums*

The Riesz means, or sometimes typical means, were introduced by M. Riesz and have been studied in connection with summability of Fourier series and of Dirichlet series [ChTyp] and [HarRie]. Given an increasing sequence $\{\lambda_k\}$ of real numbers and a sequence $\{\alpha_k\}$ of complex numbers, the Riesz sum of order $\varkappa$ is defined as in [ChTyp, p.2] and [HarRie, p.21] by

$$A^{\varkappa}(x) = A^{\varkappa}_{\lambda}(x) = \sum_{\lambda_k \leq x}{}' (x - \lambda_k)^{\varkappa} \alpha_k \qquad (6.1)$$

$$= \varkappa \int_0^x (x - t)^{\varkappa - 1} A_{\lambda}(t) \mathrm{d}t$$

$$= \varkappa \int_0^x (x - t)^{\varkappa - 1} \, \mathrm{d}A_{\lambda}(t),$$

where

$$A_{\lambda}(x) = A^0_{\lambda}(x) = \sum_{\lambda_k \leq x}{}' \alpha_k, \qquad (6.2)$$

where the prime on the summation sign means that when $\lambda_k = x$, the corresponding term is to be halved.

(6.1) or rather normalized $\frac{1}{\Gamma(\varkappa+1)} A^{\varkappa}(x)$ which appears in (6.7) is called the **Riesz sum** of order $\varkappa$ If $\frac{1}{\Gamma(\varkappa+1)} A^{\varkappa}(x)$ approaches a limit A as $x \to \infty$, the sequence $\{\alpha_k\}$ is called Riesz summable or $(R, \varkappa, \lambda)$ summable to A, which is called the **Riesz mean** of the sequence. Sometimes the negative order Riesz sum is considered, in which case the sum is taken over all n which are not equal to x.

In number-theoretic context, it is the Riesz sum rather than the Riesz mean that has been extensively studied. The Riesz sums appear as long as there appears the $G_{1,1}^{1,0}$. Cf. Remark 6.1. There is some mention on the divisor problem in [KaR] in the light of the Riesz sum and there are enormous amount of literature on the Riesz sums and we shall not dwell on well-known cases very in detail and state only unexpected use of them.

An example is given.

Recall the definition of the periodic Bernoulli polynomial etc. from §3.3.1. Then

$$B_1(x) - \psi(x) = \sum_{n \leq x}{}' 1 := A(x),$$

say, or

$$B_1(x) - \bar{B}_1(x) = \begin{cases} A(x), & x \notin \mathbb{Z}, \\ A(x) + \frac{1}{2}, & x \in \mathbb{Z}. \end{cases}$$

Integration of both sides amounts to (6.1):

$$\frac{1}{2}B_2(t) - \frac{1}{2}\bar{B}_2(t) = \int_0^x (B_1(t) - \bar{B}_1(t))\mathrm{d}t = \int_0^x A(t)\mathrm{d}t = \sum_{n \leq x}(x - n),$$

where we used $B_2(0) = \bar{B}_2(0)(= B_2 = \frac{1}{6})$.

The application of the Riesz sum comes into play through the Perron formula (6.7) below, sometimes in truncated form. The application of the truncated first order Riesz sum appears on [Dav, p.105] and a truncated general order Riesz sum is treated in [KaR] in both of which the functional equation is not assumed. Riesz sums with the functional equation can be found e.g. in [ChN3], where by differencing, the asymptotic formula for the original sum is deduced. The principle goes back to Landau [LanII] in which one can find the integral order Riesz sum and its reduction to the original partial sum by differencing.

The general formula for the **difference operator** of order $\alpha \in \mathbb{N}$ with difference $y \geq 0$ is given by

$$\Delta_y^\alpha f(x) = \sum_{\nu=0}^\alpha (-1)^{\alpha-\nu} \binom{\alpha}{\nu} f(x + \nu y). \qquad (6.3)$$

If f has the α-th derivative $f^{(\alpha)}$, then

$$\Delta_y^\alpha f(x) = \int_x^{x+y} \mathrm{d}t_1 \int_{t_1}^{t_1+y} \mathrm{d}t_2 \cdots \int_{t_{\alpha-1}}^{t_{\alpha-1}+y} f^{(\alpha)}(t_\alpha)\, \mathrm{d}t_\alpha. \qquad (6.4)$$

The Riesz kernel which produces the Riesz sum is defined by

$$G_{1,1}^{1,0}\left(z \,\middle|\, \begin{matrix} a \\ b \end{matrix}\right) = \begin{cases} \dfrac{1}{\Gamma(a-b)} z^b (1-z)^{a-b-1}, & |z| < 1 \\ \dfrac{1}{2} & z = 1, a = b+1 \\ 0, & |z| > 1. \end{cases} \qquad (6.5)$$

Remark 6.1 *Notes on (6.5). Let $\varkappa \geq 0$ denote the order of the Riesz mean and set $b = 0, a = \varkappa + 1$. Then (6.5) reads $(c > 0)$*

$$\frac{1}{2\pi i} \int_c \frac{\Gamma(s)}{\Gamma(s+\varkappa+1)} z^{-s}\, \mathrm{d}s = \begin{cases} \dfrac{1}{\Gamma(\varkappa+1)} (1-z)^\varkappa, & (|z| < 1) \\ \dfrac{1}{2}, & (\varkappa = 0, z = 1) \\ 0, & (|z| > 1). \end{cases} \qquad (6.6)$$

This can be found in Hardy-Riesz [HarRie] and Chandrasekharan and Minakshisundaram [ChTyp] and used in the context of **Perron's formula**

$$\frac{1}{\Gamma(\varkappa+1)} {\sum_{\lambda_k \leq x}}' \alpha_k (x - \lambda_n)^\varkappa = \frac{1}{2\pi i} \int_c \frac{\Gamma(s)\varphi(s) x^{s+\varkappa}}{\Gamma(s+\varkappa+1)}\, \mathrm{d}s, \qquad (6.7)$$

where the left-hand side sum is called the Riesz sum of order $\varkappa$ and denoted $A_\lambda^\varkappa(x)$ as mentioned above and $\varphi(s) = \sum_{k=1}^\infty \frac{\alpha_k}{k^s}$.

If the order $\varkappa \in \mathbb{N}$, then the right-hand side member of (6.6) is

$$\frac{1}{2\pi i} \int_c \frac{1}{s(s+1)\cdots(s+\varkappa)} z^{-s}\, \mathrm{d}s$$

and the Riesz sum amounts to the $\varkappa$ times integration of the original sum $A_\lambda(x)$. Thus Landau's differencing is an analogue of the integration and differentiation.

In view of this integration-differentiation aspect there are a number of cases in which the Riesz sum appears in disguised form. Especially, when there is a gamma factor $\frac{\Gamma(s)}{\Gamma(s+\varkappa+1)}$ or $\begin{pmatrix} \varkappa+1 \\ 0 \end{pmatrix}$ involved.

The special case of (6.6) with $\varkappa = 0$ is known as the **discontinuous integral** *whose truncated form can be found e.g. in Davenport [Dav, pp. 109-110].*

The very special case $z = 1, a - b = 1$ of (6.5) (and of the corresponding logarithmic case $(z = 1, m = 1)$) presents excessive complexities in notation, so that we follow Hardy and Riesz [HarRie] to use (6.7) by suppressing the prime on the summation. We are to bear this special case in mind although not explicitly stated.

6.1.2   *Improper modular relations as Riesz sums*

In [LWKman] we introduced the notion of an **improper modular relation** to the effect that the identity in question does not look like a modular relation because some terms are placed on the other side as part of the residues, say or for some other reasons.

Theorem 6.1   ([Kat2, Theorem 4.1]) *For all $x \geq |\operatorname{Re} \nu| + 2$ and $\nu \notin \mathbb{Z}$,*

$$\sum_{n=0}^{\infty} (-1)^n \binom{x}{n} \zeta(n - \nu) = \frac{\Gamma(x+1)\,\Gamma(-\nu-1)}{\Gamma(x-\nu)} + \mathcal{H}_\nu(x), \qquad (6.8)$$

where

$$\mathcal{H}_\nu(x) = \Gamma(x+1) \sum_{n=1}^{\infty} \Big(U(x+1, \nu+2; -2\pi i n) + U(x+1, \nu+2; 2\pi i n) \Big). \quad (6.9)$$

In the case $\nu \in \mathbb{Z}$, $\nu \geq -1$, the first term on the right of (6.8) is to be replaced by

$$(-1)^\nu \binom{x}{\nu+1} \left(\frac{\Gamma'}{\Gamma}(x - \nu) + 2\gamma - H_{\nu+1} \right), \qquad (6.10)$$

where γ is Euler's constant and $H_{\nu+1}$ is the $(\nu+1)$-th harmonic number $H_{\nu+1} = \sum_{n=1}^{\nu+1} \frac{1}{n}$.

Remark 6.2   *In the case $\operatorname{Re} \nu + 1 \geq 0$, the parameter ν affects the series and we are to shift the first $[\operatorname{Re} \nu + 1]$ partial sum to the right so as to* *indicate that these $\zeta(n-\nu)$'s are not the series but its analytic continuation, so that the right-hand side is to mean*

$$H_\nu(x) = \frac{\Gamma(x+1)\,\Gamma(-\nu-1)}{\Gamma(x-\nu)} - \sum_{n=0}^{[\text{Re}\,\nu+1]} (-1)^n \binom{x}{n} \zeta(n-\nu) + \mathcal{H}_\nu(x) \quad (6.11)$$

for $\nu \notin \mathbb{Z}$; for $\nu \in \mathbb{Z}$ we are to replace the first term in (6.11) by (6.10). On the other hand, the left-hand side sum starts from $n > [\text{Re}\,\nu + 1]$ and all the work on these problems have this type of sums.

We refer to this as the improper modular relations. They do not look like at a first glance a modular relation but even in the case $\text{Re}\,\nu + 1 > 0$, when stated in the form of our Theorem 6.1, it is a modular relation as furnished by Theorem 6.2 below.

In much the same way, we may interpret other improper modular relations ([Kat1, Theorems 3.1, 3.2]) of Katsurada as modular relations (cf. Theorem 6.2). The interpretation is based on [WW].

6.1.3 The $H_{2,1}^{1,1} \leftrightarrow H_{1,3}^{2,0}$ formula

The formula in the title asserts that the functional equation

$$\chi(s) = \begin{cases} \Gamma(Cs)\varphi(s), & \text{Re}\,s > \sigma_\varphi \\ \Gamma(C(r-s))\psi(r-s), & \text{Re}\,s < r - \sigma_\psi, \end{cases} \quad (6.12)$$

is equivalent to the modular relation

$$\sum_{k=1}^{\infty} \frac{\alpha_k}{\lambda_k^s} H_{2,1}^{1,1}\left(z\lambda_k \,\middle|\, \begin{array}{c} (1-a,A),(b,B) \\ (Cs,C) \end{array} \right)$$

$$= \sum_{k=1}^{\infty} \frac{\beta_k}{\mu_k^{r-s}} H_{0,3}^{2,0}\left(\frac{\mu_k}{z} \,\middle|\, \begin{array}{c} - \\ (C(r-s),C),(a,A),(1-b,B) \end{array} \right) \quad (6.13)$$

$$+ \sum_{k=1}^{L} \text{Res}\left(\frac{\Gamma(a+As-Aw)}{\Gamma(b-Bs+Bw)} \chi(w)\, z^{s-w}, w = s_k \right).$$

Exercise 6.1 (Katsurada's first formula [Kat1]) Setting $a = 0, A = 1$,

$b = \frac{s}{2}, B = \frac{1}{2}$ in (6.13), deduce the first generalized Katsurada formula

$$
\sum_{k=1}^{\infty} \frac{\alpha_k}{\lambda_k^s} \exp\left(-\frac{1}{z\lambda_k}\right)
$$

$$
= \frac{2}{\sqrt{\pi}} \sum_{k=1}^{\infty} \beta_k \left\{ \left(\frac{e^{\frac{\pi}{2}i}}{2z\mu_k}\right)^{\frac{1-s}{2}} K_{1-s}\left(2\sqrt{2}\, e^{-\frac{\pi}{4}i} \sqrt{\frac{\mu_k}{z}}\right) \right.
$$

$$
\left. + \left(\frac{e^{-\frac{\pi}{2}i}}{2z\mu_k}\right)^{\frac{1-s}{2}} K_{1-s}\left(2\sqrt{2}\, e^{\frac{\pi}{4}i} \sqrt{\frac{\mu_k}{z}}\right) \right\}
\qquad (6.14)
$$

$$
+ \sum_{k=1}^{L} \mathrm{Res}\left(\frac{\Gamma(s-w)}{\Gamma(\frac{w}{2})} \chi(w)\, z^{s-w}, w = s_k\right),
$$

which in the case of the Riemann zeta-function, reduces to (6.22).

6.1.4   The $H_{2,2}^{1,1} \leftrightarrow H_{1,3}^{2,0}$ formula

The formula in the title asserts that the following is equivalent to (6.12).

$$
\sum_{k=1}^{\infty} \frac{\alpha_k}{\lambda_k^s} H_{2,2}^{1,1}\left(z\lambda_k \left|\begin{array}{c}(1-a_1, A_1), (a_2, A_2) \\ (Cs, C), (1-b_1, B_1)\end{array}\right.\right)
$$

$$
= \sum_{k=1}^{\infty} \frac{\beta_k}{\mu_k^{r-s}} H_{1,3}^{2,0}\left(\frac{\mu_k}{z} \left|\begin{array}{c}(b_1, B_1) \\ (C(r-s), C), (a_1, A_1), (1-a_2, A_2)\end{array}\right.\right)
\qquad (6.15)
$$

$$
+ \sum_{k=1}^{L} \mathrm{Res}\left(\frac{\Gamma(a_1 + A_1 s - A_1 w)}{\Gamma(a_2 - A_2 s + A_2 w)\,\Gamma(b_1 + B_1 s - B_1 w)} \chi(w)\, z^{s-w}, w = s_k\right).
$$

Exercise 6.2 (Katsurada's second formula [Kat1])   Consider the case of the Riemann zeta-function $C = \frac{1}{2}$. By setting $a_1 = 0$, $a_2 = \frac{s}{2}$, $b_1 = 1 - b$, $A_1 = 1$, $A_2 = \frac{1}{2}$ and $B_1 = 1$ in (6.15), deduce the second generalized Katsurada formula

$$
\frac{1}{\pi^{\frac{s}{2}}} \sum_{k=1}^{\infty} \frac{1}{k^s} H_{2,2}^{1,1}\left(z\sqrt{\pi}\, k \left|\begin{array}{c}(1,1), \left(\frac{s}{2}, \frac{1}{2}\right) \\ \left(\frac{s}{2}, \frac{1}{2}\right), (b, 1)\end{array}\right.\right)
$$

$$
+ \mathrm{Res}\left(\frac{\Gamma(w+s-1)}{\Gamma\left(-\frac{w}{2}+\frac{1}{2}\right)\Gamma(1-b+w+s-1)}\right.
$$

$$
\left. \Gamma\left(\frac{w}{2}\right)\varphi(w)\, z^{s-1+w}, w = 1\right)
$$

$$= \frac{1}{\pi^{\frac{1-s}{2}}} \sum_{k=1}^{\infty} \frac{1}{k^{1-s}} H_{1,3}^{2,0}\!\left(\frac{\sqrt{\pi}\,k}{z} \,\middle|\, \begin{matrix} (1-b,1) \\ \left(\frac{1-s}{2},\frac{1}{2}\right),(0,1),\left(1-\frac{s}{2},\frac{1}{2}\right) \end{matrix}\right)$$

$$+ \operatorname{Res}\left(\frac{\Gamma(-w+s)}{\Gamma\!\left(\frac{w}{2}\right)\Gamma(1-b-w+s)}\,\Gamma\!\left(\tfrac{w}{2}\right)\varphi(w)\,z^{s-w}, w=1\right), \qquad (6.16)$$

which in particular entails another form of (6.23)

$$\frac{1}{\Gamma(1-b)} \sum_{n=0}^{\infty} \frac{(b)_n}{n!}\,\zeta(n-\nu)\,x^n$$

$$= x^{\nu+1} \sum_{k=1}^{\infty} \left\{ e^{-2\pi i k x}\,U\!\left(1-b,\nu+2;2\pi i k x\right) \right.$$
$$\left. + e^{2\pi i k x}\,U\!\left(1-b,\nu+2;-2\pi i k x\right)\right\} + x^{\nu+1}\frac{\Gamma(-\nu-1)}{\Gamma(-\nu-b)}. \qquad (6.17)$$

Solution On the left-hand side, $H_{1,2}^{1,1}$ reduces to the Riesz kernel while on the right-hand side

$$H_{1,3}^{2,0}\!\left(z \,\middle|\, \begin{matrix} (1-b,1) \\ (s,\frac{1}{2}),(0,1),\left(s+\frac{1}{2},\frac{1}{2}\right) \end{matrix}\right)$$

$$= \frac{z^{2s}}{\sqrt{\pi}}\left\{ e^{2iz}\,U\!\left(1-b,2s+1;-2iz\right) + e^{-2iz}\,U\!\left(1-b,2s+1;2iz\right)\right\} \qquad (6.18)$$

after some transformations.

6.1.5 *Katsurada's formula combined*

Theorem 6.2 *The Riesz sum formula, a special case of the $H_{2,2}^{1,1} \leftrightarrow H_{1,3}^{2,0}$ formula,*

$$\frac{1}{\pi^{\frac{s}{2}}} \sum_{k=1}^{\infty} \frac{1}{k^s}\,G_{1,1}^{1,0}\!\left(\frac{1}{z} \,\middle|\, \begin{matrix} 1-b \\ 0 \end{matrix}\right)$$

$$+ \operatorname{Res}\left(\frac{\Gamma(w+s-1)}{\Gamma\!\left(-\frac{w}{2}+\frac{1}{2}\right)\Gamma(1-b+w+s-1)}\right.$$

$$\left. \Gamma\!\left(\tfrac{w}{2}\right)\varphi(w)\,z^{s-1+w}, w=1\right)$$

$$= \frac{1}{\pi^{\frac{1-s}{2}}} \sum_{k=1}^{\infty} \frac{1}{k^{1-s}} H_{1,3}^{2,0}\left(\frac{\sqrt{\pi}\,k}{z}\,\middle|\,\frac{(1-b,1)}{\left(\frac{1-s}{2},\frac{1}{2}\right),(0,1),\left(1-\frac{s}{2},\frac{1}{2}\right)}\right)$$

$$+ \operatorname{Res}\left(\frac{\Gamma(-w+s)}{\Gamma\left(\frac{w}{2}\right)\Gamma(1-b-w+s)}\Gamma\left(\tfrac{w}{2}\right)\varphi(w)\,z^{s-w},\,w=1\right) \tag{6.19}$$

in the limiting case as $1 - b \to \infty$ *amounts to Katsurada's first formula* ([Kat1, Theorems 3.1, 3.2])

$$\sum_{n=0}^{\infty}(-1)^n \frac{x^n}{n!}\,\zeta(n-\nu) = \sum_{k=1}^{\infty} k^\nu e^{-\frac{x}{k}}$$

$$= 2\sum_{k=1}^{\infty}\left\{\left(\frac{x}{2\pi i k}\right)^{\frac{\nu+1}{2}} K_{\nu+1}\left(2\sqrt{2\pi i k x}\right)\right. \tag{6.20}$$

$$\left.+\left(-\frac{x}{2\pi i k}\right)^{\frac{\nu+1}{2}} K_{\nu+1}\left(2\sqrt{-2\pi i k x}\right)\right\} + x^{\nu+1}\Gamma(-\nu-1)$$

and in the case $z \to 1$ *to Katsurada's second formula* ([Kat1, Theorem 4.1])

$$\frac{1}{\Gamma(1+x)}\sum_{n=0}^{\infty}\frac{(x)_n}{n!}\,\zeta(n-\nu)$$

$$= \sum_{k=1}^{\infty}\left\{U\left(x+1,\nu+2;2\pi i k\right) + U\left(x+1,\nu+2;-2\pi i k\right)\right\} \tag{6.21}$$

$$+\frac{\Gamma(-\nu-1)}{\Gamma(x-\nu)}.$$

Proof. The first equality in (6.20) follows by expanding into the Taylor series, changing the order of summation and writing the inner sum as $\zeta(n-\nu)$.

Hence in order to prove (6.20), it suffices to prove that

$$\sum_{k=1}^{\infty} k^\nu e^{-\frac{x}{k}}$$

$$= 2\sum_{k=1}^{\infty}\left\{\left(\frac{x}{2\pi i k}\right)^{\frac{\nu+1}{2}} K_{\nu+1}\left(2\sqrt{2\pi i k x}\right)\right. \tag{6.22}$$

$$\left.+\left(-\frac{x}{2\pi i k}\right)^{\frac{\nu+1}{2}} K_{\nu+1}\left(2\sqrt{-2\pi i k x}\right)\right\} + x^{\nu+1}\Gamma(-\nu-1).$$

The special case of (6.15) with $C = \frac{1}{2}$, $a_1 = 0$, $a_2 = \frac{s}{2}$, $b_1 = 1 - b$, $A_1 = 1$, $A_2 = \frac{1}{2}$, and $B_1 = 1$ leads to (6.19) in view of the reduction formula. Although $G_{1,1}^{1,0}\left(\frac{1}{z}\,\middle|\,\begin{matrix} 1-b \\ 0 \end{matrix}\right)$ is the Riesz kernel

$$\begin{cases} \dfrac{1}{\Gamma(1-b)}\left(1-\dfrac{1}{z}\right)^{-b}, & |z| > 1, \\ 0, & |z| < 1, \end{cases}$$

and in most cases one considers the Riesz sum for $k \le x$ by taking $z = \frac{x}{k}$ and $x \ge 1$, the exceptional case $0 < x < 1$ is also considered pertaining to (6.21), in which case the left-hand side reads $\frac{1}{\Gamma(1-b)}\sum_{k=1}^{\infty} k^{\nu}\left(1-\frac{x}{k}\right)^{-b}$. Hence (6.19) amounts in the case $|z| < 1$ to

$$\frac{1}{\Gamma(1-b)}\sum_{k=1}^{\infty} k^{\nu}\left(1-\frac{z}{k}\right)^{-b} \tag{6.23}$$

$$= x^{\nu+1}\sum_{k=1}^{\infty}\left\{e^{-2\pi i k x}\,U\left(1-b,\nu+2;2\pi i k z\right) \right. \\ \left. + e^{2\pi i k z}\,U\left(1-b,\nu+2;-2\pi i k x\right)\right\} + z^{\nu+1}\frac{\Gamma(-\nu-1)}{\Gamma(-\nu-b)}.$$

Rewriting (6.23) as

$$\sum_{k=1}^{\infty} k^{\nu}\left(1-\frac{\frac{x}{k}}{1-b}\right)^{1-b} \tag{6.24}$$

$$= \frac{\Gamma(1-b)}{\Gamma(-\nu-b)(1-b)^{\nu+1}}x^{\nu+1}\sum_{k=1}^{\infty}\left\{e^{-2\pi i k x}\,U\left(1-b,\nu+2;2\pi i k\frac{x}{1-b}\right) \right. \\ \left. + e^{2\pi i k\frac{x}{1-b}}\,U\left(1-b,\nu+2;-2\pi i k x\right)\right\} + \Gamma(-\nu-1),$$

we take the limit as $1 - b \to \infty$ to conclude (6.20) for $|z| < 1$, thereby incorporating the following limit relations.

$$\lim_{a\to\infty}\left(\Gamma(a-c+1)U\left(a,c;\frac{x}{a}\right)\right) = 2x^{\frac{1}{2}(1-c)}K_{c-1}(2\sqrt{x}), \tag{6.25}$$

which is [ErdH, I,(19),p.266];

$$\lim_{1-b\to\infty}\frac{\Gamma(1-b)}{\Gamma(-\nu-b)(1-b)^{\nu+1}} = 1, \tag{6.26}$$

by (2.1) and

$$\lim_{1-b\to\infty}\left(1-\frac{\frac{x}{k}}{1-b}\right)^{1-b}=e^{-\frac{x}{k}}. \tag{6.27}$$

The restriction $|z| < 1$ may be removed by analytic continuation and the proof follows for other values of z.

Expanding into the binomial series, changing the order of summation and writing the inner sum as $\zeta(n-\nu)$ (note the similarity of the procedure as in the proof of the first equality in (6.20)), we immediately see that the left-hand side of (6.23) is

$$\frac{1}{\Gamma(1+x)}\sum_{n=0}^{\infty}\frac{(x)_n}{n!}\,\zeta(n-\nu)\,z^n,$$

so that (6.23) reads

$$\frac{1}{\Gamma(1+x)}\sum_{n=0}^{\infty}\frac{(x)_n}{n!}\,\zeta(n-\nu)\,z^n \tag{6.28}$$

$$= x^{\nu+1}\sum_{k=1}^{\infty}\Big\{e^{-2\pi ikx}\,U\big(1-b,\nu+2;2\pi ikz\big)$$
$$+ e^{2\pi ikz}\,U\big(1-b,\nu+2;-2\pi ikx\big)\Big\} + z^{\nu+1}\frac{\Gamma(-\nu-1)}{\Gamma(-\nu-b)},$$

whence (6.21) immediately follows.   □

Remark 6.3   *Another way would be to study the Riesz sum or the Riesz mean and it is well-known that the Riesz summability implies Abel summability [HarRie, Theorem 24]. There is an explicit formula known for the transition.*

Lemma 6.1   *The sum of the Abel mean $\sum_{k=1}^{\infty}a_ne^{-\lambda_k s}$ at all points of the sector $|\arg s|\le u<\frac{\pi}{2}$ other than the origin is*

$$\frac{1}{\Gamma(\varkappa+1)}\int_0^{\infty}s^{\varkappa+1}e^{-s\tau}A^{\varkappa}(\tau)\,\mathrm{d}\tau, \tag{6.29}$$

where $A^{\varkappa}(x)$ is the $\varkappa$-th Riesz sum defined in (6.1).

6.1.6   *Linearized product of two zeta-functions*

In this section we shall elucidate the results of Nakajima [Nak2] which are generalizations of identities obtained by [Wil2] and further developed by

Bellman [Bel1] on the basis of the Atkinson dissection [Atk2]. Here we mean by the **Atkinson dissection** the splitting of the double sum

$$\sum_{m,n=1}^{\infty} = \sum_{m=1}^{\infty}\sum_{n<m} + \sum_{m=n=1}^{\infty} + \sum_{n=1}^{\infty}\sum_{m<n} \qquad (6.30)$$

which is originally due to Euler. Nakajima views this as the sum of series involving discontinuous integrals rather than the Perron formula as he claims.

We consider the Dirichlet series $\varphi(s)$ and $\Phi(s)$ defined as

$$\varphi(s) = \sum_{n=1}^{\infty} \frac{\alpha_n}{\lambda_n^s}, \qquad \sigma > \sigma_\varphi, \qquad (6.31)$$

$$\Phi(s) = \sum_{n=1}^{\infty} \frac{a_n}{\gamma_n^s}, \qquad \sigma > \sigma_\Phi, \qquad (6.32)$$

where $\{\lambda_n\}$ and $\{\gamma_n\}$ are increasing sequences of real numbers and α_n and a_n are complex numbers. We assume that they are continued to meromorphic functions over the whole plane and that they satisfy the growth condition

$$\varphi(\sigma + it) << (|t| + 1)^{s_\varphi(\sigma)}, \quad \Phi(\sigma + it) << (|t| + 1)^{s_\Phi(\sigma)} \qquad (6.33)$$

in the strip $-b < \sigma < c$, where $b > 0$ and $s_\varphi(\sigma), s_\Phi(\sigma) \leq s_\zeta(\sigma)$, and where $s_\zeta(\sigma)$ is the order of the Riemann zeta-function. We use the bound $s_\zeta(-b) = \frac{1}{2} + b + \varepsilon$ for $b > 0$ for any $\varepsilon > 0$ (which we suppress in what follows).

The analytic continuation is most often supplied by the functional equation

$$\varphi(s) = G(s)\psi(r - s), \qquad (6.34)$$

which we assume in the last stage of analysis, where

$$G(s) = \frac{\Gamma(\{e_j + E_j(r - s)\}_{j=1}^{N})}{\Gamma(\{f_j + F_j(r - s)\}_{j=1}^{Q})} \frac{\Gamma(\{c_j + C_j s\}_{j=1}^{P})}{\Gamma(\{d_j + D_j s\}_{j=1}^{M})} \qquad (6.35)$$

and

$$\psi(s) = \sum_{n=1}^{\infty} \frac{\beta_n}{\mu_n^s}, \qquad \sigma > \sigma_\psi. \qquad (6.36)$$

In the beginning we consider φ and Φ just as Dirichlet series and form the integral for $\varkappa \geq 0$ and any $c > 0$.

$$
\begin{aligned}
\mathcal{F}^\varkappa_{(c)}(u, v) &= \mathcal{F}^\varkappa_c(u, v; x) \\
&:= \frac{1}{2\pi i} \int_{(c)} \frac{\Gamma(w)}{\Gamma(w + \varkappa + 1)} \varphi(u + w) \Phi(v - w) x^{w + \varkappa} \, dw,
\end{aligned} \tag{6.37}
$$

and its counterpart $\mathcal{F}^\varkappa(v, u)$ under the condition

$$
\operatorname{Re} u > \sigma_\varphi + c, \quad \operatorname{Re} v > \sigma_\Phi + c, \tag{6.38}
$$

which assures the absolute convergence of the Dirichlet series. Hence they amount to (6.39).

Theorem 6.3

(i) *The Atkinson dissection is the special case of the Riesz sum $A^\varkappa(x)$ with $\varkappa = 0$ in the sense that*

$$
\begin{aligned}
A^\varkappa(x) &= \mathcal{F}^\varkappa_{(c)}(u, v) + \mathcal{F}^\varkappa_{(c)}(v, u) \\
&= \frac{1}{\Gamma(\varkappa + 1)} \sum_{m=1}^\infty a_m \gamma_m^{-v - \varkappa} {\sum_{\lambda_n \leq \gamma_m x}}' \alpha_n \lambda_n^{-u} (\gamma_m x - \lambda_n)^\varkappa \\
&\quad + \frac{1}{\Gamma(\varkappa + 1)} \sum_{n=1}^\infty \alpha_n \lambda_n^{-u - \varkappa} {\sum_{\gamma_m \leq \lambda_n x}}' a_m \gamma_m^{-v} (\lambda_n x - \gamma_m)^\varkappa
\end{aligned} \tag{6.39}
$$

implies

$$
\begin{aligned}
\sum_{m,n=1}^\infty \alpha_m \lambda_m^{-u} a_n \gamma_n^{-v} &= \sum_{m=1}^\infty \sum_{n<m} \alpha_m \lambda_m^{-u} a_n \gamma_n^{-v} \\
&\quad + \sum_{m=n}^\infty \alpha_n \lambda_n^{-u} a_n \gamma_n^{-v} + \sum_{n=1}^\infty \sum_{m<n} \alpha_m \lambda_m^{-u} a_n \gamma_n^{-v}.
\end{aligned} \tag{6.40}
$$

(ii) *Suppose for $b > \max\{\sigma_\varphi, \sigma_\Phi\}$*

$$
\begin{aligned}
\operatorname{Re} u &> \max\{\sigma_\varphi + c, s_\zeta(-b) - \varkappa\}, \\
\operatorname{Re} v &> \max\{\sigma_\Phi + c, s_\zeta(-b) - \varkappa\}.
\end{aligned} \tag{6.41}
$$

Then

$$
\mathcal{F}^\varkappa_{(c)}(u, v) + \mathcal{F}^\varkappa_{(c)}(v, u) = \mathcal{F}^\varkappa_{(-b)}(u, v) + \mathcal{F}^\varkappa_{(-b)}(v, u) + \mathrm{P}^\varkappa(x), \tag{6.42}
$$

where

$$\mathcal{F}^{\varkappa}_{(-b)}(u,v) = x^{\varkappa} \sum_{\ell_k=1}^{\infty} \frac{\tilde{\sigma}_{r-u-v}(\ell_k)}{\ell_k^{r-u}} \tag{6.43}$$

$$\times H^{P+1,N}_{M+1,Q}\left(\frac{1}{\ell_k x} \middle| \begin{array}{l} \{(1-(e_j+E_j(r-u)),E_j)\}_{j=1}^N, \\ (0,1),\{(c_j,C_j)\}_{j=1}^P, \end{array}\right.$$

$$\left. \begin{array}{l} (\varkappa+1,1),\{(d_j,D_j)\}_{j=1}^M \\ \{(1-(f_j+F_j(r-u)),F_j)\}_{j=1}^Q \end{array}\right)$$

$$= x^{\varkappa} \sum_{\ell_k=1}^{\infty} \frac{\tilde{\sigma}_{r-u-v}(\ell_k)}{\ell_k^{r-u}}$$

$$\times H^{N,P+1}_{Q,M+1}\left(\frac{1}{\ell_k x} \middle| \begin{array}{l} (1,1),\{(1-c_j,C_j)\}_{j=1}^P, \\ \{(e_j+E_j(r-u),E_j)\}_{j=1}^N, \end{array}\right.$$

$$\left. \begin{array}{l} \{(f_j+F_j(r-u),F_j)\}_{j=1}^Q \\ (-\varkappa,1),\{(1-d_j,D_j)\}_{j=1}^M \end{array}\right)$$

and where

$$\tilde{\sigma}_{r-u-v}(\ell_k) = \sum_{\gamma_m\mu_n=\ell_k} a_m \beta_n \gamma_m^{r-u-v} \tag{6.44}$$

and $\mathrm{P}^{\varkappa}(x)$ *indicates the sum of residues of the integrand*

$$\operatorname*{Res}_{\max\{-b,-\varkappa-1\}<\operatorname{Re} w<c}\left(\frac{\Gamma(w)}{\Gamma(w+\varkappa+1)}(\varphi(u+w)\Phi(v-w)\right.$$

$$\left. + \varphi(v+w)\Phi(u-w))\right)x^{w+\varkappa}$$

in the strip $\max\{-b,-\varkappa-1\} < \operatorname{Re} w < c$.

(iii) (6.42) *entails*

$$\mathcal{F}^0_{(c)}(u,v) + \mathcal{F}^0_{(c)}(v,u) = \mathcal{F}^0_{(-b)}(u,v) + \mathcal{F}^0_{(-b)}(v,u) + \mathrm{P}^0(x). \tag{6.45}$$

Proof. (i) Substituting and changing the order of summation and integration, we obtain

$$\mathcal{F}^{\varkappa}_{(c)}(u,v)$$

$$= \sum_{m=1}^{\infty} a_m \gamma_m^{-v} \sum_{n=1}^{\infty} \beta_n \mu_n^{u-r} \frac{x^{\varkappa}}{2\pi i} \int_{(c)} \frac{\Gamma(w)}{\Gamma(w+\varkappa+1)}\left(\frac{1}{\gamma_m\mu_n x}\right)^{-w} \mathrm{d}w, \tag{6.46}$$

whence

$$\mathcal{F}_{(c)}^{\varkappa}(u,v) = \frac{1}{\Gamma(\varkappa+1)} \sum_{m=1}^{\infty} a_m \gamma_m^{-u-\varkappa} \sideset{}{'}\sum_{\lambda_n \leq \gamma_m x} \lambda_n^{-u} (\gamma_m x - \lambda_m)^{\varkappa} a_n. \quad (6.47)$$

From (6.47) and its counterpart for $\mathcal{F}(v,u)$, we deduce (6.39).

The special case of (6.39) $\varkappa = 0, x = 1$ leads to (6.40) on account of the discontinuous integral.

(ii) To shift the line of integration to $\sigma = -b$, we must assure the (absolute) convergence of the resulting integrals by Condition (6.41). However, on the line $\sigma = -b$, one of the Dirichlet series still remains a Dirichlet series and is inert in the order of magnitude, whence the name "linearized".

In order to apply the functional equation, we must have $r + b - \operatorname{Re} u > \sigma_\varphi$ which is one of the conditions in (6.41). But to make the range being consistent, we need to take $\varkappa > 0$. Under the condition (6.41), we may apply the functional equation (6.34) and we obtain

$$\mathcal{F}_{(-b)}^{\varkappa}(u,v)$$
$$= \sum_{\ell_k=1}^{\infty} \frac{\tilde{\sigma}_{r-u-v}(\ell_k)}{\ell_k^{r-u}} \frac{x^{\varkappa}}{2\pi i} \int_{(-b)} \frac{\Gamma(w)}{\Gamma(w+\varkappa+1)} G(u+w) \left(\frac{1}{\ell_k x}\right)^{-w} dw$$
$$= \sum_{\ell_k=1}^{\infty} \frac{\tilde{\sigma}_{r-u-v}(\ell_k)}{\ell_k^{r-u}} \frac{x^{\varkappa}}{2\pi i} \int_{(b)} \frac{\Gamma(-w)}{\Gamma(\varkappa+1-w)} G(u-w)(\ell_k x)^{-w} dw. \quad (6.48)$$

(6.43) is the rewriting of (6.48). (iii) This is a typical example of the $\varkappa$ times integration of the original sum and by executing $\varkappa$ times differentiation (which is legitimate in view of absolute convergence) leads to the expression for the original partial sum. $\qquad\square$

Theorem 6.4 (Generalization of Wilton's formula) *In the case of the Riemann zeta-function, we assume* $\max\{1, b + \frac{1}{2} - \varkappa\} < \operatorname{Re} u < b$ *and the same for* v. *Then* (6.42) *remains true with*

$$P^{\varkappa}(x) = 2\pi^{-\frac{u+v}{2}} \sum_{j=0}^{\varkappa} \frac{\zeta(u+j)\zeta(v+j)}{\Gamma(\varkappa+1-j)} x^{\varkappa-j} \quad (6.49)$$

$$+ \pi^{-\frac{u+v}{2}} \zeta(u+v-1) \left(\frac{x^{\varkappa+1-u}}{(1-u)\cdots(\varkappa+1-u)} + \frac{x^{\varkappa+1-v}}{(1-v)\cdots(\varkappa+1-v)} \right)$$

and (6.43) reads

$$\mathcal{F}^{\varkappa}_{(-b)}(u,v) = \frac{2^{u-1}}{\pi^{1-\frac{u-v}{2}}i} \sum_{\ell=1}^{\infty} \frac{\sigma_{1-u-v}(\ell)}{\ell^{1-u}} \tag{6.50}$$

$$\times x^{\varkappa}\left(e^{\frac{\pi i}{2}u}G^{1,1}_{1,2}\left(2i\ell x\left|\begin{matrix}1\\1-u,-\varkappa\end{matrix}\right.\right) - e^{-\frac{\pi i}{2}u}G^{1,1}_{1,2}\left(-2i\ell x\left|\begin{matrix}1\\1-u,-\varkappa\end{matrix}\right.\right)\right)$$

and more concretely

$$\mathcal{F}^{\varkappa}_{(-b)}(u,v) = \frac{x^{\varkappa+1-u}}{\pi^{1-\frac{u-v}{2}}}\frac{\Gamma(1-u)}{\Gamma(2+\varkappa-u)}\sum_{\ell=1}^{\infty}\sigma_{1-u-v}(\ell) \tag{6.51}$$

$$\times\left({}_1F_1\left(\begin{matrix}1-u\\2+\varkappa-u\end{matrix};-2i\ell x\right) - {}_1F_1\left(\begin{matrix}1-u\\2+\varkappa-u\end{matrix};2i\ell x\right)\right).$$

Proof. (6.48) reads

$$\mathcal{F}^{\varkappa}_{(-b)}(u,v) \tag{6.52}$$

$$= \frac{x^{\varkappa}}{\pi^{\frac{1-u+v}{2}}}\sum_{\ell=1}^{\infty}\frac{\sigma_{1-u-v}(\ell)}{\ell^{1-u}}H^{2,0}_{1,3}\left(\frac{1}{\ell x}\left|\begin{matrix}\left(\frac{1+u}{2},\frac{1}{2}\right),(\varkappa+1,1),\left(\frac{u}{2},\frac{1}{2}\right)\\(0,1),\end{matrix}\right.\right).$$

By the standard technique of augmentation, duplication and reciproca-
tion, we may transform the H-function in (6.52) into

$$H^{2,0}_{1,3} \tag{6.53}$$

$$= \frac{2^{u-1}}{\sqrt{\pi i}}\left(e^{\frac{\pi i}{2}u}G^{1,1}_{2,1}\left(-\frac{i}{2\ell x}\left|\begin{matrix}u,\varkappa+1\\0\end{matrix}\right.\right) - e^{-\frac{\pi i}{2}u}G^{1,1}_{2,1}\left(\frac{i}{2\ell x}\left|\begin{matrix}u,\varkappa+1\\0\end{matrix}\right.\right)\right),$$

whence by inversion we deduce (6.50).

Applying the formula

$$G^{1,1}_{1,2}\left(z\left|\begin{matrix}a\\b,c\end{matrix}\right.\right) = z^b\frac{\Gamma(1-a+b)}{\Gamma(1-c+b)}\,{}_1F_1\left(\begin{matrix}1-a+b\\1-c+b\end{matrix};-z\right), \tag{6.54}$$

leads to (6.51).   □

Corollary 6.1 (Wilton) *In the region*

$$\mathrm{Re}\,u > -1,\ \mathrm{Re}\,v > -1\ \mathrm{Re}(u+v) > 0, \tag{6.55}$$

we have

$$\zeta(u)\zeta(v) = \zeta(u+v-1)\left(\frac{1}{u-1} + \frac{1}{v-1}\right)$$

$$+ 2(2\pi)^{u-1} \sum_{\ell=1}^{\infty} \frac{\sigma_{1-u-v}(\ell)}{\ell^{1-u}} u \int_{2\pi\ell}^{\infty} x^{-u-1} \sin x \, dx \qquad (6.56)$$

$$+ 2(2\pi)^{v-1} \sum_{\ell=1}^{\infty} \frac{\sigma_{1-u-v}(\ell)}{\ell^{1-v}} v \int_{2\pi\ell}^{\infty} x^{-v-1} \sin x \, dx$$

except for poles.

Proof. We perform differentiation in the narrower domain (6.41).

$\varkappa$ times differentiation of $\mathcal{F}_{(c)}^{\varkappa}(u,v)$ by (6.1) amounts to $\zeta(u)\zeta(v)$ by the Atkinson dissection, which cancels one term in (6.57).

$\varkappa$ times differentiation of (6.49) leads to

$$P^0(x) \qquad\qquad\qquad\qquad\qquad\qquad\qquad\qquad\qquad (6.57)$$

$$= \pi^{-\frac{u+v}{2}} \left(2\zeta(u)\zeta(v) - \zeta(u+v-1)\left(\frac{1}{u-1}x^{u-1} + \frac{1}{v-1}x^{v-1}\right)\right).$$

(6.50) with $\kappa = 0$ reads

$$\mathcal{F}_{(-b)}^0(u,v) = \frac{2^{u-1}}{\pi^{1-\frac{u-v}{2}}} \sum_{\ell=1}^{\infty} \frac{\sigma_{1-u-v}(\ell)}{\ell^{1-u}} \qquad\qquad\qquad (6.58)$$

$$\times \left(e^{-\frac{\pi i}{2}(1-u)} G_{1,2}^{2,0}\left(2i\ell x \,\middle|\, \begin{matrix} 1 \\ 1-u, 0 \end{matrix}\right) + e^{\frac{\pi i}{2}(1-u)} G_{1,2}^{2,0}\left(-2i\ell x \,\middle|\, \begin{matrix} 1 \\ 1-u, 0 \end{matrix}\right)\right).$$

At the final stage we apply

$$G_{1,2}^{2,0}\left(z \,\middle|\, \begin{matrix} 1 \\ a, 0 \end{matrix}\right) = \Gamma(a, z) \qquad\qquad\qquad\qquad (6.59)$$

and also appeal to (2.17).

Substituting these in (6.42) and multiplying both sides by $\pi^{\frac{u+v}{2}}$ yields (6.56).

We need to check that the differentiated series is uniformly convergent under the condition $\mathrm{Re}(u+v) > 0$. For simplicity, we take $\varkappa = 1$ and $b = \frac{3}{2}$. We apply the known estimates $\Gamma(a, z) \sim z^{s-1}e^{-z}$ and $\sigma_{1-u-v}(\ell) << \ell^{1-u-v+\varepsilon}$ to $\mathcal{F}_{(-\frac{3}{2})}^0(u,v)$ and conclude that the series is absolutely convergent. Also the integrals on the left side of (6.56) are analytic for $\mathrm{Re}\, u > -1$, $\mathrm{Re}\, v > -1$.

By Lemma 6.2, termwise differentiation (and differentiation in the integral sign) is justified in the domain (6.55) with restriction $\operatorname{Re} u, \operatorname{Re} v < \frac{3}{2}$. By analytic continuation (6.56) holds true in the wider domain (6.55). $\square$

We note that the above reasoning entails the reduction of G-function under differentiation:

$$\frac{\mathrm{d}}{\mathrm{d}z} z G_{1,2}^{1,1}\left(z \left|\begin{array}{c} 1 \\ 1-u,-1 \end{array}\right.\right) = -G_{1,2}^{1,1}\left(z \left|\begin{array}{c} 1 \\ 1-u,0 \end{array}\right.\right). \tag{6.60}$$

6.2 Modular relations in integral form

Under the heading "integration with respect to the parameter", Koshlyakov gave a number of intriguing integral identities, which may be viewed as modular relations in integral form. In §6.2.2, we give a typical example of this which is a generalization of Ramanujan's integral formula. Then in §6.2.1 we shall interpret this and many other integral identities as manifestations of the identity between Mellin transform and its inversion.

6.2.1 *Integration in the parameter*

Theorem 6.5 *Suppose the inverse Mellin transform*

$$f(x) = \frac{1}{2\pi i} \int_{(c)} \mathfrak{F}(s) x^{-s} \,\mathrm{d}s \quad c > 0 \tag{6.61}$$

and the Mellin transform

$$b^{-s} \mathfrak{G}(s) = \int_0^\infty g(bx) x^s \frac{\mathrm{d}x}{x} \quad \sigma > 0,\, b > 0 \tag{6.62}$$

are given. Then we have the integral formulas

$$\int_0^\infty f\left(\frac{a}{x}\right) g(bx) \,\mathrm{d}x = b^{-1} \frac{1}{2\pi i} \int_{(c)} \mathfrak{F}(s) \mathfrak{G}(1+s)(ab)^{-s} \,\mathrm{d}s, \quad a > 0, \tag{6.63}$$

and

$$\int_0^\infty f(ax) g(bx) \,\mathrm{d}x = b^{-1} \frac{1}{2\pi i} \int_{(c)} \mathfrak{F}(s) \mathfrak{G}(1-s)\left(\frac{a}{b}\right)^{-s} \,\mathrm{d}s, \quad a > 0, \tag{6.64}$$

provided that the following Stirling type growth conditions are satisfied: Writing $\tilde{f}(x,t,c) = \mathfrak{F}(c+it)x^{c+it}$, *we should have*

$$\int_{x_0}^{\infty} \tilde{f}(x,t,c)g(bx)\,\mathrm{d}x << \phi_1(t), \qquad \int_{t_0}^{\infty} \tilde{f}(x,t,c)g(bx)\,\mathrm{d}t << \phi_2(x), \quad (6.65)$$

for any t_0, x_0 *big enough and* $\phi_j(t) << e^{-q_j t}$, $q_j > 0$.

Indeed, substituting (6.61) in the left-hand side of (6.64), the left-hand side becomes

$$\int_0^{\infty} g(bx)\left(\frac{1}{2\pi i}\int_{(c)} \mathfrak{F}(s)(ax)^{-s}\,\mathrm{d}s\right)\frac{\mathrm{d}x}{x}$$

$$= \frac{1}{2\pi i}\int_{(c)} \mathfrak{F}(s)a^{-s}\,\mathrm{d}s\left(\int_0^{\infty} g(bx)x^{-s}\,\mathrm{d}x\right) \qquad (6.66)$$

since the interchange of the repeated infinite integrals is permissible by absolute convergence. The inner integral is $b^{s-1}\mathfrak{G}(1-s)$ by the change of variables, whence (6.64).

Similarly for (6.63), and the proof is complete.

Recall the symbols from §3.4.1: Ω = a number field, $\zeta_\Omega(s)$ = the Dedekind zeta-function, the functional equation (1.91), $G(s)$ = the quotient of gamma factors, $K(x) = K_{r_1,r_2}(x)$ = the K-function, $\sigma(x) = \sigma_{r_1,r_2}(x)$ = the basic series.

Recall the definition (3.43) of the X-function ([KoshI, p.121]) and the Mellin transform formula for the Y-function ([KoshI, (5.4)], cf. (3.159)):

$$\int_0^{\infty} Y_{r_1,r_2}(bx)x^{s-1}\,\mathrm{d}s$$

$$= \frac{\pi}{2\sin\left(\frac{\pi}{2}s\right)\Gamma^{r_1}\left(\frac{2-s}{2}\right)\Gamma^{r_2}(2-s)}b^{-s}, \quad b > 0 \qquad (6.67)$$

for $0 < \sigma < \frac{r_2+3}{\varkappa(\varkappa-1)}$.

Example 6.1　The X-function and the Y-function are the Mellin factor of the K-function in the sense of (3.158) ([KoshI, (5.8)]):

$$\int_0^{\infty} X\left(\frac{a}{x}\right)Y(bx)\,\mathrm{d}x = \frac{1}{b}K(ab), \quad a > 0, b > 0.$$

Indeed, in this case

$$\mathfrak{F}(s)\mathfrak{G}(s+1) = \frac{\pi\Gamma^{r_1}\left(\frac{s}{2}\right)\Gamma^{r_2}(s)}{2\cos\left(\frac{\pi}{2}s\right)\Gamma^{r_1}\left(\frac{1-s}{2}\right)\Gamma^{r_2}(1-s)} = \pi\frac{G(1-s)}{2\cos\left(\frac{\pi}{2}s\right)}.$$

Example 6.2 Recall the definition of the Y-function by (3.159) ([KoshI, (5.4)]):

$$Y(x) = Y_{r_1, r_2}(x)$$
$$= \frac{1}{2\pi i} \int_{(c)} \frac{\pi}{2 \sin\left(\frac{\pi}{2} s\right) \Gamma^{r_1}\left(\frac{2-s}{2}\right) \Gamma^{r_2}(2-s)} x^{-s} \, ds, \quad c > 0, \, x > 0$$

and the Mellin transform formula for the X-function ([KoshI, (4.2)])

$$\int_0^\infty X_{r_1, r_2}(xt) x^{s-1} \, ds, = \Gamma^{r_1}\left(\frac{s}{2}\right) \Gamma^{r_2}(s) x^{-s} \quad c > 0, \quad \sigma > 0, \qquad (6.68)$$

for $x > 0$ ($\mathrm{Re}(x) > 0$). Then we have (3.158) again.

Example 6.3 Koshlyakov's [KoshI, (12.19)] is the inversion of the formula in Lemma 1.2

$$\frac{f(ix) - f(-ix)}{2} = \frac{1}{2\pi i} \int_{(c)} \mathfrak{F}(s) \sin\left(\frac{\pi s}{2}\right) x^{-s} \, ds. \qquad (6.69)$$

Koshlyakov proved [KoshI, (6.2)]

$$\zeta_\Omega(s) = 2 \sin\left(\frac{\pi s}{2}\right) \int_0^\infty \sigma(x) x^{-s} \, dx. \qquad (6.70)$$

From these we conclude [KoshI, 1.5, p.217]

$$-2 \int_0^\infty \frac{f(ix) - f(-ix)}{2} \sigma(x) x^{-s} \, dx = \frac{1}{2\pi i} \int_{(c)} \mathfrak{F}(s) \zeta_\Omega(s) \, ds. \qquad (6.71)$$

We turn to the proof of Theorem 6.6.

Koshlyakov transforms his basic series (4.58) by Cauchy's residue theorem into [KoshI, (6.12), p.124, (7.4), p.125], [KoshIII, (31.1), p.307]

$$\sigma(x) + \frac{\zeta_\Omega(0)}{\pi x} = \frac{1}{2\pi i} \int_{(c)} \frac{A^{1-2s} G(1-s) \zeta_\Omega(s)}{2 \cos\left(\frac{\pi}{2} s\right)} \frac{ds}{x^s} \qquad (6.72)$$
$$= \frac{1}{2\pi i} \int_{(c)} \frac{A^{2s-1} G(s) \zeta_\Omega(1-s)}{2 \sin\left(\frac{\pi}{2} s\right)} \frac{ds}{x^s}, \quad 0 < c < 1.$$

From (3.159) and (6.72) we deduce [KoshI, (31.3)]

$$\int_0^\infty x Y_{r_1, r_2}(ax) \left(\sigma(x) + \frac{\zeta_\Omega(0)}{x} \right) dx$$
$$= \frac{1}{2\pi i} \int_{(c)} \frac{\pi A^{2s-1} \zeta_\Omega(1-s)}{2 \sin(\pi s) \Gamma^{r_1}\left(\frac{s}{2}\right) \Gamma^{r_2}(s)} a^{-s-1} \, ds, \quad 0 < c < 1. \qquad (6.73)$$

Applying the functional equation

$$A^{2s-1}\zeta_\Omega(1-s) = G(1-s)\zeta_\Omega(s) \tag{6.74}$$

and then substituting $s = 1 - z$, we transform the right-hand side of (6.73) into

$$\frac{1}{2\pi i} \int_{(c)} \frac{\pi\zeta_\Omega(s)}{2\sin(\pi s)\,\Gamma^{r_1}\left(\frac{1-s}{2}\right)\Gamma^{r_2}(1-s)}\, a^{-s-1}\, ds \tag{6.75}$$

$$= \frac{A^3}{a^3}\frac{1}{2\pi i} \int_{(c_1)} \frac{\pi A^{2z-1}\zeta_\Omega(1-z)}{2\sin(\pi z)\,\Gamma^{r_1}\left(\frac{z}{2}\right)\Gamma^{r_2}(z)}\left(\frac{A^2}{a}\right)^{-z-1}\, dz, \quad 0 < c_1 < 1,$$

which is the corrected form of [KoshI, ll.4-5, p.308].

Theorem 6.6 ([KoshII, (19.9), p.234], [KoshIII, (31.4), p.308]) *Under the reciprocal relation*

$$ab = A^2, \tag{6.76}$$

the modular relation holds true

$$\sqrt{a^3} \int_0^\infty x Y_{r_1,r_2}(ax)\left(\sigma(x) + \frac{\zeta_\Omega(0)}{\pi}\frac{1}{x}\right)\, dx \tag{6.77}$$

$$= \sqrt{b^3} \int_0^\infty x Y_{r_1,r_2}(bx)\left(\sigma(x) + \frac{\zeta_\Omega(0)}{\pi}\frac{1}{x}\right)\, dx.$$

6.2.2 Generalization of Ramanujan's integral formula

In this section, we state consequences of Theorem 6.6.

Corollary 6.2

(i) In the rational case, (6.77) leads to Ramanujan's integral formula

$$\sqrt{a^3} \int_0^\infty x e^{-(ax)^2}\left(\frac{1}{e^{2\pi x} - 1} - \frac{1}{2\pi x}\right)\, dx \tag{6.78}$$

$$= \sqrt{b^3} \int_0^\infty x e^{-(bx)^2}\left(\frac{1}{e^{2\pi x} - 1} - \frac{1}{2\pi x}\right)\, dx,$$

where (6.76) amounts to

$$ab = \pi. \tag{6.79}$$

(ii) In the imaginary quadratic case, (6.77) leads to the modular relation

$$\sqrt{a}\left(h + w\sum_{k=1}^{\infty}F(k)e^{-na}\right) = \frac{1}{\sqrt{b}}\left(h + w\sum_{k=1}^{\infty}F(k)e^{-nb}\right), \quad (6.80)$$

which is a corrected form of [KoshIII, (31.7), p.308] and where (6.76) amounts to

$$ab = \frac{4\pi^2}{|\Delta|}. \quad (6.81)$$

(iii) In the real quadratic case, (6.77) leads to the modular relation

$$\sqrt{a^3}\int xJ_0(2ax)\sigma(x)\,\mathrm{d}x = \sqrt{b^3}\int xJ_0(2bx)\sigma(x)\,\mathrm{d}x, \quad (6.82)$$

where $\sigma(x) = \sigma_{0,2}(x)$ indicates the basic series defined in (4.58) and (6.76) amounts to

$$ab = \frac{\pi^2}{\Delta}. \quad (6.83)$$

Proof. (i) follows from

$$\sigma_{1,0}(x) = \frac{1}{e^{2\pi x} - 1}. \quad (6.84)$$

(ii) See Exercise 6.3 below. which is a corrected form of [KoshII, (8.16),p.231] (stated again on [KoshII, p.308]). This in turn follows from the generalized Plana summation formula (Theorem 1.4).

(iii) is a simple rewriting. □

In the case of the real quadratic field, it is Koshlyakov's [KoshIII, (31.10)] rather than (6.80) that leads to the modular relation.

Theorem 6.7 *Let Ω be a real quadratic field. Then the equality*

$$\int_0^{\infty}J_0(2ax)\sigma(x)\,\mathrm{d}x = \int_0^{\infty}J_0(2bx)\sigma(x)\,\mathrm{d}x, \quad (6.85)$$

valid under the reciprocal relation $ab = \frac{\pi^2}{\Delta}$ *leads to the modular relation*

$$\sqrt{\rho} \left\{ 4\zeta'_\Omega(0) - 4\sum_{k=1}^{\infty} F(k) K_0 \left(\frac{2\pi k}{\sqrt{\Delta}} \rho \right) \right\}$$

$$= \frac{1}{\sqrt{\rho}} \left\{ 4\zeta'_\Omega(0) - 4\sum_{k=1}^{\infty} F(k) K_0 \left(\frac{2\pi i}{\sqrt{\Delta}} \frac{1}{\rho} \right) \right\}. \tag{6.86}$$

Proof. Recall that (6.72) now takes the form

$$\sigma(x) \tag{6.87}$$

$$= \frac{1}{2\pi i} \int_{(c)} \left(\frac{\sqrt{\Delta}}{\pi} \right)^{1-s} \frac{\Gamma^2 \left(\frac{1-s}{2} \right)}{\Gamma^2 \left(\frac{s}{2} \right)} \frac{\pi \zeta_\Omega(1-s)}{2 \sin\left(\frac{\pi}{2} s \right)} x^{s-1} \, ds, \quad 0 < c < 1.$$

We use (6.87) and the Mellin inversion formula for the J-Bessel function defined by (2.24):

$$\int_{(c)} J_0(2ax) x^s \frac{dx}{x} = \frac{\Gamma^2 \left(\frac{s}{2} \right) \sin \left(\frac{\pi s}{2} \right)}{2\pi a^s} \quad 0 < \sigma < \frac{3}{2} \tag{6.88}$$

to deduce

$$\int_0^\infty J_0(2Ax) \sigma(x) \, dx$$

$$= \frac{1}{2\pi i} \int_{(c)} \left(\frac{\sqrt{\Delta}}{\pi} \right)^{1-s} \Gamma^2 \left(\frac{1-s}{2} \right) \zeta_\Omega(s) \frac{ds}{4 \left(\frac{\sqrt{\Delta}a}{\pi} \right)^s}. \tag{6.89}$$

Applying the functional equation (3.37), we conclude (6.85).

To deduce (6.86) from (6.85) we need one more formula which follows from a form of the Plana summation formula:

$$\sum_{k=1}^{\infty} F(k) K_0(2k) = -\frac{\pi \zeta'_\Omega(0)}{\sqrt{\Delta}a} + \pi \int_0^\infty J_0(2bx) \sigma(x) \, dx. \tag{6.90}$$

$\square$

Exercise 6.3 Prove Corollary 6.2 (ii).

Solution (6.77) leads to

$$\sqrt{a} \int_0^\infty \sin ax \left(\sigma(x) - \frac{h}{w} \frac{1}{\pi x} \right) dx \tag{6.91}$$

$$= \sqrt{b} \int_0^\infty \sin bx \left(\sigma(x) - \frac{h}{w} \frac{1}{\pi x} \right) dx,$$

which is a corrected form of [KoshIII, (31.6), p.308]. To transform (6.91) into the modular relation we need some more data. One is

$$\sum_{k=1}^{\infty} F(k)e^{-ak} = -\frac{h}{w} + \frac{h}{w}\frac{2\pi}{\sqrt{|\Delta|}}\frac{1}{a} + \int_0^{\infty} \sin(ax)\sigma(x)\,\mathrm{d}x, \qquad (6.92)$$

which is a corrected form of [KoshII, (8.16), p.231] and of the formula squeezed by [KoshIII, (31.6), p.308] and [KoshIII, (31.7), p.308]. This is proved by the generalized Plana summation formula (Theorem 1.4) and the correct form of the formula is the first formula on [KoshIII, p.256]

$$\sum_{k=1}^{\infty} F(k)e^{-ak} = \zeta_\Omega(0) - \frac{2\pi}{\sqrt{|\Delta|}}\frac{\zeta_\Omega(0)}{a} + 2\int_0^{\infty} \sin(ax)\sigma(x)\,\mathrm{d}x, \qquad (6.93)$$

and accordingly, [KoshII, (18.6), p.231] should read

$$\sum_{k=1}^{\infty} F(k)e^{-A\rho k} = \zeta_\Omega(0)f(0) - \frac{\zeta_\Omega(0)}{\rho} + 2\int_0^{\infty} \sin(A\rho x)\sigma(x)\,\mathrm{d}x, \qquad (6.94)$$

where $\zeta_\Omega(0) = -\frac{h}{w}$ and A is given by (3.35). The other is a well-known evaluation for the sinus cardinalis function

$$\int_0^{\infty} \frac{\sin x}{x}\,\mathrm{d}x = \frac{\pi}{2}. \qquad (6.95)$$

Appealing to (6.76), we deduce (6.80).

6.3 Integrated modular relations

In this section we shall give integrated modular relations which are not modular relations in a proper sense, but are their integrated form. Therefore after differentiation in the parameter, they amount to the ordinary modular relation. However, there is termwise differentiation process involved and we need to make sure that the differentiated series is uniformly convergent as given by Lemma 6.2. The first example is the case where the termwise differentiation is not possible. In §6.3.3 we shall refer to arithmetical Fourier series which may be differentiated termwise with the help of (the equivalent to) the prime number theorem.

6.3.1   *The Hardy-Littlewood sum*

The Hardy-Littlewood sum $Q(t)$ is introduced in connection with summability of Fourier series and is better understood as the series

$$\tilde{Q}(t) = \sum_{n=1}^{\infty} \frac{t}{n} \sin \frac{t}{n}. \tag{6.96}$$

This has been treated as a number-theoretic problem in [Seg] and its sequel [Kan], also referred to by Berndt [Ber75-2]. In [LWKman] we interpreted the result of Segal [Seg, Theorem 1] as an integrated modular relation.

Theorem 6.8   ([LWKman, Theorem 2]) *For $y > 0$ we have*

$$\int_0^y \tilde{Q}(x)\,dx = \sum_{k=1}^{\infty}(1 - \cos \frac{y}{k}) \tag{6.97}$$

$$= \frac{\pi}{2}y - \frac{1}{2} + \sqrt{\frac{\pi y}{2}} \sum_{k=1}^{\infty} k^{-1/2} J_1(2\sqrt{2\pi yk}),$$

which is interpreted as $G_{0,1}^{1,0} \leftrightarrow G_{0,2}^{1,0}$.

We recall the process.

The main ingredient is the integrated form of [Seg, (16)]: Let $-1 < b < 0, y > 0$. Then

$$\int_0^y \tilde{Q}(x)\,dx = \frac{\pi y}{2\pi i} \int_{(b)} \frac{\Gamma(s)}{\Gamma(2-s)} \zeta(s)(2\pi y)^{-s}\,ds \tag{6.98}$$

and we transform the right-hand side integral into the series on the right of (6.97).

Applying the asymmetric form of the functional equation

$$\zeta(s) = \pi^{-1}(2\pi)^s \sin \frac{\pi}{2}s\Gamma(1-s)\zeta(1-s), \tag{6.99}$$

we see that the right-hand side of (6.98) becomes

$$-y\frac{1}{2\pi i} \int_{(b)} \Gamma(s-1) \sin \frac{\pi}{2} s\zeta(1-s)y^{-s}\,ds.$$

Hence substituting the series expression for $\zeta(1-s)$, we conclude that

$$\int_0^y \tilde{Q}(x)\,dx = -\sum_{n=1}^{\infty} \frac{y}{n} I_n, \tag{6.100}$$

where

$$\frac{y}{n}I_n = \frac{1}{2\pi i}\int_{(b)} \Gamma(s-1)\sin\frac{\pi}{2}s\left(\frac{y}{n}\right)^{1-s}ds,$$

say. To express I_n in terms of G-functions, we assure the separation of poles by the path as above (and in [LKT]) and we move the line of integration to $\sigma = c, 1 < c < 2$, thereby encountering a simple pole of the integrand at $s = 1$ with residue 1. Hence it follows that

$$\frac{y}{n}I_n = 1 - \frac{1}{2\pi i}\int_{(c-1)} \Gamma(s)\cos\left(\frac{\pi}{2}s\right)\left(\frac{y}{n}\right)^{-s}ds \qquad (6.101)$$

$$= 1 - \left(G_{0,1}^{1,0}\left(e^{\frac{\pi i}{2}}\frac{y}{n}\middle|\begin{array}{c}-\\0\end{array}\right) + G_{0,1}^{1,0}\left(e^{\frac{-\pi i}{2}}\frac{y}{n}\middle|\begin{array}{c}-\\0\end{array}\right)\right)$$

since the path $\sigma = c - 1$ separates the poles. The last term reduces to $= 1 - \cos\frac{y}{n}$ in view of

$$G_{1,0}^{0,1}\left(x^{-1}\middle|\begin{array}{c}1\\-\end{array}\right) = G_{0,1}^{1,0}\left(x\middle|\begin{array}{c}-\\0\end{array}\right) = \frac{1}{2\pi i}\int_{(\alpha)} x^{-s}\Gamma(s)\,ds = e^{-x} \qquad (6.102)$$

valid for $1 > \alpha > 0$ and $\operatorname{Re} x \geq 0$ ([ErdH, p. 12, (33)]). Substituting (6.101) in (6.100) proves $G_{0,1}^{1,0}$ part of (6.97).

To prove the $G_{0,2}^{1,0}$ part in (6.97), we shift the line of integration to the right up to $\sigma = c > 1$, thereby encountering simple poles at $s = 0$ and $s = 1$ with residues $\frac{\pi}{2}y$ and $-\frac{1}{2}$, respectively. Hence it follows that

$$\frac{\pi y}{2\pi i}\int_{(b)} \frac{\Gamma(s)}{\Gamma(2-s)}\zeta(s)(2\pi y)^{-s}\,ds$$

$$= \frac{\pi}{2}y - \frac{1}{2} + \frac{\pi y}{2\pi i}\int_{(c)} \frac{\Gamma(s)}{\Gamma(2-s)}\zeta(s)(2\pi y)^{-s}\,ds, \qquad (6.103)$$

where we substitute the series expression in the last integral and we are left with the integral

$$\frac{1}{2\pi i}\int_{(c)} \frac{\Gamma(s)}{\Gamma(2-s)}x^{-s}\,ds = G_{0,2}^{1,0}\left(x\middle|\begin{array}{c}-\\-1,0\end{array}\right) = x^{-\frac{1}{2}}J_{-1}(2\sqrt{x}).$$

This completes the proof.

6.3.2 The $H_{2,1}^{1,1} \leftrightarrow H_{0,3}^{2,0}$ formula

We show that the $H_{2,1}^{1,1} \leftrightarrow H_{0,3}^{2,0}$ formula leads to an identity different from (6.97).

We apply (6.13) to the following case. Let

$$\varphi(s) = \psi(s) = \sum_{k=1}^{\infty} \frac{\alpha_k}{\lambda_k^s} \tag{6.104}$$

and $\alpha_k = \beta_k = \frac{1}{2\sqrt{\pi}}$ and $\lambda_k = \mu_k = \frac{\sqrt{\pi}}{2}k$. Since $\varphi(s) = \frac{2^{s-1}}{\sqrt{\pi}^{s+1}}\zeta(s)$ and $\mathrm{Res}_{s=1}\,\varphi(s) = \frac{1}{\pi}$. Then the functional equation remains of the same form as (6.12) with $C = \frac{1}{2}$ and $r = 1$. Hence

$$\frac{1}{4}\left(\frac{2}{\sqrt{\pi}}\right)^{-s-1} \sum_{k=1}^{\infty} \frac{1}{k^s} H_{2,1}^{1,1}\left(z\frac{\sqrt{\pi}}{2}k \left|\begin{matrix} (1-a,A),(b,B) \\ (\frac{s}{2},\frac{1}{2}) \end{matrix}\right.\right)$$

$$+ \mathrm{Res}\left(\frac{\Gamma(a+Aw+As-A)}{\Gamma(b-Bw-Bs+B)}\Gamma(\tfrac{w}{2})\varphi(w)\,z^{s-1+w}, w=1\right)$$

$$= \frac{1}{4}\left(\frac{2}{\sqrt{\pi}}\right)^{2-s} \sum_{k=1}^{\infty} \frac{1}{k^{1-s}} H_{0,3}^{2,0}\left(\frac{\sqrt{\pi}\,k}{2z} \left|\begin{matrix} - \\ (\frac{1-s}{2},\frac{1}{2}),(a,A),(1-b,B) \end{matrix}\right.\right)$$

$$+ \mathrm{Res}\left(\frac{\Gamma(a-Aw+As)}{\Gamma(b+Bw-Bs)}\Gamma(\tfrac{w}{2})\varphi(w)\,z^{s-w}, w=1\right). \tag{6.105}$$

With $s = 1$, $a = A = \frac{1}{2}$, $b = 2, B = 1$ and z^{-1} for z, (6.105) reads

$$\frac{\pi^2}{16} \sum_{k=1}^{\infty} \frac{1}{k} H_{2,1}^{1,1}\left(\frac{\sqrt{\pi}}{2z}k \left|\begin{matrix} (\frac{1}{2},\frac{1}{2}),(2,1) \\ (\frac{1}{2},\frac{1}{2}) \end{matrix}\right.\right)$$

$$+ \mathrm{Res}\left(\frac{\Gamma(\frac{1}{2}(w+1))}{\Gamma(2-w)}\Gamma(\tfrac{w}{2})\varphi(w)\,z^{-w}, w=1\right)$$

$$= \frac{1}{2\sqrt{\pi}} \sum_{k=1}^{\infty} H_{0,3}^{2,0}\left(z\frac{\sqrt{\pi}\,k}{2} \left|\begin{matrix} - \\ (0,\frac{1}{2}),(\frac{1}{2},\frac{1}{2}),(-1,1) \end{matrix}\right.\right) \tag{6.106}$$

$$+ \mathrm{Res}\left(\frac{\Gamma(\frac{1}{2}(2-w))}{\Gamma(1+w)}\Gamma(\tfrac{w}{2})\varphi(w)\,z^{w-1}, w=1\right).$$

Now

$$\frac{1}{2\sqrt{\pi}} H_{0,3}^{2,0}\left(\frac{\sqrt{\pi}\,k}{2}\,z \,\middle|\, \begin{matrix} - \\ (0,\frac{1}{2}),(\frac{1}{2},\frac{1}{2}),(-1,1) \end{matrix}\right) \tag{6.107}$$

$$= H_{0,2}^{1,0}\left(\sqrt{\pi}\,k\,z \,\middle|\, \begin{matrix} - \\ (0,1),(-1,1) \end{matrix}\right) = G_{0,2}^{1,0}\left(\sqrt{\pi}\,k\,z \,\middle|\, \begin{matrix} - \\ 0,-1 \end{matrix}\right)$$

$$= \frac{1}{\sqrt{\sqrt{\pi}\,k\,z}} J_1\left(2\sqrt{\sqrt{\pi}\,k\,z}\right)$$

by (2.110).

On the other hand,

$$H_{2,1}^{1,1}\left(\frac{\lambda_k}{z} \,\middle|\, \begin{matrix} (\frac{1}{2},\frac{1}{2}),(2,1) \\ (\frac{1}{2},\frac{1}{2}) \end{matrix}\right) = H_{1,2}^{1,1}\left(\frac{z}{\lambda_k} \,\middle|\, \begin{matrix} (\frac{1}{2},\frac{1}{2}) \\ (\frac{1}{2},\frac{1}{2}),(-1,1) \end{matrix}\right) \tag{6.108}$$

$$= \frac{1}{2\pi i}\int_{(c)} \frac{\Gamma\left(\frac{1}{2}+\frac{1}{2}w\right)\Gamma\left(\frac{1}{2}-\frac{1}{2}w\right)}{\Gamma(-1-w)}\left(\frac{z}{\lambda_k}\right)^{-w} \,\mathrm{d}w.$$

We transform the integrand $I(w) := \frac{\Gamma(\frac{1}{2}+\frac{1}{2}w)\Gamma(\frac{1}{2}-\frac{1}{2}w)}{\Gamma(-1-w)}$ in various ways, leading to $\Gamma(2+w)\sin\left(\frac{\pi}{2}w\right)$.

First we multiply by $\frac{\Gamma(2+w)}{\Gamma(2+w)}$, apply the difference equation and the duplication formula to rewrite it as

$$I(w) = \frac{\Gamma\left(\frac{1}{2}+\frac{1}{2}w\right)\Gamma\left(\frac{1}{2}-\frac{1}{2}w\right)\Gamma(2+w)}{\Gamma(-1-w)\Gamma(2+w)} \tag{6.109}$$

$$= -\Gamma(2+w)\frac{\Gamma\left(\frac{1}{2}+\frac{1}{2}w\right)\Gamma\left(\frac{1}{2}-\frac{1}{2}w\right)}{\Gamma(-w)\Gamma(1+w)}$$

$$= -2\pi\Gamma(2+w)\frac{\Gamma\left(\frac{1}{2}+\frac{1}{2}w\right)\Gamma\left(\frac{1}{2}-\frac{1}{2}w\right)}{\Gamma\left(-\frac{1}{2}w\right)\Gamma\left(\frac{1}{2}-\frac{1}{2}w\right)\Gamma\left(\frac{1}{2}+\frac{1}{2}w\right)\Gamma\left(1+\frac{1}{2}w\right)}$$

$$= -2\pi\Gamma(2+w)\frac{1}{\Gamma\left(-\frac{1}{2}w\right)\Gamma\left(1+\frac{1}{2}w\right)},$$

whence it follows that

$$H_{1,2}^{1,1}\left(\frac{z}{\lambda_k} \,\middle|\, \begin{matrix} (\frac{1}{2},\frac{1}{2}) \\ (\frac{1}{2},\frac{1}{2}),(-1,1) \end{matrix}\right) = -2\pi H_{1,2}^{1,0}\left(\frac{z}{\lambda_k} \,\middle|\, \begin{matrix} (1,\frac{1}{2}) \\ (2,1),(1,\frac{1}{2}) \end{matrix}\right) \tag{6.110}$$

$$= \frac{-2\pi}{2\pi i}\left(G_{0,1}^{1,0}\left(e^{\frac{\pi}{2}i}\frac{z}{\lambda_k} \,\middle|\, \begin{matrix} - \\ 2 \end{matrix}\right) - G_{0,1}^{1,0}\left(e^{-\frac{\pi}{2}i}\frac{z}{\lambda_k} \,\middle|\, \begin{matrix} - \\ 2 \end{matrix}\right)\right)$$

$$= i\left(-\frac{z^2}{\lambda_k{}^2}e^{-i\frac{z}{\lambda_k}} + \frac{z^2}{\lambda_k{}^2}e^{-i\frac{z}{\lambda_k}}\right) = 2\frac{z^2}{\lambda_k{}^2}\sin\left(\frac{z}{\lambda_k}\right).$$

Or we apply the reciprocity relation to rewrite it as

$$I(w) = \frac{\Gamma\left(\frac{1}{2} + \frac{1}{2}w\right)\Gamma\left(\frac{1}{2} - \frac{1}{2}w\right)\Gamma(2+w)}{\Gamma(-1-w)\Gamma(2+w)} \tag{6.111}$$

$$= \Gamma(2+w)\frac{\sin(\pi(2+w))}{\pi}\frac{\pi}{\sin\left(\frac{\pi}{2} - \frac{\pi}{2}w\right)}$$

$$= \Gamma(2+w)\sin\left(\frac{\pi}{2}w\right)$$

at which we also arrive from (6.109).

Theorem 6.9

$$\frac{\pi}{4}z^2 \sum_{k=1}^{\infty} \frac{1}{k^3}\sin\left(\frac{2}{\sqrt{\pi}}zk\right)$$

$$= \frac{1}{2\sqrt{\pi^3 z}} \sum_{k=1}^{\infty} \frac{1}{\sqrt{k}} J_1\left(2\sqrt{\sqrt{\pi}\,k\,z}\right) + \mathrm{P}(z), \tag{6.112}$$

where $\mathrm{P}(z) = \frac{1}{\sqrt{\pi z}} - 1$ *is the residual function.*

6.3.3   *Arithmetical Fourier series*

One of the earliest occurrences of an arithmetical Fourier series in the modular relation setting is [AZW1, p.35, l.8] (again stated in [AZW2]), i.e. the Fourier expansion for the first periodic Bernoulli polynomial (3.73) in the form:

$$x - [x] - \frac{1}{2} + \frac{1}{2}F(x)(= \psi(x)) = -\frac{1}{\pi}\sum_{m=1}^{\infty} \frac{\sin 2\pi m x}{m}, \tag{6.113}$$

where $F(x) = 1$ or 0 according as $x \in \mathbb{Z}$ or not. Recall the quotient $G(s)$ of the gamma factors (4.56) for the Dedekind zeta-function in §3.2. Let $\tau = \varkappa 2^\varkappa$ and

$$J(x) = J_{r_1, r_2}(x) = \frac{1}{2\pi i}\int_{(3/4)} G(s)\left(\frac{x}{\tau}\right)^{\varkappa s - 1} ds, \tag{6.114}$$

which is a counterpart of Koshlyakov's L-function defined by (3.177). In view of

$$J_{1,0}(x) = \frac{2}{\sqrt{\pi}}\mathrm{si}(x), \tag{6.115}$$

A. Z. Walfisz's Theorem II reduces to (6.113).

In what follows we are going to study the threshold between the equivalent assertions to the functional equation and results coming from the zero-free regions for the associated zeta-function in terms of the arithmetical Fourier series. The periodic Bernoulli polynomial case ψ has been studied rather extensively and the log gamma function was considered in [Vi2], which is the counterpart thereof in the sense that they constitute the basis of the space of Kubert functions.

H. Davenport [D1; D2] and S. Chowla were the first who established the identities given in the following corollary with $\psi(t)$ the sawtooth Fourier series in (6.113).

Corollary 6.3 (Davenport-Chowla)

$$\sum_{n=1}^{\infty} \frac{\mu(n)}{n} \psi(nx) = -\frac{1}{\pi} \sin 2\pi x \qquad \text{[Ro2, (35)]} \qquad (6.116)$$

$$\sum_{n=1}^{\infty} \frac{\Lambda(n)}{n} \psi(nx) = -\frac{1}{\pi} \sum_{n=1}^{\infty} \frac{\log n}{n} \sin 2\pi nx \qquad \text{[Ro2, (34)]} \qquad (6.117)$$

$$\sum_{n=1}^{\infty} \frac{\lambda(n)}{n} \psi(nx) = -\frac{1}{\pi} \sum_{n=1}^{\infty} \frac{\sin 2\pi n^2 x}{n^2} \qquad (6.118)$$

$$\sum_{n=1}^{\infty} \frac{\mu(n)}{n^2} \psi(n^2 x) = -\frac{1}{\pi} \sum_{n=1}^{\infty} \frac{\mu^2(n)}{n} \sin 2\pi nx. \qquad (6.119)$$

Here $\mu(n)$, $\Lambda(n)$ and $\lambda(n)$ are called the **Möbius function**, the **von Mangoldt function** and the **Liouville function** and defined respectively by

$$\mu(n) = \begin{cases} 1, & n = 1 \\ (-1)^k, & n = p_1 \cdots p_k \text{ (dictinct primes)} \\ 0, & p^2 | n, \end{cases}$$

$$\Lambda(n) = \begin{cases} \log p, & \text{if } n = p^m \text{ for some prime } p \text{ and } m \geq 1 \\ 0, & \text{otherwise}, \end{cases}$$

and

$$\lambda(n) = (-1)^{\Omega(n)},$$

where $\Omega(n)$ is the total number of prime factors of n.

We note that in view of the familiar inversion formula

$$\sum_{d|n} \mu(d) = \begin{cases} 1 & n = 1 \\ 0 & n > 1 \end{cases}, \tag{6.120}$$

(6.116) is a formal reciprocal to (6.113) and *vice versa*.

In order to show the uniform convergence of the series in (6.116) Davenport proved the estimate

$$\sum_{n \leq x} \mu(n) e^{2\pi i n t} = O(x \log^{-A} x) \tag{6.121}$$

uniformly in t as $x \to \infty$ for any $A > 0$ which is as strong as the prime number theorem. In the same year, Chowla [Ch] proved the same results but in metrical sense.

Prior to this, Chowla and Walfisz [CW] had obtained a similar result but with the divisor function. They proved that

$$-\frac{1}{\pi} \sum_{n=1}^{\infty} \frac{d(n)}{n} \sin 2\pi n x \tag{6.122}$$

converges a.e. and represents a.e.

$$\Phi_1(x) = \sum_{n=1}^{\infty} \frac{\psi(nx)}{n}, \tag{6.123}$$

for whose generalization, cf. §6.3.4. Cf. §3.3.1.

Romanoff [Ro2, p.149] proved (6.116) and (6.116) in the metrical sense.

About 40 years later S. L. Segal [S1] used the method of complex analysis and established all Davenport's identities in a unified manner (with some reservations, cf. [BT]). His method essentially depends on the integral representation

$$\psi(u) = -\frac{1}{2\pi i} \int_{(c)} \frac{\zeta(s)}{s} u^s ds, \quad -1 < c < 0, \tag{6.124}$$

in which the integral is not absolutely convergent and so the interchange of limit relations is a subtle problem.

(6.118) pointwise has been treated by Walum [Wal] proving that the right-hand side series is uniformly convergent on the basis of the estimate (6.121) with μ replaced by λ. Indeed, Walum treated the Fourier series $\sum_{n=1}^{\infty} \frac{\lambda(n)}{n} \sin 2\pi n x$ only but it implies (6.118).

Segal's results are true a.e. by the following metrical theorem

Theorem 6.10 (Davenport-Chowla-Wintner)   *If*

$$a_n = O(n^\delta), \quad \delta < \frac{1}{2} \tag{6.125}$$

we have the identity **a.e.**

$$-\sum_{n=1}^{\infty} \frac{a_n}{n} \psi(xn) = \frac{1}{\pi} \sum_{n=1}^{\infty} \frac{A_n}{n} \sin 2\pi n x, \tag{6.126}$$

where

$$A_n = \sum_{d|n} a_d. \tag{6.127}$$

This theorem is to mean that the identity (6.126) is valid **a.e.**, which follows immediately from the discrete analogue of Carlson's fundamental theorem on Fourier inversion in mathematical analysis.

Theorem 6.11   (Discrete Carlson's theorem) *Let f be periodic of period 2π and $f \in L^p(\mathbb{R})$ for some $p \in (1, \infty)$ with Fourier coefficients $\hat{f}(n)$. Then the* **Fourier expansion** *holds true:*

$$\frac{1}{\sqrt{2\pi}} \sum_{n=-\infty}^{\infty} \hat{f}(n) e^{int} = \lim_{N \to \infty} \frac{1}{\sqrt{2\pi}} \sum_{|n| \leq N} \hat{f}(n) e^{int} = f(t) \tag{6.128}$$

for **almost every** *(a.e.) $t \in \mathbb{R}$.*

An enlightening contrast may be found in the study of the Hardy-Littlewood sum $\tilde{Q}(t) = \sum_{n=1}^{\infty} \frac{t}{n} \sin \frac{t}{n}$ in [Seg] and its sequel [Kan], also referred to by Berndt [Ber75-2]. Segal's main result reads

$$\int_0^y \tilde{Q}(t) \, dt = \frac{\pi}{2} - \frac{1}{2} + \sqrt{\frac{\pi}{2} y} \sum_{n=1}^{\infty} n^{-\frac{1}{2}} J_1(2\sqrt{2\pi y n}), \tag{6.129}$$

Segal's question as to the possibility of termwise differentiation of both sides of (6.129) has been treated by Kano [Kan], who proved that it is not possible (the differentiated series being divergent for all t), but that it is $(C, 1)$-summable.

Since the integral $\int_0^y \tilde{Q}(t) \, dt$ is the Riesz sum of order 1 of $\tilde{Q}(t)$, this means that the processing gamma factor $\Delta(s) = \frac{1}{s(s+1)}$. It seems that the $(C, 1)$-summability of the differentiated series follows from a theorem of Chandrasekharan and Narasimhan [ChN2, Theorem IV] concerning the Riesz summability ($(C,1)$-summability and Riesz-summability of order 1, are equivalent). For closer analysis we refer to [KMT2].

In our specified application, we take the absolutely convergent integral ([LMZ, Proposition 1])

$$\bar{B}_\ell(u) \tag{6.130}$$

$$= -\frac{(-1)^\ell \ell!}{2\pi i} \int_{(c)} \frac{\Gamma(-s)}{\Gamma(-s+\ell)} \zeta(1-\ell+s) u^{s+\ell-1} ds, \quad -1 < c < 0$$

for $\ell \geq 2$. The case $\ell = 2$ gives the second periodic Bernoulli polynomial. We note that this is a special case of Mikolás' result [Mik].

It satisfies the functional equation in the form

$$\zeta(s) = \pi^{s-\frac{1}{2}} \frac{\Gamma\left(\frac{1-s}{2}\right)}{\Gamma\left(\frac{s}{2}\right)} \zeta(1-s). \tag{6.131}$$

We are concerned with the functional equation

$$\varphi(s) = G(s)\psi(r-s), \tag{6.132}$$

where

$$G(s) = \frac{\Gamma(\{e_j + E_j(r-s)\}_{j=1}^N)}{\Gamma(\{f_j + F_j(r-s)\}_{j=1}^Q)} \frac{\Gamma(\{c_j + C_j s\}_{j=1}^P)}{\Gamma(\{d_j + D_j s\}_{j=1}^M)}. \tag{6.133}$$

The argument is in the spirit similar to that of §6.1.6.

We consider the third Dirichlet series $\Phi(s)$ in (6.32). Let $\Psi(s)$ be the product of $\psi(s)$ and $\Phi(s)$, which can be expanded into Dirichlet series:

$$\Psi(s) = \psi(s)\Phi(s) = \sum_{n=1}^\infty \frac{A_n}{\nu_n^s}, \quad \sigma > \max\{\sigma_\psi, \sigma_\Phi\} := \sigma_\Psi, \tag{6.134}$$

where

$$A_l = \sum_{\mu_m \gamma_n = \nu_l} \beta_m a_n. \tag{6.135}$$

Let c be a constant such that $c < r - \max\{\sigma_\Phi, \sigma_\Psi\}$. We assume that the vertical line (c) separates the poles of $\Delta(s)G(s)$, where $\Delta(s)$ is the processing gamma factor which satisfies the conditions

$$\Delta(s)\varphi(s) = O\left((|t|+1)^{-1-\varepsilon}\right) \tag{6.136}$$

and

$$\Delta(s)G(s) = O\left((|t| + 1)^{-1-\varepsilon}\right), \tag{6.137}$$

in the strip which appear in the following analysis, where $\varepsilon > 0$.

Then we consider the absolutely convergent integral in two different ways $(x > 0)$

$$\mathfrak{G}(x) = \frac{1}{2\pi i} \int_{(c)} \Delta(s)\varphi(s)\Phi(r - s)x^s \mathrm{d}s. \tag{6.138}$$

During our analysis, $\Phi(r - s)$ is a Drichlet series in the first place and $\Psi(r - s) = \psi(r - s)\Phi(r - s)$ is a Drichlet series in the second place. In practice, this amounts to the cancellation of the denominator of $\Phi(r-s)$ by $\psi(r - s)$; e.g. in the case of the von Mangoldt function, $\zeta(1 - s)\frac{\zeta'(1-s)}{\zeta(1-s)} = \zeta'(1 - s)$.

Then, on one hand, we have

$$\mathfrak{G}(x) = \sum_{n=1}^{\infty} \frac{a_n}{\gamma_n^r} X(\gamma_n x), \tag{6.139}$$

where

$$X(x) = \frac{1}{2\pi i} \int_{(c)} \Delta(s)\varphi(s)x^s \mathrm{d}s. \tag{6.140}$$

On the other hand, by (6.132) and (6.134)

$$\mathfrak{G}(x) = \frac{1}{2\pi i} \int_{(c)} \Delta(s)G(s)\Psi(r - s)x^s \mathrm{d}s. \tag{6.141}$$

Hence by (6.137)

$$\begin{aligned}
\mathfrak{G}(x) &= \sum_{n=1}^{\infty} \frac{A_n}{\nu_n^r} \frac{1}{2\pi i} \int_{(c)} \Delta(s)G(s)(\nu_n x)^s \mathrm{d}s \\
&= \sum_{n=1}^{\infty} \frac{A_n}{\nu_n^r} Y(\nu_n x),
\end{aligned}$$

where

$$Y(x) = \frac{1}{2\pi i} \int_{(c)} \Delta(s)G(s)x^s \mathrm{d}s \tag{6.142}$$

is the counterpart of $X(x)$.

Theorem 6.12	*Notation being as given above, we have the modular relation under the conditions* (6.136) *and* (6.137)

$$\sum_{n=1}^{\infty} \frac{a_n}{\gamma_n^r} X(\gamma_n x) = \sum_{n=1}^{\infty} \frac{A_n}{\nu_n^r} Y(\nu_n x). \tag{6.143}$$

Condition (6.137) may be satisfied e.g. if

$$\Delta(s) = O\left((|t| + 1)^{-1-\mathrm{E}} \right), \tag{6.144}$$

and

$$\begin{aligned}
\mathrm{E} > &\sum_{j=1}^{P} \left(c_j + C_j \sigma - \frac{1}{2} \right) + \sum_{j=1}^{N} \left(e_j + E_j(\sigma - r) - \frac{1}{2} \right) \\
&- \sum_{j=1}^{M} (d_j + D_j \sigma) - \sum_{j=1}^{Q} \left(e_j + E_j(\sigma - r) - \frac{1}{2} \right).
\end{aligned} \tag{6.145}$$

As a concrete case of Theorem 6.12, we state the following theorem on point-wise convergence with the data, proof being given in Exercise 6.4:

$$\varphi(s) = \psi(s) = \pi^{-\frac{s}{2}} \zeta(s) = \sum_{n=1}^{\infty} \frac{1}{(\sqrt{\pi}n)^s}, \quad r = 1, \quad \sigma_\varphi = \sigma_\psi = 1$$

$$\Delta(s) = \frac{1}{s(s+1)}, \quad G(s) = \frac{\Gamma\left(\frac{1}{2} - \frac{1}{2}s \right)}{\Gamma\left(\frac{1}{2}s \right)}, \quad \Phi(s) = \sum_{n=1}^{\infty} \frac{a_n}{n^s}$$

$$1 \le \sigma = \Psi = \max\{\sigma_\psi, \sigma_\Phi\} \le \sigma_\Phi.$$

Theorem 6.13 (Davenport-Segal)	*Given a Dirichlet series* $\Phi(s)$ *defined by* (6.32), *where* $\sigma_\Phi < 2$ *and* $\gamma_n = n$, *we have the identity*

$$\sum_{n=1}^{\infty} \frac{a_n}{n^2} \bar{B}_2(xn) = \frac{1}{2\pi^2} \sum_{n=1}^{\infty} \frac{A_n}{n^2} \cos 2\pi nx, \tag{6.146}$$

where A_n *is given by* (6.135).

The equality (6.146) may be termwise differentiated to give rise to (6.126) if the PNT type estimate $\sum_{n \le X} a_n e^{2\pi i n x} = O(x \log^{-A} x)$ holds true.

Exercise 6.4	We assume that $\sigma_\Phi < 2$. There is a real number c such that $c < 1 - \max\{\sigma_\Phi, \sigma_\Psi\}$ and $-1 < c < 0$. Hence Condition (6.137) is satisfied.

Computation is almost verbatim to §6.1.6 and we omit the details. First, since $-1 < c < 0$, we have

$$X(x) = \frac{1}{2\pi i} \int_{(c)} \frac{\zeta(s)}{s(s+1)} \left(\frac{x}{\sqrt{\pi}}\right)^s ds = -\frac{\sqrt{\pi}}{x} \bar{B}_2\left(\frac{x}{\sqrt{\pi}}\right) \qquad (6.147)$$

by (6.130).

On the other hand,

$$Y(x) = -\frac{1}{2\sqrt{\pi x}} \cos 2x, \qquad (6.148)$$

so that

$$\mathfrak{G}(x) = -\frac{1}{2\sqrt{\pi x}} \sum_{n=1}^{\infty} \frac{A_n}{\nu_n^2} \sin 2\nu_n x = -\frac{1}{2\pi^{3/2} x} \sum_{n=1}^{\infty} \frac{A_n}{n^2} \cos 2\sqrt{\pi} n x \quad (6.149)$$

since $\nu_n = \sqrt{\pi} n$ and $r = 1$.

Substituting (6.147) in (6.139), we obtain

$$\mathfrak{G}(x) = -\frac{\sqrt{\pi}}{x} \sum_{n=1}^{\infty} \frac{a_n}{n^2} \bar{B}_2\left(\frac{x}{\sqrt{\pi}} n\right) \qquad (6.150)$$

which combined with (6.149) gives Theorem 6.13 on writing x for $\frac{x}{\sqrt{\pi}}$.

Remark 6.4 *Indeed, it is quite simple to guess the formula in Corollary 6.3. E.g. we check (6.116). Assuming the uniform convergence due to Davenport, we integrate termwise to obtain*

$$\sum_{n=1}^{\infty} \frac{\mu(n)}{n} \int_0^x \psi(nt) dt = -\frac{1}{\pi} \int_0^x \sin 2\pi t dt,$$

whence

$$\frac{1}{2} \sum_{n=1}^{\infty} \frac{\mu(n)}{n^2} (\bar{B}_2(nx) - B_2) = -\frac{1}{2\pi^2}(-\cos 2\pi x + 1). \qquad (6.151)$$

Hence, noting that $\sum_{n=1}^{\infty} \frac{\mu(n)}{n^2} = \frac{1}{\zeta(2)} = \frac{6}{\pi^2}$, we conclude

$$\sum_{n=1}^{\infty} \frac{\mu(n)}{n^2} \bar{B}_2(nx) = \frac{1}{\pi^2} \cos 2\pi x, \qquad (6.152)$$

which is a special case of Theorem 6.13 with $\Phi(s) = \frac{1}{\zeta(s)}$ and $\alpha_n = \mu(n)$.

The situation is suggestive of Ingham's treatment [Ing] of the prime number theorem, i.e. first apply the Abelian process and then Tauberian process which needs more information.

We may also deduce the following result

$$\frac{1}{(2k)!} \sum_{n=1}^{\infty} \frac{\mu(n)}{n^{2k}} \bar{B}_{2k}(nx) = \frac{2(-1)^k}{(2\pi)^{2k}} \cos 2\pi x, \qquad (6.153)$$

$$\frac{1}{(2k-1)!} \sum_{n=1}^{\infty} \frac{\mu(n)}{n^{2k-1}} \bar{B}_{2k-1}(nx) = \frac{2(-1)^k}{(2\pi)^{2k-1}} \sin 2\pi x,$$

whence we may deduce by setting $x = 0$,

$$\zeta(2k) = \frac{(-1)^{k-1} B_{2k}}{2(2k)!} (2\pi)^{2k}. \qquad (6.154)$$

Indeed, we may integrate (6.152) termwise in view of absolute convergence of the series to obtain

$$\frac{1}{(2k)!} \sum_{n=1}^{\infty} \frac{\mu(n)}{n^{2k}} \left(\bar{B}_{2k}(nx) - B_{2k} \right) = \frac{(-1)^k}{2^{2k-1}\pi^{2k}} \left(\cos 2\pi x - 1 \right). \qquad (6.155)$$

Table 6.1   Arithmetical Fourier series

type	Divisor ftns	PNT ftns	Riemann ND ftns
metric.	[Ch], [Ro2], [HW], [S1], [Wi3], [KMT1]	[S1], [LMZ], [KMT1]	[Win1]-[Wi3]
p'twise	[HL1], [HL2], [HL3], [Ko], [CW]	[D1], [D2], [Wal], [BT]	[Ri2], [I], [Kano], [HT], [Mey]

Corollary 6.4 *The Fourier series* (6.113) *is a special case of Theorem 6.13 and so, in the long run, is a consequence of the functional equation* (6.132) *and the PNT.*

This follows on choosing $\Phi(s) = 1$, i.e. $a_1 = 1$, $a_n = 0$ $(n > 1)$, so that $A_n = 1$ for all $n \geq 1$.

Exercise 6.5 Prove Corollary 6.3.

Solution (6.116) has been shown in Remark 6.4.

(6.117):

$$a_n = \Lambda(n): \qquad \Phi(s) = \sum_{n=1}^{\infty} \frac{\Lambda(n)}{n^s} = -\frac{\zeta'}{\zeta}(s),$$

$$\Psi(s) = \pi^{-\frac{s}{2}}\zeta(s)\Phi(s) = -\pi^{-\frac{s}{2}}\zeta'(s) = \sum_{n=1}^{\infty} \frac{\log n}{(\sqrt{\pi}n)^s} = \sum_{n=1}^{\infty} \frac{A_n}{\nu_n^s},$$

whence $A_n = \log n$ and $\nu_n = \sqrt{\pi}n$.

(6.118):

$$a_n = \lambda(n): \qquad \Phi(s) = \sum_{n=1}^{\infty} \frac{\lambda(n)}{n^s} = \frac{\zeta(2s)}{\zeta(s)},$$

$$\Psi(s) = \pi^{-\frac{s}{2}}\zeta(s)\Phi(s) = \pi^{-\frac{s}{2}}\zeta(2s) = \sum_{n=1}^{\infty} \frac{1}{(\sqrt{\pi}n^2)^s} = \sum_{n=1}^{\infty} \frac{A_n}{\nu_n^s},$$

so that

$$A_n = \begin{cases} 1, & n = \text{square} \\ 0, & n \neq \text{square} \end{cases} = \chi_2(n),$$

which is the characteristic function of the set of squares, and $\nu_n = \sqrt{\pi}n$. Hence (6.126) reads

$$-\sum_{n=1}^{\infty} \frac{\lambda(n)}{n}\psi(nx) = \frac{1}{\pi}\sum_{n=1}^{\infty} \frac{\chi_2(n)}{n}\sin 2\pi nx = \frac{1}{\pi}\sum_{n=1}^{\infty} \frac{1}{n^2}\sin 2\pi n^2 x, \quad (6.156)$$

i.e. (6.118).

(6.119): Since

$$\sum_{n=1}^{\infty} \frac{\mu(n)}{n^2}\psi(n^2 x) = \sum_{n=1}^{\infty} \frac{a_n}{n^2}\psi(nx), \quad a_n = \chi_2(n)\mu(\sqrt{n}),$$

where we understand $\mu(\sqrt{n}) = 0$ if n is not a square, it follows that

$$\Phi(s) = \sum_{n=1}^{\infty} \frac{\mu(n)}{n^{2s}} = \zeta(2s)^{-1}$$

and

$$\Psi(s) = \phi_2(s)\Phi(s) = \pi^{-\frac{s}{2}}\zeta(s)\frac{1}{\zeta(2s)} = \pi^{-\frac{s}{2}}\sum_{n=1}^{\infty} \frac{\mu^2(n)}{n^s},$$

whence $A_n = \mu^2(n)$, $\nu_n = \sqrt{\pi}n$.

Now (6.126) leads us to (6.119).

To establish the pointwise convergence of Fourier series, we are to make a recourse to termwise differentiation, which is assured by uniform convergence but not necessarily by absolute convergence. For example, we may appeal to the following.

Lemma 6.2   [ApoMA, Theorem 13-14,p.403] *Assume that each f_k is a real-valued function defined and differentiable at each $x \in (a,b)$. Assume that for at least one point $x_0 \in (a,b)$, the series $\sum_{k=1}^{\infty} f_k(x_0)$ is convergent. Assume further that there exists a function g such that $\sum_{k=1}^{\infty} f'_k(x) = g(x)$ uniformly on (a,b). Then there exists a function f such that $\sum_{k=1}^{\infty} f_k(x) = f(x)$ uniformly on (a,b) and that if $x \in (a,b)$, then the derivative $f'(x)$ exists and equals $\sum_{k=1}^{\infty} f_k(x) = f'(x)$.*

6.3.4   *Riemann's legacy*

In this section we shall make various remarks on Fourier series involving arithmetical functions, especially in connection with Riemann's examples. We note that the series on the right of (6.118) is Riemann's example of a continuous, "non-differentiable" function which, according to Weierstrass, he announced around 1861 ([Neu], [S2]).

It is interesting to note that the above sine series is the imaginary part of the series

$$\sum_{n=1}^{\infty} \frac{e^{2\pi i n^2 x}}{n^2} \tag{6.157}$$

considered by Itasu [I], Luther [L] et al, to prove the "non-differentiability" of its imaginary part.

The real part of (6.157) is

$$f(x) = \sum_{n=1}^{\infty} \frac{\cos 2\pi n^2 x}{n^2} \tag{6.158}$$

which is another example of Riemann discussed by Casorati and Prym in 1865 ([Neu]).

This contrast between the real and imaginary parts of (6.157) explains why Riemann was not interested in the real part in his posthumous fragment ([Rie2] (cf. [Win1, p.633], [AdR, p.59, 63]), see also below.

In §6 of his Habilitationsschrift [Ri2], Riemann gives

$$\sum_{n=1}^{\infty} \frac{\psi(nx + \frac{1}{2})}{n^2}$$

as an example of a discontinuous, nonetheless integrable function.

As remarked by [AdR, p.59], this example has its origin in the two functions considered by him in Fragment I:

$$\sum_{n=1}^{\infty} \frac{(-1)^n}{n^2} \psi\left(\frac{nx}{2\pi} + \frac{1}{2}\right), \quad \sum_{k=0}^{\infty} \frac{1}{(2k+1)^2} \psi\left(\frac{(2k+1)x}{2\pi} + \frac{1}{2}\right).$$

He also asserted [Ri2] that the trigonometrical series

$$\sum_{n=1}^{\infty} \frac{1}{n} \psi\left(nx + \frac{1}{2}\right) \tag{6.159}$$

converges to the "Fourier series"

$$\sum_{n=1}^{\infty} \frac{c(n)}{n} \sin 2\pi nx, \tag{6.160}$$

for $x \in \mathbb{Q}$, where $c(n) = \sum_{d|n}(-1)^d$.

Since

$$\psi\left(x + \frac{1}{2}\right) = \psi(2x) - \psi(x)$$

([Win1, p. 633]), this implies Riemann's assertion above.

More generally than (6.123), let $\Phi_j(s,x)$ be the series considered by Hardy and Littlewood [HL3]

$$\Phi_j(s,x) = \sum_{n=1}^{\infty} \frac{\bar{B}_j(nx)}{n^s} \quad (1 \leq j \in \mathbb{N}), \ \Phi_s(x) = P_1(s,x) \sum_{n=1}^{\infty} \frac{\psi(nx)}{n^s} \tag{6.161}$$

where

$$\bar{B}_j(x) = \sum_{k=0}^{j} \binom{j}{k} B_{j-k}\{x\}^j$$

is the j-th periodic Bernoulli polynomial.

Wintner [Win1] proved that $\Phi_1(x) = \Phi_1(1,x)$ is integrable a là Lebesgue, has the trigonometrical series (6.122) as its Fourier series and is such that $\Phi_1(x)$ is of class L^q for every $q > 0$.

The convergence of (6.122) to $\Phi_1(x)$ is a consequence of the Hausdorff-Paley extension of Fischer-Riesz theorem in §3.3.1. Hartman and Wintner [HW] note that the method of Chowla and Walfisz is not applicable to $\Phi_s(x)$ for $\sigma < 1$ and they apply the method of [Win1] to prove the a.e. convergence of the Fourier series to $\Phi_j(s,x)$:

$$\Phi_j(s,x) \sim -\frac{j!}{(2\pi i)^j} \sum_{k=-\infty}^{\infty} \frac{\sigma_{j-s}(k)}{k^j} e^{2\pi i k x} \qquad (6.162)$$

for $\frac{1}{2} < \text{Re} s \leq 1$ ([HW, p.116]).

Although Hartman and Wintner [HW] state the case of the divisor functions, their theorem is true for general arithmetic functions satisfying the order conditions in L^p-context and it reads

Theorem 6.14 (Hartman-Wintner) *Given a Dirichlet series $\Phi(s)$ defined by (6.32), where $\sigma_\Phi < 2$ and $\gamma_n = n$, we have the identity valid for $\lambda + \frac{1}{2} < \sigma < \lambda + 1$ and a.a. values of x*

$$-\sum_{n=1}^{\infty} \frac{a_n}{n^s} \psi(xn) = \frac{1}{\pi} \sum_{n=1}^{\infty} \frac{A_n}{n^s} \sin 2\pi n x, \qquad (6.163)$$

provided that

$$\sum_{d|n} |a_d| << n^{\lambda+\varepsilon} \qquad (6.164)$$

for any $\varepsilon > 0$ (6.163) means that the Fourier series $\frac{1}{\pi}\sum_{n=1}^{\infty} \frac{A_n}{n^s} \sin 2\pi n x$ converges a.e. to the L^q-class function $-\Phi_s(x)$ for

$$2 \leq q < \frac{1}{1+\lambda-\sigma}. \qquad (6.165)$$

Proof of this theorem follows almost verbatim to that of [HW, p.115]. Indeed, the Fourier series for the (Riemann integrable) partial sum $\Phi_s^N(x) = -\sum_{n=1}^{N} \frac{a_n}{n^s}\psi(xn)$ is

$$-\sum_{n=1}^{\infty} \frac{A_n^N}{n} \sin 2\pi x n,$$

where $A_n^N = \sum_{d|n, d \leq N} a_d d^{1-s}$. Hence $|A_n^N| << n^{\lambda+1-\sigma+\varepsilon}$ by (6.164). Hence it suffices to have $\frac{q}{q-1}(\sigma - \lambda) > 1$ or the condition (6.165).

Finally we state a theorem connecting (6.158) to the **Riemann Hypothesis** (RH).

Let $F_x = F_{[x]}$ denote the Farey series of order x:

$$F_x = \left\{ \rho_\nu \ \middle| \ \rho_\nu = \frac{a_\nu}{b_\nu}, \ (a_\nu, b_\nu) = 1, \ 0 \leq a_\nu < b_\nu \leq x \right\}.$$

Then

$$\#F_x = \sum_{n \leq x} \varphi(n) = \Phi(x),$$

where $\varphi(n)$ signifies the Euler function.

Define the error function

$$E_f(x) := \sum_{\nu=1}^{\Phi(x)} f(\rho_\nu) - \Phi(x) \int_0^1 f(u)\mathrm{d}u.$$

Theorem 6.15 ([KY, p. 442, Corollary]) *For every $\varepsilon > 0$,*

$$\mathrm{RH} \iff E_f(x) = O(x^{\frac{1}{2}+\varepsilon}).$$

More generally, we may formulate the equivalent condition to the RH for

$$f_{\kappa,l}(x) = \sum_{n=1}^{\infty} \frac{1}{n^\kappa} \cos 2\pi n^l x, \quad \mathrm{Re}\,\kappa > 1, \ l \in \mathbb{N},$$

whence

$$f_{2,2}(x) = f(x)$$

above.

In England there are three great analysts, Hardy, Littlewood, and Hardy-Littlewood.

— Landau

In India there are three great analytic number-theorists, Balasubramanian, Ramchandra, and Balasubramanian-Ramchandra.

Fig. 6.1   Kanakanahari Ramachandra

Chapter 7

The General Modular Relation

7.1 Definitions

In this section, we fix some notation related to Fox H-functions. We set

$$\Omega = \left(\coprod_{n=0}^{\infty} \left(\mathbb{C} \times \mathbb{R}^+ \right)^n \right)^4 \tag{7.1}$$

and for an element

$$\Lambda = \begin{pmatrix} \{(a_j, A_j)\}_{j=1}^n ; \{(a_j, A_j)\}_{j=n+1}^p \\ \{(b_j, B_j)\}_{j=1}^m ; \{(b_j, B_j)\}_{j=m+1}^q \end{pmatrix} \tag{7.2}$$

of Ω, we define

$$\Lambda^+ = \begin{pmatrix} \{(a_j, A_j)\}_{j=1}^n ; \{(a_j, A_j)\}_{j=n+1}^p \\ - \qquad ; \{(b_j, B_j)\}_{j=m+1}^q \end{pmatrix} \in \Omega \tag{7.3}$$

and

$$\Lambda^- = \begin{pmatrix} - \qquad ; \{(a_j, A_j)\}_{j=n+1}^p \\ \{(b_j, B_j)\}_{j=1}^m ; \{(b_j, B_j)\}_{j=m+1}^q \end{pmatrix} \in \Omega, \tag{7.4}$$

where "−" denotes the empty sequence. We define the Gamma factor associated to Λ by

$$
\begin{aligned}
&\Gamma(s \mid \Lambda) \\
&= \Gamma\left(s \left| \begin{array}{l} \{(a_j, A_j)\}_{j=1}^n ; \{(a_j, A_j)\}_{j=n+1}^p \\ \{(b_j, B_j)\}_{j=1}^m ; \{(b_j, B_j)\}_{j=m+1}^q \end{array} \right.\right) \\
&= \frac{\displaystyle\prod_{j=1}^m \Gamma(b_j + B_j s) \prod_{j=1}^n \Gamma(1 - a_j - A_j s)}{\displaystyle\prod_{j=n+1}^p \Gamma(a_j + A_j s) \prod_{j=m+1}^q \Gamma(1 - b_j - B_j s)} \, .
\end{aligned}
\tag{7.5}
$$

Finally, we define the Fox H-function [F; MaSa; PBM; PK] associated to Λ by

$$
\begin{aligned}
&H(z \mid \Lambda) \\
&= H_{p,q}^{m,n}\left(z \left| \begin{array}{l} \{(a_j, A_j)\}_{j=1}^n, \{(a_j, A_j)\}_{j=n+1}^p \\ \{(b_j, B_j)\}_{j=1}^m, \{(b_j, B_j)\}_{j=m+1}^q \end{array} \right.\right) \\
&= \frac{1}{2\pi i} \int_L \Gamma(s \mid \Lambda) \, z^{-s} \, ds,
\end{aligned}
\tag{7.6}
$$

provided that the integral converges absolutely, where the path $L : \gamma - i\infty \to \gamma + i\infty$ (γ : real) is taken so that the poles of

$$
\Gamma(s \mid \Lambda^+) = \frac{\displaystyle\prod_{j=1}^n \Gamma(1 - a_j - A_j s)}{\displaystyle\prod_{j=n+1}^p \Gamma(a_j + A_j s) \prod_{j=m+1}^q \Gamma(1 - b_j - B_j s)}
\tag{7.7}
$$

lie to the right of L, and those of

$$
\Gamma(s \mid \Lambda^-) = \frac{\displaystyle\prod_{j=1}^m \Gamma(b_j + B_j s)}{\displaystyle\prod_{j=n+1}^p \Gamma(a_j + A_j s) \prod_{j=m+1}^q \Gamma(1 - b_j - B_j s)}
\tag{7.8}
$$

lie to the left of L (cf. Fig. 7.1).

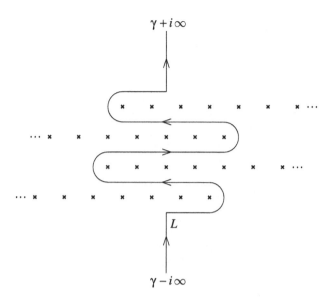

Fig. 7.1 The path L

We note that the Meijer G-functions [M1; M2] are special cases of H-functions.

$$G_{p,q}^{m,n}\left(z \left| \begin{array}{l} \{a_j\}_{j=1}^n, \{a_j\}_{j=n+1}^p \\ \{b_j\}_{j=1}^m, \{b_j\}_{j=m+1}^q \end{array} \right. \right)$$

$$= H_{p,q}^{m,n}\left(z \left| \begin{array}{l} \{(a_j,1)\}_{j=1}^n, \{(a_j,1)\}_{j=n+1}^p \\ \{(b_j,1)\}_{j=1}^m, \{(b_j,1)\}_{j=m+1}^q \end{array} \right. \right). \tag{7.9}$$

7.2 Assumptions

In this section, we state the assumptions in our main theorem.

Let H and I be two natural numbers. Let $\{\lambda_k^{(h)}\}_{k=1}^\infty$ $(1 \leq h \leq H)$, $\{\mu_k^{(i)}\}_{k=1}^\infty$ $(1 \leq i \leq I)$ be increasing sequences of positive real numbers tending to ∞, and let $\{\alpha_k^{(h)}\}_{k=1}^\infty$ $(1 \leq h \leq H)$, $\{\beta_k^{(i)}\}_{k=1}^\infty$ $(1 \leq i \leq I)$ be complex sequences. We assume that the Dirichlet series

$$\varphi_h(s) = \sum_{k=1}^\infty \frac{\alpha_k^{(h)}}{\lambda_k^{(h)s}} \quad (1 \leq h \leq H) \tag{7.10}$$

and

$$\psi_i(s) = \sum_{k=1}^{\infty} \frac{\beta_k^{(i)}}{\mu_k^{(i)s}} \quad (1 \leqq i \leqq I) \tag{7.11}$$

have finite abscissas of absolute convergence $\sigma_{\varphi_h}(1 \leqq h \leqq H)$ and $\sigma_{\psi_i}(1 \leqq i \leqq I)$, respectively. We also assume the existence of a meromorphic function χ satisfying the functional equation

$$\chi(s) = \begin{cases} \displaystyle\sum_{h=1}^{H} \Gamma\left(s\left|\Delta_1^{(h)}\right.\right) \varphi_h(s), & \mathrm{Re}(s) > \displaystyle\max_{1 \leqq h \leqq H}\left(\sigma_{\varphi_h}\right), \\[2ex] \displaystyle\sum_{i=1}^{I} \Gamma\left(r - s\left|\Delta_2^{(i)*}\right.\right) \psi_i(r-s), & \mathrm{Re}(s) < \displaystyle\min_{1 \leqq i \leqq I}\left(r - \sigma_{\psi_i}\right), \end{cases} \tag{7.12}$$

where $\Delta_1^{(h)}(1 \leqq h \leqq H)$, $\Delta_2^{(i)}(1 \leqq i \leqq I)$ are elements of Ω, and r is a real number. We further assume that among the poles of $\chi(s)$, only finitely many distinct s_k $(1 \leqq k \leqq L)$ are neither a pole of $\Gamma\left(s\left|\Delta_1^{(h)+}\right.\right)$ nor a pole of $\Gamma\left(s\left|(\Delta_2^{(i)} - r)^-\right.\right) = \Gamma\left(s - r\left|\Delta_2^{(i)-}\right.\right) = \Gamma\left(r - s\left|\Delta_2^{(i)*+}\right.\right)$. We denote the set of such poles by $S = \{s_k \mid 1 \leqq k \leqq L\}$.

Now, we introduce the processing gamma factor associated to $\Delta \in \Omega$ and suppose that for any pair of real numbers u_1, u_2 $(u_1 < u_2)$, we have the convergence

$$\lim_{|v| \to \infty} \Gamma(u + iv - s \mid \Delta) \chi(u + iv) = 0 \tag{7.13}$$

uniformly in $u_1 \leqq u \leqq u_2$. Here Ω is defined by (2.85) and means that the parameters in capitals are positive.

Let

$$L_1 = L_1(s) : \gamma_1 - i\infty \to \gamma_1 + i\infty,$$
$$L_2 = L_2(s) : \gamma_2 - i\infty \to \gamma_2 + i\infty,$$

$(\gamma_1, \gamma_2 :$ real, $\gamma_2 < \gamma_1)$ be two non-intersecting paths in the w-plane depending on s. We assume that there exists a real number Y such that L_1 and L_2 coincide with the two lines $\mathrm{Re}(w) = \gamma_1$ and $\mathrm{Re}(w) = \gamma_2$, respectively, for $|\mathrm{Im}(w)| > Y$, and that all the poles of $\Gamma\left(w\left|((\Delta - s) \oplus \Delta_1^{(h)})^+\right.\right)$ lie to the right of L_1, and those of $\Gamma\left(w\left|((\Delta - s) \oplus \Delta_1^{(h)})^-\right.\right)$ and S lie to the left of L_1, and all the poles of $\Gamma\left(w\left|((\Delta - s) \oplus (\Delta_2^{(i)} - r))^+\right.\right)$ and S lie to the right of L_2, and those of $\Gamma\left(w \left| \left((\Delta - s) \oplus (\Delta_2^{(i)} - r)\right)^-\right.\right)$ lie to the left of L_2 (cf. Fig. 7.2).

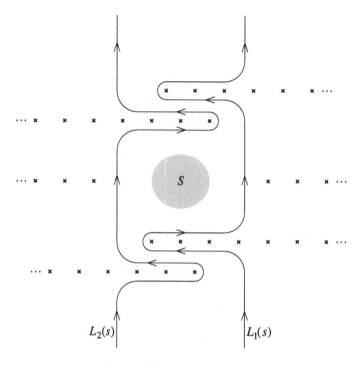

Fig. 7.2   The paths L_1 and L_2

Under these assumptions, we define the chi (key) function $\mathrm{X}(s, z \mid \Delta)$ by

$$\mathrm{X}(s, z \mid \Delta) = \frac{1}{2\pi i} \int_{L_1} \Gamma(w - s \mid \Delta) \, \chi(w) \, z^{s-w} \, dw. \tag{7.14}$$

Then, by the Cauchy residue theorem, we have

$$\begin{aligned}
\mathrm{X}(s, z \mid \Delta) &= \frac{1}{2\pi i} \int_{L_2} \Gamma(w - s \mid \Delta) \, \chi(w) \, z^{s-w} \, dw \\
&\quad + \sum_{k=1}^{L} \operatorname{Res}\left(\Gamma(w - s \mid \Delta) \, \chi(w) \, z^{s-w}, w = s_k\right).
\end{aligned} \tag{7.15}$$

7.3 Theorem

In this section we state and prove our main theorem.

Theorem 7.1   *([Ts]) For the H-function series*

$$\Phi_h(s, z \mid \Delta) = \sum_{k=1}^{\infty} \frac{\alpha_k^{(h)}}{\lambda_k^{(h)s}} H\left(z\lambda_k^{(h)} \left| \left(\Delta_1^{(h)} + s\right) \oplus \Delta\right.\right) \qquad (7.16)$$

and

$$\Psi_i(s, z \mid \Delta) = \sum_{k=1}^{\infty} \frac{\beta_k^{(i)}}{\mu_k^{(i)s}} H\left(z\mu_k^{(i)} \left| \left(\Delta_2^{(i)*} + s\right) \oplus \Delta\right.\right), \qquad (7.17)$$

the following **modular relation** *holds:*

$$\mathrm{X}(s, z \mid \Delta) \qquad\qquad\qquad\qquad\qquad (7.18)$$

$$= \begin{cases} \displaystyle\sum_{h=1}^{H} \Phi_h(s, z \mid \Delta) & \text{if } L_1 \text{ can be taken to the right of } \max_{1 \leq h \leq H}\left(\sigma_{\varphi_h}\right), \\[2em] \displaystyle\sum_{i=1}^{I} \Psi_i\left(r - s, \frac{1}{z} \,\middle|\, \Delta^*\right) + \sum_{k=1}^{L} \mathrm{Res}\left(\Gamma(w - s \mid \Delta)\, \chi(w)\, z^{s-w}, w = s_k\right) \\[1em] & \text{if } L_2 \text{ can be taken to the left of } \min_{1 \leq i \leq I}\left(r - \sigma_{\psi_i}\right), \end{cases}$$

for Δ and z such that the H-functions on the right-hand side converge absolutely.

For a traditional version of Theorem 7.1, cf. Theorem 4.3.

Proof.

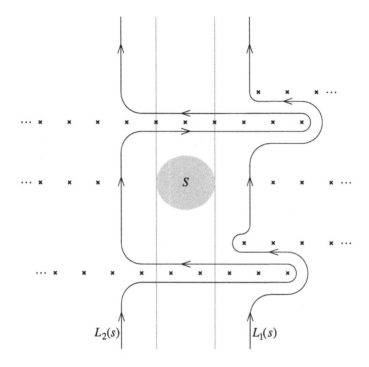

Fig. 7.3   The paths L_1 and L_2

If L_1 can be taken to the right of $\max_{1 \leq h \leq H}(\sigma_{\varphi_h})$ (cf. Fig. 7.3), by using (7.14), (7.12) and (7.10) successively, we have

$$
\begin{aligned}
\mathrm{X}(s, z \,|\, \Delta) &= \frac{1}{2\pi i} \int_{L_1} \Gamma(w - s \,|\, \Delta)\, \chi(w)\, z^{s-w}\, \mathrm{d}w \\
&= \frac{1}{2\pi i} \int_{L_1} \Gamma(w \,|\, \Delta - s) \sum_{h=1}^{H} \Gamma\!\left(w \,\Big|\, \Delta_1^{(h)}\right) \varphi_h(w)\, z^{s-w}\, \mathrm{d}w \\
&= z^s \sum_{h=1}^{H} \frac{1}{2\pi i} \int_{L_1} \Gamma(w \,|\, \Delta - s)\, \Gamma\!\left(w \,\Big|\, \Delta_1^{(h)}\right) \sum_{k=1}^{\infty} \frac{\alpha_k^{(h)}}{\left(z\lambda_k^{(h)}\right)^w}\, \mathrm{d}w \\
&= z^s \sum_{h=1}^{H} \sum_{k=1}^{\infty} \alpha_k^{(h)} \frac{1}{2\pi i} \int_{L_1} \Gamma\!\left(w \,\Big|\, (\Delta - s) \oplus \Delta_1^{(h)}\right) \left(z\lambda_k^{(h)}\right)^{-w}\, \mathrm{d}w,
\end{aligned}
$$

where the exchange of the integration and the summation is permitted by absolute convergence. Now the resulting integral is an H-function, and by (7.16), we have

$$X(s, z \mid \Delta) = z^s \sum_{h=1}^{H} \sum_{k=1}^{\infty} \alpha_k^{(h)} H\left(z\lambda_k^{(h)} \left| (\Delta - s) \oplus \Delta_1^{(h)}\right.\right)$$

$$= \sum_{h=1}^{H} \sum_{k=1}^{\infty} \frac{\alpha_k^{(h)}}{\lambda_k^{(h)s}} H\left(z\lambda_k^{(h)} \left| (\Delta_1^{(h)} + s) \oplus \Delta\right.\right)$$

$$= \sum_{h=1}^{H} \Phi_h(s, z \mid \Delta).$$

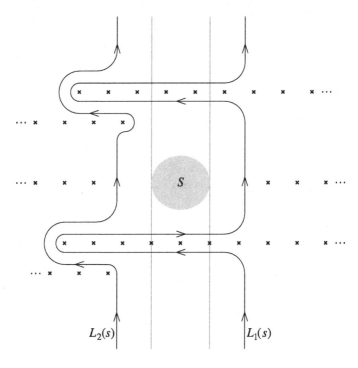

Fig. 7.4   The paths L_1 and L_2

On the other hand, if L_2 can be taken to the left of $\min_{1 \leq i \leq I}(r - \sigma_{\psi_i})$ (cf. Fig. 7.4), by using (7.15), (7.12) and (7.11) successively, we have

$$
X(s, z \mid \Delta) - \sum_{k=1}^{L} \mathrm{Res}\Big(\Gamma(w - s \mid \Delta)\, \chi(w)\, z^{s-w}, w = s_k\Big)
$$

$$
= \frac{1}{2\pi i} \int_{L_2} \Gamma(w - s \mid \Delta)\, \chi(w)\, z^{s-w}\, \mathrm{d}w
$$

$$
= \frac{1}{2\pi i} \int_{L_2} \Gamma(w \mid \Delta - s) \sum_{i=1}^{I} \Gamma\Big(r - w \Big| \Delta_2^{(i)*}\Big)\, \psi_i(r - w)\, z^{s-w}\, \mathrm{d}w
$$

$$
= z^{s-r} \sum_{i=1}^{I} \frac{1}{2\pi i} \int_{L_2} \Gamma(w \mid \Delta - s)\, \Gamma\Big(w - r \Big| \Delta_2^{(i)}\Big) \sum_{k=1}^{\infty} \frac{\beta_k^{(i)}}{\left(\frac{\mu_k^{(i)}}{z}\right)^{-w}}\, \mathrm{d}w \left(\frac{z}{\mu_k^{(i)}}\right)^r
$$

$$
= z^{s-r} \sum_{i=1}^{I} \sum_{k=1}^{\infty} \beta_k^{(i)} \frac{1}{2\pi i} \int_{L_2} \Gamma(w \mid \Delta - s)\Gamma\Big(w \Big| \Delta_2^{(i)} - r\Big) \left(\frac{z}{\mu_k^{(i)}}\right)^{-w}\, \mathrm{d}w \left(\frac{z}{\mu_k^{(i)}}\right)^r,
$$

where we changed the order of integration and summation. Now the resulting integral is an H-function which can be transformed as follows by the properties (2.87) and (2.89):

$$
\frac{1}{2\pi i} \int_{L_2} \Gamma\Big(w \Big| (\Delta - s) \oplus (\Delta_2^{(i)} - r)\Big) \left(\frac{z}{\mu_k^{(i)}}\right)^{-w}\, \mathrm{d}w \left(\frac{z}{\mu_k^{(i)}}\right)^r
$$

$$
= H\left(\frac{z}{\mu_k^{(i)}} \Big| (\Delta - s) \oplus (\Delta_2^{(i)} - r)\right) \left(\frac{z}{\mu_k^{(i)}}\right)^r
$$

$$
= H\left(\frac{z}{\mu_k^{(i)}} \Big| (\Delta_2^{(i)} - (r - s)) \oplus \Delta\right) \left(\frac{z}{\mu_k^{(i)}}\right)^{r-s}
$$

$$
= H\left(\frac{\mu_k^{(i)}}{z} \Big| (\Delta_2^{(i)*} + (r - s)) \oplus \Delta^*\right) \left(\frac{z}{\mu_k^{(i)}}\right)^{r-s}.
$$

Therefore, we have, by (7.17),

$$X(s, z \mid \Delta) - \sum_{k=1}^{L} \mathrm{Res}\Big(\Gamma(w - s \mid \Delta)\, \chi(w)\, z^{s-w}, w = s_k\Big)$$

$$= \sum_{i=1}^{I} \sum_{k=1}^{\infty} \frac{\beta_k^{(i)}}{\mu_k^{(i)r-s}} H\left(\frac{\mu_k^{(i)}}{z} \,\bigg|\, (\Delta_2^{(i)*} + (r - s)) \oplus \Delta^* \right)$$

$$= \sum_{i=1}^{I} \Psi_i\left(r - s, \frac{1}{z} \,\bigg|\, \Delta^* \right).$$

This completes the proof of (7.18). □

7.4   The Main Formula (basic version)

Let $\{\lambda_k\}_{k=1}^{\infty}, \{\mu_k\}_{k=1}^{\infty}$ be increasing sequences of positive real numbers tending to ∞, and let $\{\alpha_k\}_{k=1}^{\infty}, \{\beta_k\}_{k=1}^{\infty}$ be complex sequences. We form the Dirichlet series

$$\varphi(s) = \sum_{k=1}^{\infty} \frac{\alpha_k}{\lambda_k^s}, \tag{7.19}$$

and

$$\psi(s) = \sum_{k=1}^{\infty} \frac{\beta_k}{\mu_k^s} \tag{7.20}$$

and suppose that they have finite abscissas of absolute convergence σ_φ and σ_ψ, respectively.

We suppose the existence of the meromorphic function χ satisfying the functional equation

$$
\chi(s) = \begin{cases}
\dfrac{\displaystyle\prod_{j=1}^{M} \Gamma(d_j + D_j s)}{\displaystyle\prod_{j=1}^{P} \Gamma(c_j + C_j s)}\, \varphi(s), & \mathrm{Re}(s) > \sigma_\varphi \\[3em]
\dfrac{\displaystyle\prod_{j=1}^{N} \Gamma(e_j + E_j(r - s))}{\displaystyle\prod_{j=1}^{Q} \Gamma(f_j + F_j(r - s))}\, \psi(r - s), & \mathrm{Re}(s) < r - \sigma_\psi
\end{cases}
\tag{7.21}
$$

that is

$$
\chi(s) \tag{7.22}
$$

$$
= \begin{cases}
\Gamma\left(s \,\middle|\, \begin{matrix} - & ; \{(c_j, C_j)\}_{j=1}^{P} \\ \{(d_j, D_j)\}_{j=1}^{M}; & - \end{matrix} \right) \varphi(s), & \mathrm{Re}(s) > \sigma_\varphi \\[2em]
\Gamma\left(r - s \,\middle|\, \begin{matrix} - & ; \{(f_j, F_j)\}_{j=1}^{Q} \\ \{(e_j, E_j)\}_{j=1}^{N}; & - \end{matrix} \right) \psi(r - s), & \mathrm{Re}(s) < r - \sigma_\psi
\end{cases}
$$

with r a real number, and $C_j, D_j, E_j, F_j > 0$, and having a finite number of poles s_k ($1 \le k \le L$). We set $S = \{s_k | 1 \le k \le L\}$.

We introduce the gamma factor

$$
\Delta = \begin{pmatrix} \{(1 - a_j, A_j)\}_{j=1}^{n}; & \{(a_j, A_j)\}_{j=n+1}^{p} \\ \{(b_j, B_j)\}_{j=1}^{m} & ; \{(1 - b_j, B_j)\}_{j=m+1}^{q} \end{pmatrix} \in \Omega
\tag{7.23}
$$

with its associated Gamma factor

$$
\Gamma(w \,|\, \Delta) = \frac{\displaystyle\prod_{j=1}^{m} \Gamma(b_j + B_j w) \prod_{j=1}^{n} \Gamma(a_j - A_j w)}{\displaystyle\prod_{j=n+1}^{p} \Gamma(a_j + A_j w) \prod_{j=m+1}^{q} \Gamma(b_j - B_j w)},
\tag{7.24}
$$

and suppose that for any real numbers u_1, u_2 ($u_1 < u_2$)

$$
\lim_{|v| \to \infty} \Gamma(u + iv - s \,|\, \Delta)\, \chi(u + iv) = 0,
\tag{7.25}
$$

uniformly in $u_1 \le u \le u_2$.

In the w-plane we take two deformed Bromwich paths

$$L_1(s) : \gamma_1 - i\infty \to \gamma_1 + i\infty,$$
$$L_2(s) : \gamma_2 - i\infty \to \gamma_2 + i\infty \quad (\gamma_2 < \gamma_1)$$

such that all the poles of

$$\Gamma(w \,|\, \Delta^+ - s) = \frac{\displaystyle\prod_{j=1}^{n} \Gamma(a_j + A_j s - A_j w)}{\displaystyle\prod_{j=n+1}^{p} \Gamma(a_j - A_j s + A_j w) \prod_{j=m+1}^{q} \Gamma(b_j + B_j s - B_j w)}$$

lie to the right of $L_1(s)$ and $L_2(s)$, and all the poles of

$$\Gamma(w \,|\, \Delta^- - s) = \frac{\displaystyle\prod_{j=1}^{m} \Gamma(b_j - B_j s + B_j w)}{\displaystyle\prod_{j=n+1}^{p} \Gamma(a_j - A_j s + A_j w) \prod_{j=m+1}^{q} \Gamma(b_j + B_j s - B_j w)}$$

lie to the left of $L_1(s)$ and $L_2(s)$, and S lies to the left of $L_1(s)$ and to the right of $L_2(s)$ (cf. Fig. 7.2).

Under these conditions we define $\mathrm{X}(s, z \,|\, \Delta)$ by

$$\mathrm{X}(s, z \,|\, \Delta) = \frac{1}{2\pi i} \int_{L_1(s)} \Gamma(w - s \,|\, \Delta) \chi(w) \, z^{s-w} dw$$

$$= \frac{1}{2\pi i} \int_{L_2(s)} \Gamma(w - s \,|\, \Delta) \, \chi(w) \, z^{s-w} \, dw \qquad (7.26)$$

$$+ \sum_{k=1}^{L} \mathrm{Res}\Big(\Gamma(w - s \,|\, \Delta) \, \chi(w) \, z^{s-w}, w = s_k\Big).$$

Then by Theorem 7.1, we have a general modular relation:

$$X(s, z \mid \Delta)$$

$$
= \begin{cases}
\displaystyle\sum_{k=1}^{\infty} \frac{\alpha_k}{\lambda_k^s} \, H\!\left(z\lambda_k \left| \left(\begin{array}{c} - \; ; \{(c_j + C_j s, C_j)\}_{j=1}^{P} \\ \{(d_j + D_j s, D_j)\}_{j=1}^{M}; \; - \end{array}\right) \oplus \Delta \right.\right) \\
\qquad\qquad\qquad \text{if } L_1(s) \text{ can be taken to the right of } \sigma_\varphi, \\[2em]
\displaystyle\sum_{k=1}^{\infty} \frac{\beta_k}{\mu_k^{r-s}} \, H\!\left(\frac{\mu_k}{z} \left| \left(\begin{array}{c} - \; ; \{(f_j + F_j(r - s), F_j)\}_{j=1}^{Q} \\ \{(e_j + E_j(r - s), E_j)\}_{j=1}^{N}; \; - \end{array}\right) \oplus \Delta^* \right.\right) \\
\qquad + \displaystyle\sum_{k=1}^{L} \mathrm{Res}\!\left(\Gamma(w - s \mid \Delta)\, \chi(w)\, z^{s-w}, w = s_k\right) \\
\qquad\qquad\qquad \text{if } L_2(s) \text{ can be taken to the left of } r - \sigma_\psi.
\end{cases}
$$

In the traditional notation,

$$X(s, z \mid \Delta) \tag{7.27}$$

$$
= \begin{cases}
\displaystyle\sum_{k=1}^{\infty} \frac{\alpha_k}{\lambda_k^s} \, H_{p+P,\,q+M}^{m+M,\,n}\!\left(z\lambda_k \left| \begin{array}{c} \{(1 - a_j, A_j)\}_{j=1}^{n}, \\ \{(d_j + D_j s, D_j)\}_{j=1}^{M}, \{(b_j, B_j)\}_{j=1}^{m}, \\ \{(c_j + C_j s, C_j)\}_{j=1}^{P}, \{(a_j, A_j)\}_{j=n+1}^{p} \\ \{(1 - b_j, B_j)\}_{j=m+1}^{q} \end{array}\right.\right) \\
\qquad\qquad\qquad \text{if } L_1(s) \text{ can be taken to the right of } \sigma_\varphi, \\[3em]
\displaystyle\sum_{k=1}^{\infty} \frac{\beta_k}{\mu_k^{r-s}} \, H_{q+Q,\,p+N}^{n+N,\,m}\!\left(\frac{\mu_k}{z} \left| \begin{array}{c} \{(1 - b_j, B_j)\}_{j=1}^{m}, \\ \{(e_j + E_j(r - s), E_j)\}_{j=1}^{N}, \{(a_j, A_j)\}_{j=1}^{n}, \\ \{(f_j + F_j(r - s), F_j)\}_{j=1}^{Q}, \{(b_j, B_j)\}_{j=m+1}^{q} \\ \{(1 - a_j, A_j)\}_{j=n+1}^{p} \end{array}\right.\right) \\
\qquad + \displaystyle\sum_{k=1}^{L} \mathrm{Res}\!\left(\Gamma(w - s \mid \Delta)\, \chi(w)\, z^{s-w}, w = s_k\right) \\
\qquad\qquad\qquad \text{if } L_2(s) \text{ can be taken to the left of } r - \sigma_\psi.
\end{cases}
$$

Fig. 7.5   Erich Hecke

Chapter 8

The Hecke Type Zeta-functions

In this chapter we consider the special case of the Main Formula (Chapter 7) in which the Dirichlet series $\phi(s), \psi(s)$ satisfy the **Hecke type functional equation** (8.1). The modular relation (8.6) includes the Bochner modular relation (cf. Theorem 3.4 etc.), the Fourier-Bessel expansion (Chapter 4), the Ewald expansion (Chapter 5) which includes the Bochner-Chandrasekharan formula (§5.1.2), the Riesz sum (Chapter 6). We shall state special cases of (8.7), which is to work as a database. Examples of non-Hecke type were studied in [BeI; BeII; BeIII; BeIV], and also in [KTT1; KTT2] many of which may be reduced to Hecke type and the corresponding results can be readily guessed from the database.

8.1 Statement of the formula

Given two Dirichlet series, we assume the functional equation of Hecke type

$$\chi(s) = \begin{cases} \Gamma(s)\, \varphi(s), & \mathrm{Re}(s) > \sigma_\varphi \\ \Gamma(r - s)\, \psi(r - s), & \mathrm{Re}(s) < r - \sigma_\psi, \end{cases} \tag{8.1}$$

or

$$\chi(s) = \begin{cases} \Gamma\!\left(s \left|\begin{array}{c} - \; ; \; - \\ (0,1) \; ; \; - \end{array}\right.\right)\varphi(s), & \mathrm{Re}(s) > \sigma_\varphi \\ \Gamma\!\left(r - s \left|\begin{array}{c} - \; ; \; - \\ (0,1) \; ; \; - \end{array}\right.\right)\psi(r - s), & \mathrm{Re}(s) < r - \sigma_\psi. \end{cases} \tag{8.2}$$

By Theorem 7.1, we have the following modular relation:

$$\sum_{k=1}^{\infty} \alpha_k H\left(z\lambda_k \,\left|\, (\Delta - s) \oplus \left(\begin{matrix} - \\ (0,1) \end{matrix} ; \begin{matrix} - \\ - \end{matrix} \right) \right. \right)$$

$$= z^{-r} \sum_{k=1}^{\infty} \beta_k H\left(\frac{\mu_k}{z} \,\left|\, (\Delta^* - (r - s)) \oplus \left(\begin{matrix} - \\ (0,1) \end{matrix} ; \begin{matrix} - \\ - \end{matrix} \right) \right. \right) \qquad (8.3)$$

$$+ \sum_{k=1}^{L} \mathrm{Res}\left(\Gamma(w - s \,|\, \Delta) \chi(w) z^{-w}, w = s_k \right),$$

which can be rewritten as

$$\sum_{k=1}^{\infty} \frac{\alpha_k}{\lambda_k^s} H\left(z\lambda_k \,\left|\, \Delta \oplus \left(\begin{matrix} - \\ (s,1) \end{matrix} ; \begin{matrix} - \\ - \end{matrix} \right) \right. \right)$$

$$= \sum_{k=1}^{\infty} \frac{\beta_k}{\mu_k^{r-s}} H\left(\frac{\mu_k}{z} \,\left|\, \Delta^* \oplus \left(\begin{matrix} - \\ (r-s,1) \end{matrix} ; \begin{matrix} - \\ - \end{matrix} \right) \right. \right)$$

$$+ \sum_{k=1}^{L} \mathrm{Res}\left(\Gamma(w - s \,|\, \Delta) \chi(w) z^{s-w}, w = s_k \right).$$

In the traditional notation, if

$$\Delta = \begin{pmatrix} \{(1 - a_j, A_j)\}_{j=1}^{n}; & \{(a_j, A_j)\}_{j=n+1}^{p} \\ \{(b_j, B_j)\}_{j=1}^{m} & ; \{(1 - b_j, B_j)\}_{j=m+1}^{q} \end{pmatrix} \in \Omega \qquad (8.4)$$

with its associated Gamma factor

$$\Gamma(s \,|\, \Delta) = \frac{\displaystyle\prod_{j=1}^{m} \Gamma(b_j + B_j s) \prod_{j=1}^{n} \Gamma(a_j - A_j s)}{\displaystyle\prod_{j=n+1}^{p} \Gamma(a_j + A_j s) \prod_{j=m+1}^{q} \Gamma(b_j - B_j s)}, \qquad (8.5)$$

we have

$$\sum_{k=1}^{\infty} \frac{\alpha_k}{\lambda_k^s} H_{p,q+1}^{m+1,n}\left(z\lambda_k \middle| \begin{array}{c} \{(1-a_j,A_j)\}_{j=1}^n, \{(a_j,A_j)\}_{j=n+1}^p \\ (s,1), \{(b_j,B_j)\}_{j=1}^m, \{(1-b_j,B_j)\}_{j=m+1}^q \end{array} \right) \quad (8.6)$$

$$= \sum_{k=1}^{\infty} \frac{\beta_k}{\mu_k^{r-s}} H_{q,p+1}^{n+1,m}\left(\frac{\mu_k}{z} \middle| \begin{array}{c} \{(1-b_j,B_j)\}_{j=1}^m, \{(b_j,B_j)\}_{j=m+1}^q \\ (r-s,1), \{(a_j,A_j)\}_{j=1}^n, \{(1-a_j,A_j)\}_{j=n+1}^p \end{array} \right)$$

$$+ \sum_{k=1}^{L} \mathrm{Res}\Big(\Gamma(w-s\,|\,\Delta)\,\chi(w)\,z^{s-w}, w=s_k \Big).$$

In the special case of $A_j = B_j = 1$, we have

$$\sum_{k=1}^{\infty} \frac{\alpha_k}{\lambda_k^s} G_{p,q+1}^{m+1,n}\left(z\lambda_k \middle| \begin{array}{c} 1-a_1,\ldots,1-a_n,a_{n+1},\ldots,a_p \\ s,b_1,\ldots,b_m,1-b_{m+1},\ldots,1-b_q \end{array} \right) \quad (8.7)$$

$$= \sum_{k=1}^{\infty} \frac{\beta_k}{\mu_k^{r-s}} G_{q,p+1}^{n+1,m}\left(\frac{\mu_k}{z} \middle| \begin{array}{c} 1-b_1,\ldots,1-b_m,b_{m+1},\ldots,b_q \\ r-s,a_1,\ldots,a_n,1-a_{n+1},\ldots,1-a_p \end{array} \right)$$

$$+ \sum_{k=1}^{L} \mathrm{Res}\Big(\Gamma(w-s\,|\,\Delta)\,\chi(w)\,z^{s-w}, w=s_k \Big).$$

8.1.1 *The bilateral form*

The special case of (8.7) with

$$m = n,\ p = q,\ a_1 = b_1 - c, \ldots, a_n = b_n - c,\ b_{n+1} = a_{n+1} - c, \ldots, b_p = a_p - c$$

leads to

$$\Delta = \left(\begin{array}{cc} \{(1-b_j+c,1)\}_{j=1}^n; & \{(a_j,1)\}_{j=n+1}^p \\ \{(b_j,1)\}_{j=1}^n & ;\{(1-a_j+c,1)\}_{j=m+1}^p \end{array} \right)$$

and

$$\Gamma(s\,|\,\Delta) = \frac{\displaystyle\prod_{j=1}^{n}\Gamma(b_j+s)\prod_{j=1}^{n}\Gamma(b_j-c-s)}{\displaystyle\prod_{j=n+1}^{p}\Gamma(a_j+s)\prod_{j=n+1}^{p}\Gamma(a_j-c-s)}.$$

$$\sum_{k=1}^{\infty} \frac{\alpha_k}{\lambda_k^s} \, G_{p,p+1}^{n+1,n} \left(z\lambda_k \left| \begin{array}{c} 1 - b_1 + c, \ldots, 1 - b_n + c, a_{n+1}, \ldots, a_p \\ s, b_1, \ldots, b_n, 1 - a_{n+1} + c, \ldots, 1 - a_p + c \end{array} \right. \right)$$

$$= \sum_{k=1}^{\infty} \frac{\beta_k}{\mu_k^{r-s}} \, G_{p,p+1}^{n+1,n} \left(\frac{\mu_k}{z} \left| \begin{array}{c} 1 - b_1, \ldots, 1 - b_n, a_{n+1} - c, \ldots, a_p - c \\ r - s, b_1 - c, \ldots, b_n - c, 1 - a_{n+1}, \ldots, 1 - a_p \end{array} \right. \right)$$

$$+ \sum_{k=1}^{L} \mathrm{Res}\Big(\Gamma(w - s \,|\, \Delta) \, \chi(w) \, z^{s-w}, w = s_k \Big).$$

Since the G-function coefficients on the right-side becomes

$$\frac{\beta_k}{\mu_k^{r-s}} \left(\frac{\mu_k}{z} \right)^{-c} G_{p,p+1}^{n+1,n} \left(\frac{\mu_k}{z} \left| \begin{array}{c} 1 - b_1 + c, \ldots, 1 - b_n + c, \\ r - s + c, b_1, \ldots, b_n, \\ a_{n+1}, \ldots, a_p \\ 1 - a_{n+1} + c, \ldots, 1 - a_p + c \end{array} \right. \right),$$

we have the bilateral formula

$$\sum_{k=1}^{\infty} \frac{\alpha_k}{\lambda_k^s} \, G_{p,p+1}^{n+1,n} \left(z\lambda_k \left| \begin{array}{c} 1 - b_1 + c, \ldots, 1 - b_n + c, a_{n+1}, \ldots, a_p \\ s, b_1, \ldots, b_n, 1 - a_{n+1} + c, \ldots, 1 - a_p + c \end{array} \right. \right)$$

$$= z^c \sum_{k=1}^{\infty} \frac{\beta_k}{\mu_k^{r+c-s}} \, G_{p,p+1}^{n+1,n} \left(\frac{\mu_k}{z} \left| \begin{array}{c} 1 - b_1 + c, \ldots, 1 - b_n + c, \\ r - s + c, b_1, \ldots, b_n, \\ a_{n+1}, \ldots, a_p \\ 1 - a_{n+1} + c, \ldots, 1 - a_p + c \end{array} \right. \right) \qquad (8.8)$$

$$+ \sum_{k=1}^{L} \mathrm{Res}\Big(\Gamma(w - s \,|\, \Delta) \, \chi(w) \, z^{s-w}, w = s_k \Big).$$

For each $p \geq 0$, we may write down the formula similar to (4.84) corresponding to $n, \frac{p}{2} \leq n \leq p$.

Convention

In almost what follows we shall state specific form of the general formula (8.7) described in the section titles and we omit this passage. Also in most case $z > 0$ or more generally $\mathrm{Re}\, z > 0$. It would be instructive to check the validity of various identities among G-functions that occur subsequently.

8.1.2 The Bochner modular relation: $G^{1,0}_{0,1} \leftrightarrow G^{1,0}_{0,1}$

This is the genesis of all subsequent studies on the modular relations and has been fully discussed in general form in Chapter 3 above.

8.2 The Riesz sums or the first J-Bessel expansion: $G^{1,0}_{1,1} \leftrightarrow G^{1,0}_{0,2}$

We have

$$
\sum_{k=1}^{\infty} \frac{\alpha_k}{\lambda_k^s} G^{1,0}_{1,1}\left(z\,\lambda_k \,\middle|\, \begin{matrix} a \\ s \end{matrix} \right)
$$

$$
= \sum_{k=1}^{\infty} \frac{\beta_k}{\mu_k^{r-s}} G^{1,0}_{0,2}\left(\frac{\mu_k}{z} \,\middle|\, \begin{matrix} - \\ r-s, 1-a \end{matrix} \right) \tag{8.9}
$$

$$
+ \sum_{k=1}^{L} \mathrm{Res}\left(\frac{\chi(w)\, z^{s-w}}{\Gamma(a-s+w)}, w = s_k \right) \quad (z > 0),
$$

which by (6.5) and (2.110) amounts to

$$
\frac{z^s}{\Gamma(a-s)} \sum_{\lambda_k < \frac{1}{z}} \alpha_k (1 - z\lambda_k)^{a-s-1}
$$

$$
= z^{\frac{1}{2}(s-r-1+a)} \sum_{k=1}^{\infty} \frac{\beta_k}{\mu_k^{\frac{1}{2}(r-s-1+a)}} J_{r-s-1+a}\left(2\sqrt{\frac{\mu_k}{z}} \right) \tag{8.10}
$$

$$
+ \sum_{k=1}^{L} \mathrm{Res}\left(\frac{\chi(w)\, z^{s-w}}{\Gamma(a-s+w)}, w = s_k \right) \quad (z > 0).
$$

The special case of (8.10) with $a = 1$, after slight changes of notation gives rise to the formula for the Riesz sum of order κ of the Idealfunktion of an imaginary quadratic field due to Landau [LanII], Hecke [Hecl], Chandrasekharan and Narasimhan [ChN2, Lemma 5] et al:

Theorem 8.1

$$
\frac{1}{\Gamma(\kappa+1)} \sum_{\lambda_k < z}{}' \alpha_k \big(z - \lambda_k\big)^{\kappa}
$$

$$= z^{\frac{r+\kappa}{2}} \sum_{k=1}^{\infty} \frac{\beta_k}{\mu_k^{\frac{r+\kappa}{2}}} J_{r+\kappa}\left(2\sqrt{\mu_k z}\right)$$

$$+ \sum_{k=1}^{L} \operatorname{Res}\left(\frac{\chi(w)\,z^{\rho+w}}{\Gamma(1+\kappa+w)}, w = s_k\right). \tag{8.11}$$

By **differencing** due to Landau [LanII] (cf. §6.1.1 and [ChN3]) applied to (8.11), the formula for the summatory function of the coefficients of the imaginary quadratic field is deduced.

Corollary 8.1

$$\sideset{}{'}\sum_{k\leq x} F(k) = -\zeta_\Omega(0)x + \zeta_\Omega(0) + \sqrt{x}\sum_{k=1}^{\infty} \frac{F(k)}{\sqrt{k}} J_1\left(4\pi\sqrt{\frac{kx}{\sqrt{|\Delta|}}}\right) \tag{8.12}$$

by (3.151). The series on the right of (8.12) is convergent for all $x > 0$, boundedly convergent on any closed interval and uniformly convergent on any closed interval free from any jump discontinuities of the left side.

(8.12) was first deduced formally by Voronoĭ [Vor1] and rigorously proved by Hardy [HarCP]. Then it was generalized by Arnold Walfisz [AZW1], [AZW2] and further simplified by Anna Walfisz [AAW1].

Since a very wide class of zeta-functions (including those which are associated to cusp forms and positive definite quadratic forms) satisfies the function equation (8.1), Corollary 8.1 holds true for such zeta-functions. For the summatory function for a real quadratic field, cf. §9.1.2, especially (9.36).

8.3   The partial sum formula: $G_{2,2}^{2,0} \leftrightarrow G_{1,3}^{1,1}$

$$\sum_{k=1}^{\infty} \frac{\alpha_k}{\lambda_k^s} G_{2,2}^{2,0}\left(z\lambda_k \left| \begin{matrix} a,c \\ s,b \end{matrix} \right.\right)$$

$$= \sum_{k=1}^{\infty} \frac{\beta_k}{\mu_k^{r-s}} G_{1,3}^{1,1}\left(\frac{\mu_k}{z} \left| \begin{matrix} 1-b \\ r-s, 1-a, 1-c \end{matrix} \right.\right) \tag{8.13}$$

$$+ \sum_{k=1}^{L} \operatorname{Res}\left(\frac{\Gamma(a-s+w)}{\Gamma(b-s+w)\,\Gamma(c-s+w)} \chi(w)\,z^{s-w}, w = s_k\right).$$

Equation (8.9) with $c = b$ amounts to (8.9), whence to Theorem 8.1, the Riesz sum formula. We remark that (8.13) with $c = s$ leads to another

Riesz sum:

$$\sum_{k=1}^{\infty} \frac{\alpha_k}{\lambda_k^s} G_{1,1}^{1,0}\left(z \left| \begin{matrix} a \\ b \end{matrix} \right.\right) = \sum_{k=1}^{\infty} \frac{\beta_k}{\mu_k^{r-s}} G_{1,3}^{1,1}\left(\frac{\mu_k}{z} \left| \begin{matrix} 1-b \\ r-s, 1-s, 1-a \end{matrix} \right.\right)$$
$$+ \sum_{k=1}^{L} \mathrm{Res}\left(\frac{\Gamma(b-s+w)}{\Gamma(a-s+w)\Gamma(w)} \chi(w) z^{s-w}, w = s_k\right).$$

Now, on appealing to (2.76) with $p = 1, q = 2$

$$G_{1,3}^{1,1}\left(z \left| \begin{matrix} 1-b \\ r-s, 1-s, 1-a \end{matrix} \right.\right)$$
$$= \frac{\Gamma(b+r-s)}{\Gamma(r)\Gamma(a+r-s)} z^{r-s} \, {}_1F_2\left(\begin{matrix} b+r-s \\ r, a+r-s \end{matrix} ; -z\right),$$

and slightly changing notation, we obtain the Riesz sum formula

$$\frac{1}{\Gamma(\varkappa+1)} \sideset{}{'}\sum_{\lambda_k \le z} \frac{\alpha_k}{\lambda_k^{s-a}} (z-\lambda_k)^{\varkappa} \tag{8.14}$$
$$= \frac{\Gamma(a+r-s) z^{\varkappa+a+r-s}}{\Gamma(r)\Gamma(\varkappa+1+a+r-s)} \sum_{k=1}^{\infty} \beta_k \, {}_1F_2\left(\begin{matrix} a+r-s \\ r, \varkappa+1+a+r-s \end{matrix} ; -z\mu_k\right)$$
$$+ \sum_{k=1}^{L} \mathrm{Res}\left(\frac{\Gamma(a-s+w)}{\Gamma(\varkappa+1+a-s+w)\Gamma(w)} \chi(w) z^{\varkappa+a-s+w}, w = s_k\right).$$

From (8.14) we may readily read off a partial sum or the Dirichlet polynomial for the zeta-function.

8.4 The Fourier-Bessel expansion: $G_{1,1}^{1,1} \leftrightarrow G_{0,2}^{2,0}$

This is expounded in more general form in Chapter 4.

8.5 The Ewald expansion: $G_{1,2}^{2,0} \leftrightarrow G_{1,2}^{1,1}$

This is expounded in more general form in Chapter 5. Cf. especially, Theorems 5.2 and 5.3. The latter implies the incomplete gamma expansion (Ewald expansion) from the confluent hypergeometric function expansion.

8.6   The Bochner-Chandrasekharan formula: $H_{1,2}^{2,0} \leftrightarrow H_{1,2}^{1,1}$

This has been expounded as §5.1.2.

8.7   The $G_{1,2}^{2,1} \leftrightarrow G_{1,2}^{2,1}$ formula

This being a special case of (8.8), we find it more appropriate to state the counterpart of (8.6) in the case of the Riemann type functional equation, i.e. we assume that the gamma factor in (8.1) is $\Gamma\left(\frac{s}{2}\right)$

$$\sum_{k=1}^{\infty} \frac{\alpha_k}{\lambda_k^s} H_{p,q+1}^{m+1,n}\left(z\lambda_k \left| \begin{array}{c} \{(1-a_j, A_j)\}_{j=1}^n, \{(a_j, A_j)\}_{j=n+1}^p \\ \left(\frac{s}{2}, \frac{1}{2}\right), \{(b_j, B_j)\}_{j=1}^m, \{(1-b_j, B_j)\}_{j=m+1}^q \end{array} \right. \right) \quad (8.15)$$

$$= \sum_{k=1}^{\infty} \frac{\beta_k}{\mu_k^{r-s}} H_{q,p+1}^{n+1,m}\left(\frac{\mu_k}{z} \left| \begin{array}{c} \{(1-b_j, B_j)\}_{j=1}^m, \{(b_j, B_j)\}_{j=m+1}^q \\ \left(\frac{r-s}{2}, \frac{1}{2}\right), \{(a_j, A_j)\}_{j=1}^n, \{(1-a_j, A_j)\}_{j=n+1}^p \end{array} \right. \right)$$

$$+ \sum_{k=1}^{L} \mathrm{Res}\Big(\Gamma(w - s \mid \Delta)\, \chi(w)\, z^{s-w}, w = s_k\Big).$$

This leads to the confluent hypergeometrc expansion and thence to incomplete gamma expansion similar to the one in Chapter 5. It further boils down to the Ueno-Nishizawa formula [UN] ([Lav1]), which as has been enunciated in several papers of ours leads to the Hurwitz formula [Vista].

8.8   The second J-Bessel expansion: $G_{2,2}^{1,1} \leftrightarrow G_{1,3}^{2,0}$

$$\sum_{k=1}^{\infty} \frac{\alpha_k}{\lambda_k^s} G_{2,2}^{1,1}\left(z\lambda_k \left| \begin{array}{c} 1-b, c \\ s, 1-a \end{array} \right. \right)$$

$$= \sum_{k=1}^{\infty} \frac{\beta_k}{\mu_k^{r-s}} G_{1,3}^{2,0}\left(\frac{\mu_k}{z} \left| \begin{array}{c} a \\ r-s, b, 1-c \end{array} \right. \right)$$

$$+ \sum_{k=1}^{L} \mathrm{Res}\left(\frac{\Gamma(b+s-w)}{\Gamma(a+s-w)\,\Gamma(c-s+w)}\, \chi(w)\, z^{s-w}, w = s_k\right).$$

We do not have a closed form for $G_{1,3}^{2,0}$ in general and only partial results are known including (9.11), cf. Remark 9.1. Here we refer to the special case $a = r - s$, in which $G_{1,3}^{2,0}$ reduces to $G_{0,2}^{1,0}\left(\frac{\mu_k}{z} \left| \begin{array}{c} - \\ b, 1-c \end{array} \right. \right)$, the J-Bessel function by (2.110), which is also a consequence of (9.11).

On setting $z \leftarrow \frac{1}{z}$, we obtain

$$\frac{\Gamma(b+s)z^{-s}}{\Gamma(c-s)\Gamma(r)} \sum_{\lambda_k < z} \alpha_k \, {}_2F_1 \left(\begin{matrix} b+s, 1-c+s \\ r \end{matrix} ; \frac{\lambda_k}{z} \right) \tag{8.16}$$

$$+ \frac{\Gamma(b+s)z^b}{\Gamma(-b-s+r)\Gamma(b+c)} \sum_{\lambda_k > z} \frac{\alpha_k}{\lambda_k^{s+b}} \, {}_2F_1 \left(\begin{matrix} b+s, b+1-r+s \\ b+c \end{matrix} ; \frac{z}{\lambda_k} \right)$$

$$= \sum_{k=1}^{\infty} \frac{\beta_k}{\mu_k^{r-s}} \sqrt{z\mu_k}^{1+b-c} J_{b+c-1}(2\sqrt{z\mu_k})$$

$$+ \sum_{k=1}^{L} \operatorname{Res}\left(\frac{\Gamma(b+s-w)}{\Gamma(r-w)\Gamma(c-s+w)} \chi(w) \, z^{w-s}, w = s_k \right).$$

Here to the left-hand side, we applied the special case of (2.74):

$$G_{2,2}^{1,1}\left(z \left| \begin{matrix} a_1, a_2 \\ b_1, b_2 \end{matrix} \right. \right) \tag{8.17}$$

$$= \begin{cases} \dfrac{\Gamma(1-a_1+b_1)}{\Gamma(a_2-b_1)\Gamma(1-b_2+b_1)} z^{b_1} \\ \qquad \times {}_2F_1\left(\begin{matrix} 1-a_1+b_1, 1-a_2+b_1 \\ 1-b_2+b_1 \end{matrix} ; z \right), & |z| < 1 \\[2em] \dfrac{\Gamma(1-a_1+b_1)}{\Gamma(a_1-b_2)\Gamma(1-a_1+a_2)} z^{a_1-1} \\ \qquad \times {}_2F_1\left(\begin{matrix} 1-a_1+b_1, 1-a_1+b_2 \\ 1-a_1+a_2 \end{matrix} ; \frac{1}{z} \right), & |z| > 1 \end{cases}$$

and for curiosity we note its special case.

$$G_{2,2}^{1,1}\left(z \left| \begin{matrix} a, c \\ b, c \end{matrix} \right. \right) = \begin{cases} \dfrac{\Gamma(1-a+b)}{\pi} \sin((c-b)\pi) \, z^b \, (1-z)^{a-b-1}, & |z| < 1 \\[1.5em] \dfrac{\Gamma(1-a+b)}{\pi} \sin((a-c)\pi) \, z^b \, (z-1)^{a-b-1}, & |z| > 1 \end{cases}.$$

8.9   The $H_{2,2}^{2,1} \leftrightarrow H_{1,3}^{2,1}$ formula

$$\sum_{k=1}^{\infty} \frac{\alpha_k}{\lambda_k^s} H_{2,2}^{2,1}\left(z\lambda_k \,\middle|\, \begin{array}{c} (1-b,B),(c,C) \\ (s,1),(a,A) \end{array}\right)$$

$$= \sum_{k=1}^{\infty} \frac{\beta_k}{\mu_k^{r-s}} H_{1,3}^{2,1}\left(\frac{\mu_k}{z} \,\middle|\, \begin{array}{c} (1-a,A) \\ (r-s,1),(b,B),(1-c,C) \end{array}\right)$$

$$+ \sum_{k=1}^{L} \mathrm{Res}\left(\frac{\Gamma(a-A(s-w))\Gamma(b+B(s-w))}{\Gamma(c-C(s-w))}\chi(w)\,z^{s-w}, w=s_k\right).$$

We state its special case with $c=s, C=1$, $A=B=2, b=0$, so that on the left we have $H_{1,1}^{1,1}\left(z\,\lambda_k \,\middle|\, \begin{array}{c} (1,2) \\ (a,2) \end{array}\right)$, a reminiscent of the Fourier-Bessel expansion in Chapter 4.

$$\frac{\Gamma(2s)}{2}\sum_{k=1}^{\infty} \frac{\alpha_k}{\left(\sqrt{\lambda_k}+\sqrt{z}^{-1}\right)^{2s}} \tag{8.18}$$

$$= \sum_{k=1}^{\infty} \frac{\beta_k}{\mu_k^{r-s}} H_{1,3}^{2,1}\left(\frac{\mu_k}{z} \,\middle|\, \begin{array}{c} (1-a,2) \\ (r-s,1),(0,2),(1-s,1) \end{array}\right)$$

$$+ \sum_{k=1}^{L} \mathrm{Res}\left(\frac{\Gamma(a-2s+2w))\Gamma(2s-2w)}{\Gamma(w)}\chi(w)\,z^{s-w}, w=s_k\right)$$

because

$$H_{1,1}^{1,1}\left(z \,\middle|\, \begin{array}{c} (1,2) \\ (a,2) \end{array}\right) = \frac{1}{2} G_{1,1}^{1,1}\left(\sqrt{z} \,\middle|\, \begin{array}{c} 1 \\ a \end{array}\right) = \frac{\Gamma(a)}{2} \frac{1}{\left(1+\sqrt{z}^{-1}\right)^a}.$$

We state the results for $r=\frac{1}{2}$ with $a=2s$ and $r=\frac{3}{2}$ with $a=2s-1$ for curiosity.

For the H-function on the right of (8.18), we apply the formula (2.11), which reads in the case $a=2s$

$$H_{1,3}^{2,1}\left(z \,\middle|\, \begin{array}{c} (1-2s,2) \\ (\frac{1}{2}-s,1),(0,2),(1-s,1) \end{array}\right)$$

$$= \frac{\Gamma(2s)}{2^{2-2s}\sqrt{\pi}}\left(e^{-\frac{1-2s}{2}\pi\, i+2\sqrt{z}\, i}\Gamma\left(1-2s,2\sqrt{z}\, i\right)\right. \tag{8.19}$$

$$\left. + e^{\frac{1-2s}{2}\pi\, i-2\sqrt{z}\, i}\Gamma\left(1-2s,-2\sqrt{z}\, i\right)\right).$$

Substituting (8.19) and making the change of variable $z \leftarrow \frac{1}{z^2}$, we obtain

$$
\frac{\Gamma(2s)}{2} \sum_{k=1}^{\infty} \frac{\alpha_k}{\left(\sqrt{\lambda_k} + z\right)^{2s}}
$$
$$
= \frac{\Gamma(2s)}{2^{2-2s}\sqrt{\pi}} \sum_{k=1}^{\infty} \frac{\beta_k}{\mu_k^{\frac{1}{2}-s}} \left(e^{-\frac{1-2s}{2}\pi\, i + 2z\sqrt{\mu_k}\, i}\, \Gamma(1 - 2s, 2z\sqrt{\mu_k}\, i) \right.
$$
$$
\left. + e^{\frac{1-2s}{2}\pi\, i - 2z\sqrt{\mu_k}\, i}\, \Gamma(1 - 2s, -2z\sqrt{\mu_k}\, i) \right) \tag{8.20}
$$
$$
+ \sum_{k=1}^{L} \mathrm{Res}\left(\frac{\Gamma(2w)\Gamma(2s - 2w)}{\Gamma(w)} \chi(w)\, z^{2w-2s},\, w = s_k \right).
$$

In exactly the same way, (8.18) with $r = \frac{3}{2}, a = 2s - 1$ amounts to

$$
\frac{\Gamma(2s - 1)}{2} \sum_{k=1}^{\infty} \frac{\alpha_k}{\sqrt{\lambda_k}\left(\sqrt{\lambda_k} + z\right)^{2s-1}}
$$
$$
= \frac{\Gamma(2s - 1)}{2^{3-2s}\sqrt{\pi}} \sum_{k=1}^{\infty} \frac{\beta_k}{\mu_k^{\frac{3}{2}-s}} \left(i e^{-(1-s)\pi i + 2z\sqrt{\mu_k}\, i}\, \Gamma(2 - 2s, 2z\sqrt{\mu_k}\, i) \right.
$$
$$
\left. - i e^{(1-s)\pi i - 2z\sqrt{\mu_k}\, i}\, \Gamma(2 - 2s, -2z\sqrt{\mu_k}\, i) \right) \tag{8.21}
$$
$$
+ \sum_{k=1}^{L} \mathrm{Res}\left(\frac{\Gamma(2w - 1)\Gamma(2s - 2w)}{\Gamma(w)} \chi(w)\, z^{2w-2s},\, w = s_k \right).
$$

8.10 The second K-Bessel expansion: $G_{1,3}^{3,0} \leftrightarrow G_{2,2}^{1,2}$

$$
\sum_{k=1}^{\infty} \frac{\alpha_k}{\lambda_k^s}\, G_{1,3}^{3,0}\left(z\lambda_k \,\middle|\, \begin{matrix} c \\ s, a, b \end{matrix} \right)
$$
$$
= \sum_{k=1}^{\infty} \frac{\beta_k}{\mu_k^{r-s}}\, G_{2,2}^{1,2}\left(\frac{\mu_k}{z} \,\middle|\, \begin{matrix} 1 - a, 1 - b \\ r - s, 1 - c \end{matrix} \right)
$$
$$
+ \sum_{k=1}^{L} \mathrm{Res}\left(\frac{\Gamma(a - s + w)\,\Gamma(b - s + w)}{\Gamma(c - s + w)} \chi(w)\, z^{s-w},\, w = s_k \right),
$$

which, for $b = c = s$, reduces to

$$
G_{0,2}^{2,0}\left(z\lambda_k \,\middle|\, \begin{matrix} - \\ a, b \end{matrix} \right) \leftrightarrow G_{1,1}^{1,1}\left(\frac{\mu_k}{z} \,\middle|\, \begin{matrix} 1 - a \\ r - s \end{matrix} \right),
$$

the Fourier-Bessel expansion in Chapter 5.

For $c = s$, the right-hand side member becomes the K-Bessel function by (2.109) and we have

$$2 \sum_{k=1}^{\infty} \frac{\alpha_k}{\lambda_k^s} \sqrt{z\lambda_k}^{-a+b} K_{b-a}\left(2\sqrt{z\lambda_k}\right) \tag{8.22}$$

$$= \frac{\Gamma(a+r-s)\,\Gamma(b+r-s)}{\Gamma(r)} z^{s-r} \sum_{k=1}^{\infty} \beta_k \, {}_2F_1\left(\begin{array}{c} a+r-s, b+r-s \\ r \end{array}; -\frac{\mu_k}{z}\right)$$

$$+ \sum_{k=1}^{L} \mathrm{Res}\left(\frac{\Gamma(a-s+w)\,\Gamma(b-s+w)}{\Gamma(w)} \chi(w)\, z^{s-w}, w = s_k\right),$$

on using (2.76).

8.11 The $G_{1,3}^{3,1} \leftrightarrow G_{2,2}^{2,2}$ formula

$$\sum_{k=1}^{\infty} \frac{\alpha_k}{\lambda_k^s} \, G_{1,3}^{3,1}\left(z\lambda_k \,\middle|\, \begin{array}{c} 1 \\ s, a, b \end{array}\right)$$

$$= \sum_{k=1}^{\infty} \frac{\beta_k}{\mu_k^{r-s}} \, G_{2,2}^{2,2}\left(\frac{\mu_k}{z} \,\middle|\, \begin{array}{c} 1-a, 1-b \\ 0, r-s \end{array}\right)$$

$$+ \sum_{k=1}^{L} \mathrm{Res}\left(\Gamma(a-s+w)\,\Gamma(b-s+w)\,\Gamma(s-w)\,\chi(w)\, z^{s-w}, w = s_k\right),$$

which leads to a formula involving ${}_1F_2$ and ${}_2F_2$ by (2.74).

8.12 The $G_{2,3}^{2,1} \leftrightarrow G_{2,3}^{2,1}$ formula

This is another special case of (8.8):

$$\sum_{k=1}^{\infty} \frac{\alpha_k}{\lambda_k^s} \, G_{2,3}^{2,1}\left(z\lambda_k \,\middle|\, \begin{array}{c} 1-a+c, b \\ s, a, 1-b+c \end{array}\right)$$

$$= z^c \sum_{k=1}^{\infty} \frac{\beta_k}{\mu^{r+c-s}} \, G_{2,3}^{2,1}\left(\frac{\mu_k}{z} \,\middle|\, \begin{array}{c} 1-a+c, b \\ r-s+c, a, 1-b+c \end{array}\right) \tag{8.23}$$

$$+ \sum_{k=1}^{L} \mathrm{Res}\left(\frac{\Gamma(a-s+w)\Gamma(a-c+s-w)}{\Gamma(b-s+w)\Gamma(b-c+s-w)} \chi(w)\, z^{s-w}, w = s_k\right),$$

which leads to a formula involving ${}_2F_2$ by (2.74).

8.13 The $G_{2,3}^{3,0} \leftrightarrow G_{2,3}^{1,2}$ formula

We have

$$
\sum_{k=1}^{\infty} \frac{\alpha_k}{\lambda_k^s} \, G_{2,3}^{3,0}\left(z\lambda_k \, \middle| \, \begin{matrix} b_1, b_2 \\ a, a_1, a_2 \end{matrix} \right)
$$

$$
= \sum_{k=1}^{\infty} \frac{\beta_k}{\mu^{r-s}} \, G_{2,3}^{1,2}\left(\frac{\mu_k}{z} \, \middle| \, \begin{matrix} 1-a_1, 1-a_2 \\ r-s, 1-b_1, 1-b_2 \end{matrix} \right) \tag{8.24}
$$

$$
+ \sum_{k=1}^{L} \operatorname{Res}\left(\frac{\Gamma(a_1 - s + w)\,\Gamma(a_2 - s + w)}{\Gamma(b_1 - s + w)\,\Gamma(b_2 - s + w)} \, \chi(w)\, z^{s-w}, w = s_k \right),
$$

which leads to a formula involving $_2F_2$ by (2.74).

8.14 The $G_{2,3}^{3,1} \leftrightarrow G_{2,3}^{2,2}$ formula

We have

$$
\sum_{k=1}^{\infty} \frac{\alpha_k}{\lambda_k^s} \, G_{2,3}^{3,1}\left(z\lambda_k \, \middle| \, \begin{matrix} 1, c \\ s, a, b \end{matrix} \right)
$$

$$
= \sum_{k=1}^{\infty} \frac{\beta_k}{\mu^{r-s}} \, G_{2,3}^{2,2}\left(\frac{\mu_k}{z} \, \middle| \, \begin{matrix} 1-a, 1-b \\ 0, r-s, 1-c \end{matrix} \right) \tag{8.25}
$$

$$
+ \sum_{k=1}^{L} \operatorname{Res}\left(\frac{\Gamma(a - s + w)\,\Gamma(b - s + w)\,\Gamma(s - w)}{\Gamma(c - s + w)} \, \chi(w)\, z^{s-w}, w = s_k \right),
$$

which by (2.74) leads to the $_2F_2$ formula adopted in the comic [KH]:

$$
\sum_{k=1}^{\infty} \alpha_k \left\{ \frac{\Gamma(s-a)\Gamma(b-a)\Gamma(a)}{\Gamma(c-a)} \frac{z^a}{\lambda_k^{s-a}} \, {}_2F_2\left(\begin{matrix} a, 1+a-c \\ 1+a-s, 1+a-b \end{matrix} ; z\lambda_k \right) \right.
$$

$$
+ \frac{\Gamma(s-b)\Gamma(a-b)\Gamma(b)}{\Gamma(c-b)} \frac{z^b}{\lambda_k^{s-b}} \, {}_2F_2\left(\begin{matrix} b, 1+b-c \\ 1+b-s, 1+b-a \end{matrix} ; z\lambda_k \right)
$$

$$
\left. + \frac{\Gamma(a-s)\Gamma(b-s)\Gamma(s)}{\Gamma(c-s)} z^s \, {}_2F_2\left(\begin{matrix} s, 1+s-c \\ 1+s-a, 1+s-b \end{matrix} ; z\lambda_k \right) \right\}
$$

$$= \sum_{k=1}^{\infty} \beta_k \left\{ \frac{\Gamma(r-s)\Gamma(a)\Gamma(b)}{\Gamma(c)} \frac{1}{\mu_k^{r-s}} \, {}_2F_2\left(\begin{array}{c} a,b \\ 1+s-r,c \end{array}; \frac{\mu_k}{z} \right) \right. \tag{8.26}$$

$$+ \frac{\Gamma(s-r)\Gamma(a+r-s)\Gamma(b+r-s)}{\Gamma(c+r-s)} \frac{1}{z^{r-s}}$$

$$\times \, {}_2F_2\left(\begin{array}{c} a+r-s, b+r-s \\ 1+r-s, c+r-s \end{array}; \frac{\mu_k}{z} \right) \right\}$$

$$+ \sum_{k=1}^{L} \mathrm{Res}\left(\frac{\Gamma(a-s+w)\,\Gamma(b-s+w)\,\Gamma(s-w)}{\Gamma(c-s+w)} \chi(w)\, z^{s-w}, w=s_k \right).$$

The left-hand side member, in the case $c=s$, reduces to ${}_1F_1$.

$$\sum_{k=1}^{\infty} \frac{\alpha_k}{\lambda_k^s} \, G_{1,2}^{2,1}\left(z\lambda_k \left| \begin{array}{c} 1 \\ a,b \end{array} \right. \right)$$

$$= \sum_{k=1}^{\infty} \frac{\beta_k}{\mu_k^{r-s}} \, G_{2,3}^{2,2}\left(\frac{\mu_k}{z} \left| \begin{array}{c} 1-a,1-b \\ 0,r-s,1-s \end{array} \right. \right)$$

$$+ \sum_{k=1}^{L} \mathrm{Res}\left(\frac{\Gamma(a-s+w)\,\Gamma(b-s+w)\,\Gamma(s-w)}{\Gamma(w)} \chi(w)\, z^{s-w}, w=s_k \right).$$

8.15   The $G_{2,3}^{3,2} \leftrightarrow G_{2,3}^{3,2}$ formula

This is another example of the bilateral formula.

$$\sum_{k=1}^{\infty} \frac{\alpha_k}{\lambda_k^s} \, G_{2,3}^{3,2}\left(z\lambda_k \left| \begin{array}{c} 1-a+c,1-b+c \\ s,a,b \end{array} \right. \right)$$

$$= z^c \sum_{k=1}^{\infty} \frac{\beta_k}{\mu_k^{r+c-s}} \, G_{2,3}^{3,2}\left(\frac{\mu_k}{z} \left| \begin{array}{c} 1-a+c,1-b+c \\ r-s+c,a,b \end{array} \right. \right) \tag{8.27}$$

$$+ \sum_{k=1}^{L} \mathrm{Res}\Big(\Gamma(a-s+w)\,\Gamma(b-s+w)\,\Gamma(a-c+s-w)\,\Gamma(b-c+s-w)$$

$$\chi(w)\, z^{s-w}, w=s_k \Big),$$

which leads to a formula similar to but more general than (8.26). We state it for curiosity's sake.

$$
\sum_{k=1}^{\infty} \alpha_k \Bigg\{ \Gamma(a-s)\Gamma(b-s)\Gamma(a-c+s)\Gamma(b-c+s)
$$

$$
\times z^s \, {}_2F_2 \left(\begin{matrix} a-c+s, b-c+s \\ 1-a+s, 1-b+s \end{matrix} ; -z\lambda_k \right)
$$

$$
+ \Gamma(s-a)\Gamma(b-a)\Gamma(2a-c)\Gamma(a+b-c)
$$

$$
\times \frac{z^a}{\lambda_k^{s-a}} \, {}_2F_2 \left(\begin{matrix} 2a-c, a+b-c \\ 1+a-s, 1+a-b \end{matrix} ; -z\lambda_k \right)
$$

$$
+ \Gamma(s-b)\Gamma(a-b)\Gamma(2b-c)\Gamma(a+b-c)
$$

$$
\times \frac{z^b}{\lambda_k^{s-b}} \, {}_2F_2 \left(\begin{matrix} 2b-c, a+b-c \\ 1+b-s, 1+b-a \end{matrix} ; -z\lambda_k \right) \Bigg\}
$$

$$
= \sum_{k=1}^{\infty} \beta_k \Bigg\{ \Gamma(a-c-r+s)\Gamma(b-c-r+s)\Gamma(a+r-s)\Gamma(b+r-s)
$$

$$
\times \frac{1}{z^{r-s}} \, {}_2F_2 \left(\begin{matrix} a+r-s, b+r-s \\ 1-a+c+r-s, 1-b+c+r-s \end{matrix} ; -\frac{\mu_k}{z} \right)
$$

$$
+ \Gamma(r-s-a+c)\Gamma(b-a)\Gamma(2a-c)\Gamma(a+b-c)
$$

$$
\times \frac{z^{c-a}}{\mu_k^{r-s+c-a}} \, {}_2F_2 \left(\begin{matrix} 2a-c, a+b-c \\ 1+a-c-r+s, 1+a-b \end{matrix} ; -\frac{\mu_k}{z} \right)
$$

$$
+ \Gamma(r-s-b+c)\Gamma(a-b)\Gamma(2b-c)\Gamma(a+b-c)
$$

$$
\times \frac{z^{c-b}}{\mu_k^{r-s+c-b}} \, {}_2F_2 \left(\begin{matrix} 2b-c, a+b-c \\ 1+b-c-r+s, 1+b-a \end{matrix} ; -\frac{\mu_k}{z} \right) \Bigg\}
$$

$$
+ \sum_{k=1}^{L} \operatorname{Res} \Big(\Gamma(a-s+w)\,\Gamma(b-s+w)
$$

$$
\times \Gamma(a-c+s-w)\,\Gamma(b-c+s-w)\,\chi(w)\,z^{s-w}, \atop w=s_k \Big). \quad (8.28)
$$

8.16 The $G_{1,p+1}^{p+1,0} \leftrightarrow G_{p,2}^{1,p}$ formula

This has some relevance to §3.1 and §9.2.

$$
\sum_{k=1}^{\infty} \frac{\alpha_k}{\lambda_k^s} \, G_{1,p+1}^{p+1,0} \left(z\lambda_k \, \middle| \, \begin{matrix} b \\ s, a_1, \ldots, a_p \end{matrix} \right)
$$

$$= \sum_{k=1}^{\infty} \frac{\beta_k}{\mu_k^{r-s}} \, G_{p,2}^{1,p}\left(\frac{\mu_k}{z} \,\middle|\, \begin{array}{c} 1-a_1,\ldots,1-a_p \\ r-s,1-b \end{array} \right)$$

$$+ \sum_{k=1}^{L} \mathrm{Res}\left(\frac{\displaystyle\prod_{j=1}^{p} \Gamma(a_j - s + w)}{\Gamma(b - s + w)} \, \chi(w) \, z^{s-w}, w = s_k \right), \tag{8.29}$$

which reduces, in the special case $b = s$, to

$$\sum_{k=1}^{\infty} \frac{\alpha_k}{\lambda_k^s} \, V(z\lambda_k; a_1, \cdots, a_p) \tag{8.30}$$

$$= \frac{\displaystyle\prod_{j=1}^{p} \Gamma(a_j + r - s)}{\Gamma(r)} \, z^{s-r}$$

$$\times \sum_{k=1}^{\infty} \beta_k \, {}_pF_1\left(\begin{array}{c} a_1 + r - s, \ldots, a_p + r - s \\ r \end{array} ; -\frac{\mu_k}{z} \right)$$

$$+ \sum_{k=1}^{L} \mathrm{Res}\left(\frac{\displaystyle\prod_{j=1}^{p} \Gamma(a_j - s + w)}{\Gamma(w)} \, \chi(w) \, z^{s-w}, w = s_k \right),$$

where V is Steen's function defined by (3.9) and we applied (2.76) on the right.

In the special case $a_1 = 0$, $a_2 = \frac{1}{p}, \ldots, a_p = \frac{p-1}{p}$ we have the formula

$$G_{0,p}^{p,0}\left(z\lambda_k \,\middle|\, \begin{array}{c} - \\ 0, \frac{1}{p}, \ldots, \frac{p-1}{p} \end{array} \right) = \frac{\sqrt{2\pi}^{p-1}}{\sqrt{p}} e^{-p\sqrt[p]{z\lambda_k}} \tag{8.31}$$

and so the left-hand side member of (8.30) reduces to the exponential function.

Chapter 9

The Product of Zeta-functions

In this chapter, we consider the product of zeta-functions. The most general case of the product of several different zeta-functions is not treated and will be touched upon in later volumes. We restrict to the case of the product of two zeta-functions and the (arbitrary) powers of zeta-functions. We shall not state all the cases and confine to the Riesz sums only. The main examples are due to Oppenheim-Wilton in the case of the product of zeta-functions and in the case of powers of zeta-functions, they are due to Berndt (and [Ric] for the estimate on the resulting Bessel series).

9.1 The product of zeta-functions

We consider the ΓΓ-type functional equation. §6.1.6 also falls in this section but we treat it independently since it is a degenerate case. Needless to say, this section overlaps §9.2.

9.1.1 *Statement of the Main Formula*

Given two Dirichlet series, we make the following
Convention.
In the case of the product of the Riemann zeta-functions, we consider $\varphi(s) = \psi(s) = \pi^{-s}\zeta(2s+\nu)\zeta(2s-\nu)$ and so $\lambda_k = \mu_k = \pi k^2$, $\alpha_k = \beta_k = k^{-\nu}\sigma_{2\nu}(k)$. Then they satisfy the **functional equation of ΓΓ-type** of the form:

$$\chi(s) = \begin{cases} \Gamma\!\left(s + \frac{\nu}{2}\right)\Gamma\!\left(s - \frac{\nu}{2}\right)\varphi(s), & \mathrm{Re}(s) > \sigma_\varphi \\ \Gamma\!\left(r - s + \frac{\nu}{2}\right)\Gamma\!\left(r - s - \frac{\nu}{2}\right)\psi(r - s), & \mathrm{Re}(s) < r - \sigma_\psi \end{cases} \tag{9.1}$$

or

$$\chi(s) = \begin{cases} \Gamma\left(s \left| \begin{matrix} - & ; & - \\ (\frac{\nu}{2},1),(-\frac{\nu}{2},1) & ; & - \end{matrix} \right. \right) \varphi(s), & \mathrm{Re}(s) > \sigma_\varphi \\ \\ \Gamma\left(r - s \left| \begin{matrix} - & ; & - \\ (\frac{\nu}{2},1),(-\frac{\nu}{2},1) & ; & - \end{matrix} \right. \right) \psi(r-s), & \mathrm{Re}(s) < r - \sigma_\psi. \end{cases}$$

By Theorem 7.1, we have the following modular relation:

$$\sum_{k=1}^{\infty} \frac{\alpha_k}{\lambda_k^s} H\left(z\lambda_k \left| \left(\begin{matrix} - & ; & - \\ (s+\frac{\nu}{2},1),(s-\frac{\nu}{2},1) & ; & - \end{matrix} \right) \oplus \Delta \right. \right)$$

$$= \sum_{k=1}^{\infty} \frac{\beta_k}{\mu_k^{r-s}} H\left(\frac{\mu_k}{z} \left| \left(\begin{matrix} - & ; & - \\ (r-s+\frac{\nu}{2},1),(r-s-\frac{\nu}{2},1) & ; & - \end{matrix} \right) \oplus \Delta^* \right. \right) \quad (9.2)$$

$$+ \sum_{k=1}^{L} \mathrm{Res}\Big(\Gamma(w - s \,|\, \Delta)\, \chi(w)\, z^{s-w}, w = s_k \Big).$$

In the traditional notation, if

$$\Delta = \begin{pmatrix} \{(1 - a_j, A_j)\}_{j=1}^{n}; & \{(a_j, A_j)\}_{j=n+1}^{p} \\ \{(b_j, B_j)\}_{j=1}^{m} & ; \{(1 - b_j, B_j)\}_{j=m+1}^{q} \end{pmatrix} \in \Omega \quad (9.3)$$

with its associated Gamma factor:

$$\Gamma(s \,|\, \Delta) = \frac{\displaystyle\prod_{j=1}^{m} \Gamma(b_j + B_j s) \prod_{j=1}^{n} \Gamma(a_j - A_j s)}{\displaystyle\prod_{j=n+1}^{p} \Gamma(a_j + A_j s) \prod_{j=m+1}^{q} \Gamma(b_j - B_j s)}, \quad (9.4)$$

we have

$$
\sum_{k=1}^{\infty} \frac{\alpha_k}{\lambda_k^s} H_{p,q+2}^{m+2,n}\left(z\lambda_k \,\middle|\, \begin{array}{c} \{(1-a_j, A_j)\}_{j=1}^n, \\ (s+\frac{\nu}{2}, 1), (s-\frac{\nu}{2}, 1), \{(b_j, B_j)\}_{j=1}^m, \end{array}\right.
$$
$$
\left.\begin{array}{c} \{(a_j, A_j)\}_{j=n+1}^p \\ \{(1-b_j, B_j)\}_{j=m+1}^q \end{array}\right)
$$
$$
= \sum_{k=1}^{\infty} \frac{\beta_k}{\mu_k^{r-s}} H_{q,p+2}^{n+2,m}\left(\frac{\mu_k}{z} \,\middle|\, \begin{array}{c} \{(1-b_j, B_j)\}_{j=1}^m, \\ (r-s+\frac{\nu}{2}, 1), (r-s-\frac{\nu}{2}, 1), \{(a_j, A_j)\}_{j=1}^n, \end{array}\right.
$$
$$
\left.\begin{array}{c} \{(b_j, B_j)\}_{j=m+1}^q \\ \{(1-a_j, A_j)\}_{j=n+1}^p \end{array}\right)
$$
$$
+ \sum_{k=1}^{L} \mathrm{Res}\Big(\Gamma(w-s \,|\, \Delta)\,\chi(w)\,z^{s-w}, w=s_k\Big). \tag{9.5}
$$

In the special case where $A_j = B_j = 1$, we have

$$
\sum_{k=1}^{\infty} \frac{\alpha_k}{\lambda_k^s} G_{p,q+2}^{m+2,n}\left(z\lambda_k \,\middle|\, \begin{array}{c} 1-a_1, \ldots, 1-a_n, a_{n+1}, \ldots, a_p \\ s+\frac{\nu}{2}, s-\frac{\nu}{2}, b_1, \ldots, b_m, 1-b_{m+1}, \ldots, 1-b_q \end{array}\right)
$$
$$
= \sum_{k=1}^{\infty} \frac{\beta_k}{\mu_k^{r-s}} G_{q,p+2}^{n+2,m}\left(\frac{\mu_k}{z} \,\middle|\, \begin{array}{c} 1-b_1, \ldots, 1-b_m, \\ r-s+\frac{\nu}{2}, r-s-\frac{\nu}{2}, a_1, \ldots, a_n, \end{array}\right.
$$
$$
\left.\begin{array}{c} b_{m+1}, \ldots, b_q \\ 1-a_{n+1}, \ldots, 1-a_p \end{array}\right)
$$
$$
+ \sum_{k=1}^{L} \mathrm{Res}\Big(\Gamma(w-s \,|\, \Delta)\,\chi(w)\,z^{s-w}, w=s_k\Big). \tag{9.6}
$$

9.1.2   Wilton's Riesz sum: $G_{4,4}^{2,2} \leftrightarrow G_{2,6}^{4,0}$

In this subsection we deduce Wilton's Riesz sum formula, Theorem 9.2, which is the most far-reaching example of (9.6). We begin by stating the formulas for G, H-functions.

By a similar reasoning to the proof of Theorem 2.4, we may prove

Proposition 9.1   *For $c, d \in \mathbb{C}$, $C, D \in \mathbb{R}^+$ such that $C > D$, and for any*

$\Delta \in \Omega$, *we have*

$$H\left(z \,\middle|\, \Delta \oplus \begin{pmatrix} -;(c,C),(d,D) \\ -;(c,C),(d,D) \end{pmatrix}\right)$$
$$= \frac{1}{2\pi}\left\{ H\left(z \,\middle|\, \Delta \oplus \begin{pmatrix} -;(c-d+\frac{1}{2},C-D) \\ -;(c-d+\frac{1}{2},C-D) \end{pmatrix}\right) \right. \tag{9.7}$$
$$\left. + H\left(z \,\middle|\, \Delta \oplus \begin{pmatrix} -;(c+d-\frac{1}{2},C+D) \\ -;(c+d-\frac{1}{2},C+D) \end{pmatrix}\right) \right\}.$$

Proof. Let $A = c - d + \frac{1}{2} + (C-D)w$ and $B = c + d - \frac{1}{2} + (C+D)w$. Then by the reciprocity formula

$$\frac{1}{\Gamma(c+Cw)\Gamma(1-(c+Cw)\Gamma(d+Dw)\Gamma(1-(d+Dw)}$$
$$= \frac{1}{\pi^2} \sin\pi(c+Cw)\sin\pi(d+Dw)$$
$$= \frac{1}{2\pi^2}\left(\cos\pi(c+Cw-(d+Dw)) - \cos\pi(c+Cw+(d+Dw))\right),$$

which may be written as

$$\frac{1}{2\pi^2}\left(\sin\pi A + \sin\pi B\right) = \frac{1}{2\pi}\left(\frac{1}{\Gamma(A)\Gamma(1-A)} + \frac{1}{\Gamma(B)\Gamma(1-B)}\right),$$

i.e. (9.7). $\qquad\square$

If $C = D$ in (9.8) we have

$$\frac{1}{\Gamma(c+Cw)\Gamma(1-(c+Cw)\Gamma(d+Cw)\Gamma(1-(d+Cw)}$$
$$= \frac{1}{2\pi^2}\left(\cos\pi(c-d) - \cos\pi(c+d+2Cw)\right), \tag{9.8}$$

whence we deduce the following

Corollary 9.1 *For $c, d \in \mathbb{C}$, $C \in \mathbb{R}^+$, and for any $\Delta \in \Omega$, we have*

$$H\left(z \,\middle|\, \Delta \oplus \begin{pmatrix} -;(c,C),(d,C) \\ -;(c,C),(d,C) \end{pmatrix}\right) \tag{9.9}$$
$$= \frac{1}{2\pi^2}\cos\left((c-d)\pi\right) H\left(z \,\middle|\, \Delta\right) + \frac{1}{2\pi} H\left(z \,\middle|\, \Delta \oplus \begin{pmatrix} -;(c+d-\frac{1}{2},2C) \\ -;(c+d-\frac{1}{2},2C) \end{pmatrix}\right),$$

or

$$H^{m,n}_{p+2,q+2}\left(z \left| \begin{array}{l} \{(a_j, A_j)\}_{j=1}^n, \{(a_j, A_j)\}_{j=n+1}^p, (c, C), (d, C) \\ \{(b_j, B_j)\}_{j=1}^m, \{(b_j, B_j)\}_{j=m+1}^q, (c, C), (d, C) \end{array} \right. \right) \tag{9.10}$$

$$= \frac{1}{2\pi^2} \cos((c-d)\pi) \, H^{m,n}_{p,q}\left(z \left| \begin{array}{l} \{(a_j, A_j)\}_{j=1}^n, \{(a_j, A_j)\}_{j=n+1}^p \\ \{(b_j, B_j)\}_{j=1}^m, \{(b_j, B_j)\}_{j=m+1}^q \end{array} \right. \right)$$

$$+ \frac{1}{2\pi} H^{m,n}_{p+1,q+1}\left(z \left| \begin{array}{l} \{(a_j, A_j)\}_{j=1}^n, \{(a_j, A_j)\}_{j=n+1}^p, \left(c+d-\frac{1}{2}, 2C\right) \\ \{(b_j, B_j)\}_{j=1}^m, \{(b_j, B_j)\}_{j=m+1}^q, \left(c+d-\frac{1}{2}, 2C\right) \end{array} \right. \right).$$

Theorem 9.1 *We have*

$$G^{2,0}_{1,3}\left(z \left| \begin{array}{c} c \\ a, b, c \end{array} \right. \right) \tag{9.11}$$

$$= z^{\frac{1}{2}(a+b)} \left\{ -\sin((c-b)\pi) \, Y_{a-b}\big(2\sqrt{z}\big) + \cos((c-b)\pi) \, J_{a-b}\big(2\sqrt{z}\big) \right\}$$

and

$$G^{4,0}_{2,6}\left(z \left| \begin{array}{c} c, d \\ a, b, a+\frac{1}{2}, b+\frac{1}{2}, c, d \end{array} \right. \right) \tag{9.12}$$

$$= \frac{1}{4^{a+b}} \left\{ \frac{\cos((c-d)\pi)}{\pi} \, G^{2,0}_{0,2}\left(4\sqrt{z} \left| \begin{array}{c} - \\ 2a, 2b \end{array} \right. \right) \right.$$

$$\left. + G^{2,0}_{1,3}\left(4\sqrt{z} \left| \begin{array}{c} c+d-\frac{1}{2} \\ 2a, 2b, c+d-\frac{1}{2} \end{array} \right. \right) \right\}$$

$$= z^{\frac{1}{2}(a+b)} \left\{ \frac{2}{\pi} \cos((c-d)\pi) \, K_{2a-2b}\big(4\sqrt[4]{z}\big) \right.$$

$$\left. + \cos((c+d-2b)\pi) \, Y_{2a-2b}\big(4\sqrt[4]{z}\big) + \sin((c+d-2b)\pi) \, J_{2a-2b}\big(4\sqrt[4]{z}\big) \right\}.$$

Proof. By (9.10)

$$G^{4,0}_{2,6}\left(z \left| \begin{array}{c} c, d \\ a, b, a+\frac{1}{2}, b+\frac{1}{2}, c, d \end{array} \right. \right) \tag{9.13}$$

$$= \frac{1}{2\pi^2} \cos((c-d)\pi) \, G^{4,0}_{0,4}\left(z \left| \begin{array}{c} - \\ a, b, a+\frac{1}{2}, b+\frac{1}{2} \end{array} \right. \right)$$

$$+ \frac{1}{2\pi} H^{4,0}_{1,5}\left(z \left| \begin{array}{c} \left(c+d-\frac{1}{2}, 2\right) \\ (a,1), (b,1), \left(a+\frac{1}{2},1\right), \left(b+\frac{1}{2},1\right), \left(c+d-\frac{1}{2},2\right) \end{array} \right. \right).$$

By Theorem 2.3 we have

$$G_{0,4}^{4,0}\left(z \,\middle|\, \begin{matrix} - \\ a, b, a + \frac{1}{2}, b + \frac{1}{2} \end{matrix}\right) = \frac{2\pi}{4^{a+b}\pi} G_{0,2}^{2,0}\left(4\sqrt{z} \,\middle|\, \begin{matrix} - \\ 2a, 2b \end{matrix}\right)$$

and

$$H_{1,5}^{4,0}\left(z \,\middle|\, \begin{matrix} \left(c + d - \frac{1}{2}, 2\right) \\ (a, 1), (b, 1), \left(a + \frac{1}{2}, 1\right), \left(b + \frac{1}{2}, 1\right), \left(c + d - \frac{1}{2}, 2\right) \end{matrix}\right)$$
$$= \frac{\pi}{4^{a+b}} H_{1,3}^{2,0}\left(16z \,\middle|\, \begin{matrix} \left(c + d - \frac{1}{2}, 2\right) \\ (2a, 2), (2b, 2), \left(c + d - \frac{1}{2}, 2\right) \end{matrix}\right),$$

and so the first equality of (9.12) follows from (9.13).

The second follows from (9.11).   □

Corollary 9.2

(1) By setting $c = a + \frac{1}{2}$, $d = b + \frac{1}{2}$, we have

$$G_{0,4}^{2,0}\left(z \,\middle|\, \begin{matrix} - \\ a, b, a + \frac{1}{2}, b + \frac{1}{2} \end{matrix}\right) \tag{9.14}$$
$$= z^{\frac{1}{2}(a+b)}\left\{ \frac{2}{\pi}\cos((a-b)\pi) K_{2a-2b}\left(4\sqrt[4]{z}\right) - \cos((a-b)\pi) Y_{2a-2b}\left(4\sqrt[4]{z}\right) \right.$$
$$\left. - \sin((a-b)\pi) J_{2a-2b}\left(4\sqrt[4]{z}\right) \right\}$$

and in particular

$$G_{0,4}^{2,0}\left(z \,\middle|\, \begin{matrix} - \\ a, a, a + \frac{1}{2}, a + \frac{1}{2} \end{matrix}\right) = z^a\left\{ \frac{2}{\pi} K_0\left(4\sqrt[4]{z}\right) - Y_0\left(4\sqrt[4]{z}\right) \right\}. \tag{9.15}$$

(2) By setting $c = a + \frac{1}{2}$, $d = b$, we have

$$G_{0,4}^{2,0}\left(z \,\middle|\, \begin{matrix} - \\ a, b + \frac{1}{2}, a + \frac{1}{2}, b \end{matrix}\right) \tag{9.16}$$
$$= z^{\frac{1}{2}(a+b)}\left\{ -\frac{2}{\pi}\sin((a-b)\pi) K_{2a-2b}\left(4\sqrt[4]{z}\right) - \sin((a-b)\pi) Y_{2a-2b}\left(4\sqrt[4]{z}\right) \right.$$
$$\left. + \cos((a-b)\pi) J_{2a-2b}\left(4\sqrt[4]{z}\right) \right\}.$$

(3) By setting $c = a$, $d = b$, we have

$$G_{0,4}^{2,0}\left(z \left|_{a + \frac{1}{2}, b + \frac{1}{2}, a, b}^{-}\right.\right) \tag{9.17}$$

$$= z^{\frac{1}{2}(a+b)}\left\{\frac{2}{\pi}\cos((a - b)\pi)\,K_{2a-2b}\left(4\sqrt[4]{z}\right) + \cos((a - b)\pi)\,Y_{2a-2b}\left(4\sqrt[4]{z}\right)\right.$$

$$\left. + \sin((a - b)\pi)\,J_{2a-2b}\left(4\sqrt[4]{z}\right)\right\}$$

and in particular

$$G_{0,4}^{2,0}\left(z \left|_{a + \frac{1}{2}, a + \frac{1}{2}, a, a}^{-}\right.\right) = z^{a}\left\{\frac{2}{\pi}K_{0}\left(4\sqrt[4]{z}\right) + Y_{0}\left(4\sqrt[4]{z}\right)\right\}. \tag{9.18}$$

Remark 9.1 $G_{1,3}^{2,0}\left(z \left|_{0,\, s_2 - s_1,\, 1 - s_1}^{s_2}\right.\right)$ *appears as* $W(2z; s_1.s_2)$ *on [Mot, (2.1.3), p.44] ([Mot, (2.6.15), p.77]) and its special case is computed [Mot, (2.4.3), p.63].*

$$\frac{\Gamma(s)}{\Gamma(1 - s)}W(z; 1, s) = \left(\frac{z}{2}\right)^2 - \frac{2}{z}\sum_{k=1}^{\infty}(2k - 1)\frac{k - (1 - s)}{k + 1 - s}J_{2k-1}(z), \tag{9.19}$$

which provides an evaluation of $G_{1,3}^{2,0}\left(z \left|_{0,\, 1 - s,\, 0}^{s}\right.\right)$. *He introduced it in evaluating the inner product of two Poincaré series.* $\mathfrak{p}(x, t)$ *on [Mot, (2.3.5), p.75] is another form for the same G-function. [Mot, (2.6.15), p.77] is another form for the W-function. As he states on pp.91-92, his Lemma 2.1 replaces Kuznetsov's heavy use of Bessel functions.*

Another G-function that appears is [Mot, (2.6.11), p.75] with half-unit argument though, which is known as the Barnes integral.

$$G_{1,4}^{4,0}\left(\frac{1}{2} \left|_{\left(\frac{1}{2}(\mu + \nu), \frac{1}{2}\right),\, \left(\frac{1}{2}(\mu - \nu), \frac{1}{2}\right),\, \left(\frac{1}{2}(-\mu + \nu), \frac{1}{2}\right),\, \left(-\frac{1}{2}(\mu + \nu), \frac{1}{2}\right)}^{(0, 1)}\right.\right)$$

$$= K_{\mu}\left(\frac{1}{2}\right)K_{\nu}\left(\frac{1}{2}\right), \tag{9.20}$$

which is then applied in deriving [Mot, Lemma 3.5, p.107] for the functional equation for the Rankin L-function associated with a pair of cusp forms.

(9.12) with $c = b, d = b + 1/2$ leads to

$$
G_{0,4}^{2,0}\left(z \left|\begin{array}{c} - \\ a, a + \frac{1}{2}, b, b + \frac{1}{2} \end{array}\right.\right)
$$

$$
= \frac{1}{4^{a+b}} G_{0,2}^{1,0}\left(4\sqrt{z} \left|\begin{array}{c} - \\ 2a, 2b \end{array}\right.\right) = z^{\frac{1}{2}(a+b)} J_{2a-2b}\left(4\sqrt[4]{z}\right).
$$

$$
\sum_{k=1}^{\infty} \frac{\alpha_k}{\lambda_k^s} G_{4,4}^{2,2}\left(z\lambda_k \left|\begin{array}{c} 1 - a, 1 - b, c, d \\ s + \frac{\nu}{2}, s - \frac{\nu}{2}, 1 - e, 1 - f \end{array}\right.\right) \tag{9.21}
$$

$$
= \sum_{k=1}^{\infty} \frac{\beta_k}{\mu_k^{r-s}} G_{2,6}^{4,0}\left(\frac{\mu_k}{z} \left|\begin{array}{c} e, f \\ r - s + \frac{\nu}{2}, r - s - \frac{\nu}{2}, a, b, 1 - c, 1 - d \end{array}\right.\right)
$$

$$
+ \sum_{k=1}^{L} \mathrm{Res}\left(\frac{\Gamma(a + s - w)\,\Gamma(b + s - w)}{\Gamma(c - s + w)\,\Gamma(d - s + w)\,\Gamma(e + s - w)\,\Gamma(f + s - w)}\right.
$$

$$
\left. \chi(w)\, z^{s-w}, w = s_k\right).
$$

We treat the case $r = \frac{1}{2}$. Assuming $\varkappa$ is a non-negative integer, we put $a = -s - \frac{\nu}{2} - \frac{\varkappa}{2} + \frac{1}{2}, b = -s - \frac{\nu}{2} - \frac{\varkappa}{2}, c = s - \frac{\nu}{2}, d = s + \frac{\nu}{2} + \varkappa + 1, e = -s - \frac{\nu}{2} + \frac{1}{2}, f = -s - \frac{\nu}{2} - \varkappa$. Then (9.21) becomes

$$
\sum_{k=1}^{\infty} \alpha_k G_{4,4}^{2,2}\left(z\lambda_k \left|\begin{array}{c} \frac{\nu}{2} + \frac{\varkappa}{2} + \frac{1}{2}, \frac{\nu}{2} + \frac{\varkappa}{2} + 1, -\frac{\nu}{2}, \frac{\nu}{2} + \varkappa + 1 \\ \frac{\nu}{2}, -\frac{\nu}{2}, \frac{\nu}{2} + \frac{1}{2}, \frac{\nu}{2} + \varkappa + 1 \end{array}\right.\right) \tag{9.22}
$$

$$
= z^{-\frac{1}{2}} \sum_{k=1}^{\infty} \beta_k G_{2,6}^{4,0}\left(\frac{\mu_k}{z} \left|\begin{array}{c} -\frac{\nu}{2}, -\frac{\nu}{2} - \varkappa - \frac{1}{2} \\ \frac{\nu}{2}, -\frac{\nu}{2}, -\frac{\nu}{2} - \frac{\varkappa}{2}, -\frac{\nu}{2} - \frac{\varkappa}{2} - \frac{1}{2}, \frac{\nu}{2} + \frac{1}{2}, -\frac{\nu}{2} - \varkappa - \frac{1}{2} \end{array}\right.\right)
$$

$$
+ \sum_{k=1}^{L} \mathrm{Res}\left(\frac{\Gamma\left(-\frac{\nu}{2} - \frac{\varkappa}{2} + \frac{1}{2} - w\right)\Gamma\left(-\frac{\nu}{2} - \frac{\varkappa}{2} - w\right)}{\Gamma\left(-\frac{\nu}{2} + w\right)\Gamma\left(\frac{\nu}{2} + \varkappa + 1 + w\right)\Gamma\left(-\frac{\nu}{2} + \frac{1}{2} - w\right)\Gamma\left(-\frac{\nu}{2} - \varkappa - w\right)}\right.
$$

$$
\left. \chi(w)\, z^{-w}, w = s_k\right).
$$

We transform the G-functions in (9.22). By (2.96), $-\frac{\nu}{2}$'s in the first and the second lines of $G_{4,4}^{2,2}$ may be deleted and then $\frac{\nu}{2} + \frac{1}{2}$'s may be added, i.e. replacing $-\frac{\nu}{2}$'s by $\frac{\nu}{2} + \frac{1}{2}$'s, we apply (9.10) to obtain

$$G_{4,4}^{2,2} = \frac{1}{2\pi^2} \cos\left(\left(\varkappa + \tfrac{1}{2}\right)\pi\right) G_{2,2}^{2,2}\left(z \left| \begin{array}{c} \frac{\nu}{2} + \frac{\varkappa}{2} + \frac{1}{2}, \frac{\nu}{2} + \frac{\varkappa}{2} + 1 \\ \frac{\nu}{2}, \frac{\nu}{2} + \frac{1}{2} \end{array}\right.\right) \qquad (9.23)$$
$$+ \frac{1}{2\pi} H_{3,3}^{2,2}\left(z \left| \begin{array}{c} \left(\frac{\nu}{2} + \frac{\lambda}{2} + \frac{1}{2}, 1\right), \left(\frac{\nu}{2} + \frac{\lambda}{2} + 1, 1\right), (\nu + \lambda + 1, 2) \\ \left(\frac{\nu}{2}, 1\right), \left(\frac{\nu}{2} + \frac{1}{2}, 1\right), (\nu + \lambda + 1, 2) \end{array}\right.\right).$$

Hence only the second term remains, which becomes

$$2^{\lambda+1} H_{2,2}^{1,1}\left(z \left| \begin{array}{c} (\nu + \lambda + 1, 2), (\nu + \lambda + 1, 2) \\ (\nu, 2), (\nu + \lambda + 1, 2) \end{array}\right.\right)$$

by (2.99). This in turn reduces by (2.96) to $2^\lambda G_{1,1}^{1,0}\left(\sqrt{z} \left| \begin{array}{c} \nu + \lambda + 1 \\ \nu \end{array}\right.\right)$, the Riesz kernel.

Similarly, replacing $-\frac{\nu}{2}$'s by $\frac{\nu}{2} + \frac{1}{2}$'s, we apply (9.10) to obtain

$$G_{2,6}^{4,0} = z^{-\frac{\lambda}{4} - \frac{1}{4}} G_{2\nu+\lambda+1}^{\lambda}\left(4\sqrt[4]{z}\right) \qquad (9.24)$$

(by (9.12)) where, slightly more general than Wilton's (1.22) [Wil4], we introduce **Wilton's generalized Bessel function**

$$G_\nu^\lambda(z) = -(-1)^\lambda \frac{2}{\pi} \sin\left(\frac{\nu - \lambda}{2}\pi\right) K_\nu(z)$$
$$- \sin\left(\frac{\nu - \lambda}{2}\pi\right) Y_\nu(z) + \cos\left(\frac{\nu - \lambda}{2}\pi\right) J_\nu(z). \qquad (9.25)$$

Therefore we obtain

$$\frac{z^{\frac{\nu}{2}} 2^\lambda}{\Gamma(\lambda + 1)} \sum_{\lambda_k < \frac{1}{z}} \alpha_k \lambda_k^{\frac{\nu}{2}} \left(1 - \sqrt{z\lambda_k}\right)^\lambda$$
$$= z^{\frac{\lambda}{4} - \frac{1}{4}} \sum_{k=1}^{\infty} \frac{\beta_k}{\mu_k^{\frac{\lambda}{4} + \frac{1}{4}}} G_{2\nu+\lambda+1}^{\lambda}\left(4\sqrt[4]{\frac{\mu_k}{z}}\right)$$
$$+ \sum_{k=1}^{L} \mathrm{Res}\left(\frac{\Gamma\left(-\frac{\nu}{2} - \frac{\lambda}{2} + \frac{1}{2} - w\right)\Gamma\left(-\frac{\nu}{2} - \frac{\lambda}{2} - w\right)}{\Gamma\left(-\frac{\nu}{2} + w\right)\Gamma\left(\frac{\nu}{2} + \lambda + 1 + w\right)\Gamma\left(-\frac{\nu}{2} + \frac{1}{2} - w\right)\Gamma\left(-\frac{\nu}{2} - \lambda - w\right)}\right.$$
$$\left. \chi(w) z^{-w}, w = s_k\right),$$

and by replacing z by $\frac{1}{z}$, we have

$$
\frac{z^{-\frac{\nu}{2}}2^{\lambda}}{\Gamma(\lambda+1)} \sum_{\lambda_k<z} \alpha_k \lambda_k^{\frac{\nu}{2}} \left(1 - \sqrt{\frac{\lambda_k}{z}}\right)^{\lambda} \tag{9.26}
$$

$$
= z^{-\frac{\lambda}{4}+\frac{1}{4}} \sum_{k=1}^{\infty} \frac{\beta_k}{\mu_k^{\frac{\lambda}{4}+\frac{1}{4}}} G_{2\nu+\lambda+1}^{\lambda}\left(4\sqrt[4]{z\mu_k}\right)
$$

$$
+ \sum_{k=1}^{L} \operatorname{Res}\left(\frac{\Gamma\left(-\frac{\nu}{2}-\frac{\lambda}{2}+\frac{1}{2}-w\right)\Gamma\left(-\frac{\nu}{2}-\frac{\lambda}{2}-w\right)}{\Gamma\left(-\frac{\nu}{2}+w\right)\Gamma\left(\frac{\nu}{2}+\lambda+1+w\right)\Gamma\left(-\frac{\nu}{2}+\frac{1}{2}-w\right)\Gamma\left(-\frac{\nu}{2}-\lambda-w\right)} \right.
$$

$$
\left. \vphantom{\frac{\Gamma}{\Gamma}} \chi(w)\,z^w, w=s_k \right),
$$

which leads to the following generalization of Wilton's Theorem 1 in [Wil4]:

Theorem 9.2 (Generalization of Wilton's Riesz sum) *For a non-negative integer $\varkappa$ and $z>0$, we have the Riesz sum formula*

$$
\frac{2^{\varkappa+1}z^{-\nu-\varkappa}}{\Gamma(\varkappa+1)} \sum_{\lambda_k<z} \alpha_k \lambda_k^{\nu}(z-\lambda_k)^{\varkappa} \tag{9.27}
$$

$$
= 2\,z^{-\frac{\lambda-1}{2}} \sum_{k=1}^{\infty} \frac{\beta_k}{\mu_k^{\frac{\varkappa+1}{2}}} G_{2\nu+\lambda+1}^{\lambda}\left(4\sqrt{z\mu_k}\right)
$$

$$
+ \sum_{k=1}^{L} \operatorname{Res}\left(\frac{\Gamma\left(-\frac{\nu}{2}-\frac{\varkappa}{2}+\frac{1}{2}-\frac{w}{2}\right)\Gamma\left(-\frac{\nu}{2}-\frac{\varkappa}{2}-\frac{w}{2}\right)}{\Gamma\left(-\frac{\nu}{2}+\frac{w}{2}\right)\Gamma\left(\frac{\nu}{2}+\varkappa+1+\frac{w}{2}\right)\Gamma\left(-\frac{\nu}{2}+\frac{1}{2}-\frac{w}{2}\right)\Gamma\left(-\frac{\nu}{2}-\varkappa-\frac{w}{2}\right)} \right.
$$

$$
\left. \vphantom{\frac{\Gamma}{\Gamma}} \chi(w)\,z^w, w=s_k \right),
$$

where $G_{\nu}^{\varkappa}(z)$ is Wilton's generalized Bessel function in (9.25).

The special case of $\varkappa=0$ of (9.27) reads

Corollary 9.3

$$
2\,z^{-\nu} \sum_{\lambda_k\le z}' \alpha_k \lambda_k^{\nu}
$$

$$
= 2\,z^{\frac{1}{2}} \sum_{k=1}^{\infty} \frac{\beta_k}{\sqrt{\mu_k}} F_{2\nu+1}\left(4\sqrt{z\mu_k}\right) \tag{9.28}
$$

$$
+ \sum_{k=1}^{L} \operatorname{Res}\left(\frac{1}{\Gamma\left(\frac{\nu}{2}+\frac{w}{2}+1\right)\Gamma\left(-\frac{\nu}{2}+\frac{w}{2}\right)} \chi(w)\,z^w, w=s_k \right),
$$

where

$$F_\nu(z) = G_\nu^0(z) \tag{9.29}$$

$$= -\frac{2}{\pi} \sin\left(\frac{\nu}{2}\pi\right) K_\nu(z) - \sin\left(\frac{\nu}{2}\pi\right) Y_\nu(z) + \cos\left(\frac{\nu}{2}\pi\right) J_\nu(z).$$

By Convention, Theorem 9.2 reduces to the well-known Oppenheim-Wilton's formula (for $x > 0$)

$$\frac{2^{\varkappa+1}}{\Gamma(\varkappa+1)} {\sum_{k\leq x}}' \sigma_{2\nu}(k)(x-k)^\varkappa \tag{9.30}$$

$$= 2\pi^{-\varkappa} x^{\nu+\frac{\varkappa}{2}+\frac{1}{2}} \sum_{k=1}^\infty \frac{\sigma_{2\nu}(k)}{k^{\nu+\frac{\varkappa}{2}+\frac{1}{2}}} G_{2\nu+\varkappa+1}^\lambda\left(4\pi\sqrt{kz}\right)$$

$$+ \sum_{v\in S_\nu} \mathrm{Res}\left(\frac{\Gamma\left(-\frac{\nu}{2}-\frac{\varkappa}{2}+\frac{1}{2}-\frac{w}{2}\right)\Gamma\left(-\frac{\nu}{2}-\frac{\varkappa}{2}-\frac{w}{2}\right)}{\Gamma\left(\frac{\nu}{2}+\varkappa+1+\frac{w}{2}\right)\Gamma\left(-\frac{\nu}{2}+\frac{1}{2}-\frac{w}{2}\right)\Gamma\left(-\frac{\nu}{2}-\varkappa-\frac{w}{2}\right)}\right.$$

$$\left.\Gamma\left(\tfrac{w+\nu}{2}\right)\zeta(w+\nu)\zeta(w-\nu)x^{\nu+\varkappa+w}, w=v\right),$$

while Corollary 9.3 reduces to

$$\begin{aligned}{\sum_{k\leq x}}'\sigma_{2\nu}(k) &= x^{\frac{2\nu+1}{2}} \sum_{k=1}^\infty \frac{\sigma_{2\nu}(k)}{k^{\nu+\frac{1}{2}}} F_{2\nu+1}\left(4\pi\sqrt{xk}\right) \\ &+ x^\nu \sum_{v\in S_\nu} \mathrm{Res}\left(\frac{\zeta(w+\nu)\zeta(w-\nu)}{w+\nu}x^w, w=v\right).\end{aligned} \tag{9.31}$$

We may easily calculate the residues and we have

$$\mathrm{P}_\nu(x) = \zeta(1-2\nu)x + \frac{\zeta(2\nu+1)}{2v+1}x^{2\nu+1} - \frac{1}{2}\zeta(-2\nu) \tag{9.32}$$

for $\nu \neq 0, \pm\frac{1}{2}$,

$$\mathrm{P}_0(x) = x\log x + (2\gamma-1)x + \frac{1}{4} \tag{9.33}$$

and

$$\mathrm{P}_{1/2}(x) = -\frac{1}{12}\left(\sqrt{x}+\frac{1}{\sqrt{x}}\right), \mathrm{P}_{1/2}(x) = \frac{\pi^2}{6}x^{\frac{3}{2}} - \frac{1}{12}\sqrt{z}(\log x + \gamma). \tag{9.34}$$

(9.31) together with (9.32) coincide with [AAW4, pp.9-10] and especially, the function $F(w)$ on [AAW4, ll.1-2, p.10] is $w^{\frac{\nu}{2}}F_\nu(4\sqrt{w})$.

(9.31) with $\nu = 0$ together with (9.33) coincide with [AAW1, (16)] and especially, the function $F(w)$ on [AAW1, (17)] is $w^{\frac{1}{2}} F_\nu(4\sqrt{w})$. This is the celebrated Voronoï formula ([Vor1])

$$\sideset{}{'}\sum_{k\leq x} d(k) = x \log x + (2\gamma - 1)x + \frac{1}{4}$$
$$- \sqrt{x} \sum_{k=1}^{\infty} \frac{d(k)}{\sqrt{k}} \left(Y_1\left(4\pi\sqrt{xk}\right) + \frac{2}{\pi} K_1\left(4\pi\sqrt{xk}\right) \right). \tag{9.35}$$

In exactly the same way, we obtain the formula for the summatory function of the coefficients of the Dedekind zeta-function of a real quadratic field [AAW1, (18)]

$$\sideset{}{'}\sum_{k\leq x} F(k)$$
$$= \rho x - \sqrt{x} \sum_{k=1}^{\infty} \frac{F(k)}{\sqrt{k}} \left(Y_1\left(4\pi\sqrt{\frac{xk}{|\Delta|}}\right) + \frac{2}{\pi} K_1\left(4\pi\sqrt{\frac{xk}{|\Delta|}}\right) \right), \tag{9.36}$$

where the residue is defined by (3.38).

Remark 9.2 *Corollary 9.3 follows also readily from the* $G_{2,2}^{2,0} \leftrightarrow G_{0,4}^{2,0}$ *formula, which we state as another example of (9.6):*

$$\sum_{k=1}^{\infty} \frac{\alpha_k}{\lambda_k^s} G_{2,2}^{2,0}\left(z\lambda_k \,\Big|\, \begin{matrix} a, b \\ s + \frac{\nu}{2}, s - \frac{\nu}{2} \end{matrix} \right)$$
$$= \sum_{k=1}^{\infty} \frac{\beta_k}{\mu_k^{r-s}} G_{0,4}^{2,0}\left(\frac{\mu_k}{z} \,\Big|\, \begin{matrix} - \\ r - s + \frac{\nu}{2}, r - s - \frac{\nu}{2}, 1 - a, 1 - b \end{matrix} \right) \tag{9.37}$$
$$+ \sum_{k=1}^{L} \mathrm{Res}\left(\frac{1}{\Gamma(a + w - s)\,\Gamma(b + w - s)} \chi(w)\, z^{s-w}, w = s_k \right).$$

In the case $r = \frac{1}{2}$, $a = s + \frac{\nu}{2} + 1$, $b = s - \frac{\nu}{2}$, *(9.37) becomes*

$$\sum_{k=1}^{\infty} \alpha_k\, G_{2,2}^{2,0}\left(z\lambda_k \,\Big|\, \begin{matrix} \frac{\nu}{2} + 1, -\frac{\nu}{2} \\ \frac{\nu}{2}, -\frac{\nu}{2} \end{matrix} \right)$$
$$= z^{-\frac{1}{2}} \sum_{k=1}^{\infty} \beta_k\, G_{0,4}^{2,0}\left(\frac{\mu_k}{z} \,\Big|\, \begin{matrix} - \\ \frac{\nu}{2}, -\frac{\nu}{2}, -\frac{\nu}{2} - \frac{1}{2}, \frac{\nu}{2} + \frac{1}{2} \end{matrix} \right) \tag{9.38}$$
$$+ \sum_{k=1}^{L} \mathrm{Res}\left(\frac{1}{\Gamma(\frac{\nu}{2} + w + 1)\,\Gamma(-\frac{\nu}{2} + w)} \chi(w)\, z^{-w}, w = s_k \right).$$

$G_{2,2}^{2,0}$ amounts to the Riesz kernel while $G_{0,4}^{2,0}$ amounts to $F_\nu(z) = G_\nu^0(z)$ in (9.29), thus leading to

$$z^{-\frac{\nu}{2}} \sum_{\lambda_k < z} \alpha_k \lambda_k^{\frac{\mu}{2}} \tag{9.39}$$

$$= z^{\frac{1}{4}} \sum_{k=1}^{\infty} \frac{\beta_k}{\mu_k^{\frac{1}{4}}} F_{2\nu+1}\left(4\sqrt[4]{z\mu_k}\right)$$

$$+ \sum_{k=1}^{L} \operatorname{Res}\left(\frac{1}{\Gamma(\frac{\nu}{2} + w + 1)\Gamma(-\frac{\nu}{2} + w)} \chi(w) z^w, w = s_k\right) \quad (z > 0).$$

9.2 Powers of zeta-functions

In this section we consider the functional equation of Γ^N-type (9.40) and derive the corresponding modular relation. Naturally, there are some results overlapping those in Chapter 3. We provide a few specific examples theorems of Berndt in [BeI] and [BeII], the square case overlapping some of the results in § 9.1.

9.2.1 *Statement of the Main Formula*

We assume that the given Dirichlet series $\varphi(s), \psi(s)$ satisfy the **functional equation of Γ^N-type**

$$\chi(s) = \begin{cases} \Gamma(s)^N \, \varphi(s), & \operatorname{Re}(s) > \sigma_\varphi \\ \Gamma(r - s)^N \, \psi(r - s), & \operatorname{Re}(s) < r - \sigma_\psi, \end{cases} \tag{9.40}$$

or

$$\chi(s) = \begin{cases} \Gamma\left(s \, \middle| \, \begin{matrix} - & ; & - \\ \underbrace{(0,1),\ldots,(0,1)}_{N} & ; & - \end{matrix}\right) \varphi(s), & \operatorname{Re}(s) > \sigma_\varphi \\[3em] \Gamma\left(r - s \, \middle| \, \begin{matrix} - & ; & - \\ \underbrace{(0,1),\ldots,(0,1)}_{N} & ; & - \end{matrix}\right) \psi(r - s), & \operatorname{Re}(s) < r - \sigma_\psi, \end{cases}$$

where $N = 1, 2, 3, \ldots$.

By Theorem 7.1, we have the following modular relation:

$$\sum_{k=1}^{\infty} \frac{\alpha_k}{\lambda_k^s} H\left(z\lambda_k \left| \left(\begin{array}{c} - \quad ; \quad - \\ \underbrace{(s,1),\ldots,(s,1)}_{N} ; \quad - \end{array} \right) \oplus \Delta \right. \right)$$

$$= \sum_{k=1}^{\infty} \frac{\beta_k}{\mu_k^{r-s}} H\left(\frac{\mu_k}{z} \left| \left(\begin{array}{c} - \quad ; \quad - \\ \underbrace{(r-s,1),\ldots,(r-s,1)}_{N} ; \quad - \end{array} \right) \oplus \Delta^* \right. \right) \quad (9.41)$$

$$+ \sum_{k=1}^{L} \operatorname{Res}\left(\Gamma(w-s \mid \Delta) \chi(w) z^{s-w}, w = s_k \right).$$

In the traditional notation, if

$$\Delta = \left(\begin{array}{cc} \{(1-a_j, A_j)\}_{j=1}^n; & \{(a_j, A_j)\}_{j=n+1}^p \\ \{(b_j, B_j)\}_{j=1}^m & ; \{(1-b_j, B_j)\}_{j=m+1}^q \end{array} \right) \in \Omega \quad (9.42)$$

then the associated Gamma factor is

$$\Gamma(s \mid \Delta) = \frac{\displaystyle\prod_{j=1}^{m} \Gamma(b_j + B_j s) \prod_{j=1}^{n} \Gamma(a_j - A_j s)}{\displaystyle\prod_{j=n+1}^{p} \Gamma(a_j + A_j s) \prod_{j=m+1}^{q} \Gamma(b_j - B_j s)}, \quad (9.43)$$

and we have

$$\sum_{k=1}^{\infty} \frac{\alpha_k}{\lambda_k^s} H_{p,q+N}^{m+N,n} \left(z\lambda_k \left| \begin{array}{c} \{(1-a_j,A_j)\}_{j=1}^{n}, \\ \underbrace{(s,1),\ldots,(s,1)}_{N}, \{(b_j,B_j)\}_{j=1}^{m}, \\ \{(a_j,A_j)\}_{j=n+1}^{p} \\ \{(1-b_j,B_j)\}_{j=m+1}^{q} \end{array} \right. \right)$$

$$= \sum_{k=1}^{\infty} \frac{\beta_k}{\mu_k^{r-s}} H_{q,p+N}^{n+N,m} \left(\frac{\mu_k}{z} \left| \begin{array}{c} \{(1-b_j,B_j)\}_{j=1}^{m}, \\ \underbrace{(r-s,1),\ldots,(r-s,1)}_{N}, \{(a_j,A_j)\}_{j=1}^{n}, \\ \{(b_j,B_j)\}_{j=m+1}^{q} \\ \{(1-a_j,A_j)\}_{j=n+1}^{p} \end{array} \right. \right)$$

$$+ \sum_{k=1}^{L} \mathrm{Res}\Big(\Gamma(w-s\,|\,\Delta)\,\chi(w)\,z^{s-w}, w=s_k \Big). \qquad (9.44)$$

In the special case where $A_j = B_j = 1$, we have

$$\sum_{k=1}^{\infty} \frac{\alpha_k}{\lambda_k^s} G_{p,q+N}^{m+N,n} \left(z\lambda_k \left| \begin{array}{c} 1-a_1,\ldots,1-a_n,a_{n+1},\ldots,a_p \\ \underbrace{s,\ldots,s}_{N},b_1,\ldots,b_m,1-b_{m+1},\ldots,1-b_q \end{array} \right. \right)$$

$$= \sum_{k=1}^{\infty} \frac{\beta_k}{\mu_k^{r-s}} G_{q,p+N}^{n+N,m} \left(\frac{\mu_k}{z} \left| \begin{array}{c} 1-b_1,\ldots,1-b_m,b_{m+1},\ldots,b_q \\ \underbrace{r-s,\ldots,r-s}_{N},a_1,\ldots,a_n,1-a_{n+1},\ldots,1-a_p \end{array} \right. \right)$$

$$+ \sum_{k=1}^{L} \mathrm{Res}\Big(\Gamma(w-s\,|\,\Delta)\,\chi(w)\,z^{s-w}, w=s_k \Big). \qquad (9.45)$$

For the $G_{0,N}^{N,0} \leftrightarrow G_{0,N}^{N,0}$ formula, cf. Exercise 3.1 and (3.5) in Chapter 3. We add a formula for the Voronoĭ function $E_N(x)$ defined by (3.6).

Exercise 9.1   By

$$\int_0^\infty G_{p,q}^{m,n}\left(tz\,\middle|\,\begin{matrix} a_1,\ldots,a_n,a_{n+1},\ldots,a_p \\ b_1,\ldots,b_m,b_{m+1},\ldots,b_q \end{matrix}\right)$$

$$\times\, G_{p',q'}^{m',n'}\left(\frac{1}{tz'}\,\middle|\,\begin{matrix} a'_1,\ldots,a'_{n'},a'_{n'+1},\ldots,a'_{p'} \\ b'_1,\ldots,b'_{m'},b'_{m'+1},\ldots,b'_{q'} \end{matrix}\right)\frac{dt}{t}$$

$$= G_{p+p',q+q'}^{m+m',n+n'}\left(\frac{z}{z'}\,\middle|\,\begin{matrix} a_1,\ldots,a_n,a'_1,\ldots,a'_{n'}, \\ b_1,\ldots,b_m,b'_1,\ldots,b'_{m'}, \end{matrix}\right.$$

$$\left.\begin{matrix} a_{n+1},\ldots,a_p,a'_{n'+1},\ldots,a'_{p'} \\ b_{m+1},\ldots,b_q,b'_{m'+1},\ldots,b'_{q'} \end{matrix}\right) \tag{9.46}$$

deduce

$$E_N(z) \tag{9.47}$$

$$= \int_0^\infty e^{-u_{N-1}}\left(\int_0^\infty e^{-u_{N-2}}\cdots\right.$$

$$\left(\int_0^\infty e^{-u_1}e^{-\frac{z}{u_1\cdots u_{N-2}u_{N-1}}}\frac{du_1}{u_1}\right)\cdots\left.\frac{du_{N-2}}{u_{N-2}}\right)\frac{du_{N-1}}{u_{N-1}}$$

for $N = 2,3,\ldots$, which is [BeI, Lemma 5].

Solution.   By (9.46), we have

$$E_N(z) = \int_0^\infty G_{0,1}^{1,0}\left(u\,\middle|\,\begin{matrix} - \\ 0 \end{matrix}\right) G_{0,N-1}^{N-1,0}\left(\frac{z}{u}\,\middle|\,\begin{matrix} - \\ \underbrace{0,\ldots,0}_{N-1} \end{matrix}\right)\frac{du}{u}$$

$$= \int_0^\infty e^{-u}E_{N-1}\left(\frac{z}{u}\right)\frac{du}{u}$$

by

$$E_1(z) = G_{0,1}^{1,0}\left(z\,\middle|\,\begin{matrix} - \\ 0 \end{matrix}\right) = e^{-z}.$$

Hence inductively,

$$E_N(z)$$

$$= \int_0^\infty e^{-u_{N-1}}\left(\int_0^\infty e^{-u_{N-2}}\cdots\right.$$

$$\left(\int_0^\infty e^{-u_1}E_1\left(\frac{z}{u_1\cdots u_{N-2}u_{N-1}}\right)\frac{du_1}{u_1}\right)\cdots\left.\frac{du_{N-2}}{u_{N-2}}\right)\frac{du_{N-1}}{u_{N-1}},$$

whence (9.47).

9.2.2 The $G_{N,N}^{N,0} \leftrightarrow G_{0,2N}^{N,0}$ formula

As an example of (9.45), we have

$$
\sum_{k=1}^{\infty} \frac{\alpha_k}{\lambda_k^s} \, G_{N,N}^{N,0}\left(z\lambda_k \, \middle| \, \begin{matrix} a_1, \ldots, a_N \\ \underbrace{s, \ldots, s}_{N} \end{matrix} \right)
$$

$$
= \sum_{k=1}^{\infty} \frac{\beta_k}{\mu_k^{r-s}} \, G_{0,2N}^{N,0}\left(\frac{\mu_k}{z} \, \middle| \, \begin{matrix} - \\ \underbrace{r-s, \ldots, r-s}_{N}, 1-a_1, \ldots, 1-a_N \end{matrix} \right) \tag{9.48}
$$

$$
+ \sum_{k=1}^{L} \operatorname{Res}\left(\frac{\chi(w)\, z^{s-w}}{\displaystyle\prod_{j=1}^{N} \Gamma\left(a_j - s + w\right)}, w = s_k \right).
$$

Example 9.1 (logarithmic Riesz sum) For $q = 0, 1, \ldots, N-1$, by setting

$$
a_1 = \cdots = a_{q+1} = s+1
$$

and

$$
a_{q+2} = \cdots = a_N = s
$$

in (9.48), we have

$$
\sum_{k=1}^{\infty} \frac{\alpha_k}{\lambda_k^s} \, G_{N,N}^{N,0}\left(z\lambda_k \, \middle| \, \begin{matrix} \overbrace{s+1, \ldots, s+1}^{q+1}, \overbrace{s, \ldots, s}^{N-q-1} \\ \underbrace{s, \ldots, s}_{N} \end{matrix} \right)
$$

$$
= \sum_{k=1}^{\infty} \frac{\beta_k}{\mu_k^{r-s}} \, G_{0,2N}^{N,0}\left(\frac{\mu_k}{z} \, \middle| \, \begin{matrix} - \\ \underbrace{r-s, \ldots, r-s}_{N}, \underbrace{-s, \ldots, -s}_{q+1}, \underbrace{1-s, \ldots, 1-s}_{N-q-1} \end{matrix} \right)
$$

$$
+ \sum_{k=1}^{L} \operatorname{Res}\left(\frac{\chi(w)\, z^{s-w}}{\Gamma(w+1)^{q+1}\Gamma(w)^{N-q-1}}, w = s_k \right) \quad (z > 0). \tag{9.49}
$$

To express this in a more concrete form we use

$$G_{m,m}^{m,0}\left(z\left|\begin{array}{c}s+1,\ldots,s+1\\s,\ldots,s\end{array}\right.\right) \tag{9.50}$$

$$=\begin{cases}\dfrac{z^s}{(m-1)!}\log^{m-1}\left(\dfrac{1}{z}\right), & |z|<1\\[2mm]\dfrac{1}{2}, & z=1, m=1\\[2mm]0, & |z|>1.\end{cases}$$

Then the left-hand side becomes the logarithmic Riesz sum

$$G_{N,N}^{N,0}\left(z\left|\begin{array}{c}\overbrace{s+1,\ldots,s+1}^{q+1},\overbrace{s,\ldots,s}^{N-q-1}\\\underbrace{s,\ldots,s}_{N}\end{array}\right.\right)$$

$$=G_{q+1,q+1}^{q+1,0}\left(z\left|\begin{array}{c}\overbrace{s+1,\ldots,s+1}^{q+1}\\\underbrace{s,\ldots,s}_{q+1}\end{array}\right.\right)=\begin{cases}\dfrac{z^s}{q!}\log^q\left(\dfrac{1}{z}\right), & |z|<1\\[2mm]0, & |z|>1.\end{cases} \tag{9.51}$$

The right-hand side becomes

$$G_{0,2N}^{N,0}\left(z\left|\begin{array}{c}-\\\underbrace{a,\ldots,a}_{N},\underbrace{b,\ldots,b}_{n},\underbrace{c,\ldots,c}_{N-n}\end{array}\right.\right)$$

$$=\sqrt{z}^{a+b}G_{0,2N}^{N,0}\left(z\left|\begin{array}{c}-\\\underbrace{\dfrac{a-b}{2},\ldots,\dfrac{a-b}{2}}_{N},\underbrace{-\dfrac{a-b}{2},\ldots,-\dfrac{a-b}{2}}_{n},\end{array}\right.\right.$$

$$\left.\underbrace{\dfrac{a-b}{2}-(a-c),\ldots,\dfrac{a-b}{2}-(a-c)}_{N-n}\right)$$

$$=2^{(N-n)(b-c+1)-n+1}\sqrt{z}^{a+b}K_{a-b}\left(2^N\sqrt{z};a-c;N,n\right), \tag{9.52}$$

by (9.54) below.

By writing z by $\frac{1}{z}$ in the resulting formula, we deduce

$$\frac{1}{q!}\sum_{\lambda_k<z}\alpha_k\log^q\left(\frac{z}{\lambda_k}\right) = 2^q\sum_{k=1}^{\infty}\beta_k\left(\frac{z}{\mu_k}\right)^{\frac{r}{2}}K_r\left(2^N\sqrt{z\mu_k}\,;r-1;N,q+1\right)$$

$$+\sum_{k=1}^{L}\mathrm{Res}\left(\frac{\chi(w)\,z^w}{w^{q+1}\Gamma(w)^N},w=s_k\right), \qquad (9.53)$$

which is [BeI, Theorem 1 (i)].

Exercise 9.2 Recall Berndt's generalized Bessel function defined by

$$K_\nu\bigl(z;\mu;N,n\bigr)$$
$$= 2^{-(N-n)(\mu-\nu+1)-n+1}$$
$$\times G_{0,2N}^{N,0}\left(\left(\frac{z}{2^N}\right)^2\;\middle|\;\underbrace{\frac{\nu}{2},\dots,\frac{\nu}{2}}_{N},\overset{\displaystyle-}{\underbrace{-\frac{\nu}{2},\dots,-\frac{\nu}{2}}_{n}},\underbrace{\frac{\nu}{2}-\mu,\dots,\frac{\nu}{2}-\mu}_{N-n}\right). \qquad (9.54)$$

Cf. (3.7) in Chapter 3. Similarly to Exercise 9.1 prove the following.

$$K_\nu(z;\mu;N,n)$$
$$= \int_0^\infty u_{N-1}^{\nu-\mu}J_\mu(u_{N-1})\left(\int_0^\infty u_{N-2}^{\nu-\mu}J_\mu(u_{N-2})\cdots\right.$$
$$\times\left(\int_0^\infty u_n^{\nu-\mu}J_\mu(u_n)K_\nu\left(\frac{z}{u_n\cdots u_{N-2}u_{N-1}};\mu;n,n\right)\frac{du_n}{u_n}\right) \qquad (9.55)$$
$$\left.\cdots\frac{du_{N-2}}{u_{N-2}}\right)\frac{du_{N-1}}{u_{N-1}}.$$

Solution By (2.110), we have

$$K_\nu\bigl(z;\mu;1,0\bigr) = G_{0,2}^{1,0}\left(\left(\frac{z}{2}\right)^2\;\middle|\;\overset{\displaystyle-}{\tfrac{\nu}{2},\tfrac{\nu}{2}-\mu}\right) = \left(\frac{z}{2}\right)^{\nu-\mu}J_\mu(z), \qquad (9.56)$$

$$K_\nu\bigl(z;\mu;1,1\bigr) = G_{0,2}^{1,0}\left(\left(\frac{z}{2}\right)^2\;\middle|\;\overset{\displaystyle-}{\tfrac{\nu}{2},-\tfrac{\nu}{2}}\right) = J_\nu(z). \qquad (9.57)$$

As in Exercise 9.1, it follows from (9.46) that for $N \geq n+1$

$$
K_\nu\big(z; \mu; N, n\big)
$$

$$
= 2^{-(N-n)(\mu-\nu+1)-n+1} \int_0^\infty G_{0,2}^{1,0}\left(\frac{t}{2^2} \,\middle|\, \begin{matrix} - \\ \frac{\nu}{2}, \frac{\nu}{2} - \mu \end{matrix} \right)
$$

$$
\times G_{0,2N-2}^{N-1,0}\left(\frac{1}{t}\left(\frac{z}{2^{N-1}}\right)^2 \,\middle|\, \begin{matrix} - \\ \underbrace{\frac{\nu}{2}, \dots, \frac{\nu}{2}}_{N-1}, \underbrace{-\frac{\nu}{2}, \dots, -\frac{\nu}{2}}_{n}, \underbrace{\frac{\nu}{2} - \mu, \dots, \frac{\nu}{2} - \mu}_{N-n-1} \end{matrix} \right) \frac{dt}{t}
$$

$$
= \int_0^\infty 2^{\nu-\mu} G_{0,2}^{1,0}\left(\left(\frac{u}{2}\right)^2 \,\middle|\, \begin{matrix} - \\ \frac{\nu}{2}, \frac{\nu}{2} - \mu \end{matrix} \right) 2^{-(N-n-1)(\mu-\nu+1)-n+1}
$$

$$
\times G_{0,2N-2}^{N-1,0}\left(\left(\frac{z}{2^{N-1}u}\right)^2 \,\middle|\, \begin{matrix} - \\ \underbrace{\frac{\nu}{2}, \dots, \frac{\nu}{2}}_{N-1}, \underbrace{-\frac{\nu}{2}, \dots, -\frac{\nu}{2}}_{n}, \underbrace{\frac{\nu}{2} - \mu, \dots, \frac{\nu}{2} - \mu}_{N-n-1} \end{matrix} \right) \frac{du}{u}
$$

$(t = u^2)$. Using (9.56), we get

$$
K_\nu\big(z; \mu; N, n\big) = \int_0^\infty u^{\nu-\mu} J_\mu(u)\, K_\nu\left(\frac{z}{u}; \mu; N-1, n\right) \frac{du}{u}.
$$

Exercise 9.3   Now, by the same reasoning as above, using (9.46) and (9.57), we have

$$
K_\nu\big(z; \mu; n, n\big) = 2^{-n+1} G_{0,2n}^{n,0}\left(\left(\frac{z}{2^n}\right)^2 \,\middle|\, \begin{matrix} - \\ \underbrace{\frac{\nu}{2}, \dots, \frac{\nu}{2}}_{n}, \underbrace{-\frac{\nu}{2}, \dots, -\frac{\nu}{2}}_{n} \end{matrix} \right)
$$

$$
= \int_0^\infty G_{0,2}^{1,0}\left(\left(\frac{u}{2}\right)^2 \,\middle|\, \begin{matrix} - \\ \frac{\nu}{2}, -\frac{\nu}{2} \end{matrix} \right)
$$

$$
\times 2^{-n+2} G_{0,2n-2}^{n-1,0}\left(\left(\frac{z}{2^{n-1}u}\right)^2 \,\middle|\, \begin{matrix} - \\ \underbrace{\frac{\nu}{2}, \dots, \frac{\nu}{2}}_{n-1}, \underbrace{-\frac{\nu}{2}, \dots, -\frac{\nu}{2}}_{n-1} \end{matrix} \right) \frac{du}{u}
$$

$$
= \int_0^\infty J_\nu(u)\, K_\nu\left(\frac{z}{u}; \mu; n-1, n-1\right) \frac{du}{u}.
$$

Hence, inductively,

$$
K_\nu\big(z; \mu; n, n\big)
$$

$$
= \int_0^\infty J_\nu(u_{n-1}) \bigg(\int_0^\infty J_\nu(u_{n-2}) \cdots
$$

$$
\times \bigg(\int_0^\infty J_\nu(u_1) K_\nu\bigg(\frac{z}{u_1 \cdots u_{n-2} u_{n-1}}; \mu; 1, 1 \bigg) \frac{\mathrm{d}u_1}{u_1} \bigg) \cdots \frac{\mathrm{d}u_{n-2}}{u_{n-2}} \bigg) \frac{\mathrm{d}u_{n-1}}{u_{n-1}}
$$

$$
= \int_0^\infty J_\nu(u_{n-1}) \bigg(\int_0^\infty J_\nu(u_{n-2}) \cdots
$$

$$
\times \bigg(\int_0^\infty J_\nu(u_1) J_\nu\bigg(\frac{z}{u_1 \cdots u_{n-2} u_{n-1}} \bigg) \frac{\mathrm{d}u_1}{u_1} \bigg) \cdots \frac{\mathrm{d}u_{n-2}}{u_{n-2}} \bigg) \frac{\mathrm{d}u_{n-1}}{u_{n-1}} \qquad (9.58)
$$

on appealing to (9.57). Substitution of (9.58) in (9.55) leads to Berndt's Definition 4 ([BeI]).

Example 9.2 By setting $a_1 = 1$ and $a_2 = \cdots = a_N = s$ in (9.48), we have

$$
\sum_{k=1}^\infty \frac{\alpha_k}{\lambda_k^s} \, G_{N,N}^{N,0}\left(z\lambda_k \,\middle|\, \begin{matrix} \overbrace{1, s, \ldots, s}^{N-1} \\ \underbrace{s, \ldots, s}_{N} \end{matrix} \right)
$$

$$
= \sum_{k=1}^\infty \frac{\beta_k}{\mu_k^{r-s}} \, G_{0,2N}^{N,0}\left(\frac{\mu_k}{z} \,\middle|\, \begin{matrix} - \\ \underbrace{r-s, \ldots, r-s}_{N}, 0, \underbrace{1-s, \ldots, 1-s}_{N-1} \end{matrix} \right) \qquad (9.59)
$$

$$
+ \sum_{k=1}^L \mathrm{Res}\bigg(\frac{\chi(w) \, z^{s-w}}{\Gamma(1-s+w)\,\Gamma(w)^{N-1}}, w = s_k \bigg) \qquad (z > 0).
$$

We have

$$
G_{N,N}^{N,0}\left(z \,\middle|\, \begin{matrix} \overbrace{1, s, \ldots, s}^{N-1} \\ \underbrace{s, \ldots, s}_{N} \end{matrix} \right) = G_{1,1}^{1,0}\left(z \,\middle|\, \begin{matrix} 1 \\ s \end{matrix} \right)
$$

$$
(9.60)
$$

$$
= \begin{cases} \dfrac{1}{\Gamma(1-s)} \, z^s (1-z)^{-s}, & |z| < 1 \\[2mm] 0, & |z| > 1 \end{cases}
$$

by (6.5), and

$$
G_{0,2N}^{N,0}\left(z \left| \underbrace{a,\ldots,a}_{N}, \overset{-}{b}, \underbrace{c,\ldots,c}_{N-1} \right. \right)
$$
$$
= 2^{(N-1)(b-c+1)} \sqrt{z}^{\,a+b}\, K_{a-b}\left(2^N \sqrt{z}\,; a-c; N; 1\right) \tag{9.61}
$$

by (9.52). Therefore, setting

$$
K_\nu\big(z; \mu; N\big) = K_\nu\big(z; \mu; N, 1\big), \tag{9.62}
$$

we have

$$
\frac{z^s}{\Gamma(1-s)} \sum_{\lambda_k < \frac{1}{z}} \alpha_k \big(1 - z\,\lambda_k\big)^{-s}
$$
$$
= 2^{(N-1)s} \sum_{k=1}^{\infty} \frac{\beta_k}{\mu_k^{r-s}} \sqrt{\frac{\mu_k}{z}}^{\,r-s} K_{r-s}\left(2^N \sqrt{\frac{\mu_k}{z}}\,; r-1; N\right)
$$
$$
+ \sum_{k=1}^{L} \mathrm{Res}\left(\frac{\chi(w)\, z^{s-w}}{\Gamma(1-s+w)\,\Gamma(w)^{N-1}}, w = s_k\right).
$$

Replacing z by $\frac{1}{z}$, we get Berndt's Theorem 2 in [Be1]:

$$
\frac{1}{\Gamma(1-s)} \sum_{\lambda_k < z} \alpha_k \big(z - \lambda_k\big)^{-s}
$$
$$
= 2^{(N-1)s} \sum_{k=1}^{\infty} \beta_k \sqrt{\frac{z}{\mu_k}}^{\,r-s} K_{r-s}\left(2^N \sqrt{z\mu_k}\,; r-1; N\right) \tag{9.63}
$$
$$
+ \sum_{k=1}^{L} \mathrm{Res}\left(\frac{\chi(w)\, z^{w-s}}{\Gamma(1-s+w)\,\Gamma(w)^{N-1}}, w = s_k\right).
$$

9.2.3 The $G_{q+1,q+1}^{q+1,0} \leftrightarrow G_{q-N+1,q+N+1}^{N,q-N+1}$ formula

For $q = N, N+1, \ldots,$ as an example of (9.45), we have

$$\sum_{k=1}^{\infty} \frac{\alpha_k}{\lambda_k^s} G_{q+1,q+1}^{q+1,0} \left(z\lambda_k \left| \begin{array}{c} a_1, \ldots, a_{q+1} \\ \underbrace{s, \ldots, s}_{N}, b_1, \ldots, b_{q-N+1} \end{array} \right. \right)$$

$$= \sum_{k=1}^{\infty} \frac{\beta_k}{\mu_k^{r-s}} G_{q-N+1,q+N+1}^{N,q-N+1} \left(\frac{\mu_k}{z} \left| \begin{array}{c} 1-b_1, \ldots, 1-b_{q-N+1} \\ \underbrace{r-s, \ldots, r-s}_{N}, 1-a_1, \ldots, 1-a_{q+1} \end{array} \right. \right)$$

$$+ \sum_{k=1}^{L} \operatorname{Res}\left(\frac{\prod\limits_{j=1}^{q-N+1} \Gamma(b_j + w - s)}{\prod\limits_{j=1}^{q+1} \Gamma(a_j + w - s)} \chi(w) \, z^{s-w}, w = s_k \right). \qquad (9.64)$$

By setting $a_1 = \cdots = a_{q+1} = s+1$ and $b_1 = \cdots = b_{q-N+1} = s$, we have

$$\sum_{k=1}^{\infty} \frac{\alpha_k}{\lambda_k^s} G_{q+1,q+1}^{q+1,0} \left(z\lambda_k \left| \begin{array}{c} \overbrace{s+1, \ldots, s+1}^{q+1} \\ \underbrace{s, \ldots, s}_{q+1} \end{array} \right. \right)$$

$$= \sum_{k=1}^{\infty} \frac{\beta_k}{\mu_k^{r-s}} G_{q-N+1,q+N+1}^{N,q-N+1} \left(\frac{\mu_k}{z} \left| \begin{array}{c} \overbrace{1-s, \ldots, 1-s}^{q-N+1} \\ \underbrace{r-s, \ldots, r-s}_{N}, \underbrace{-s, \ldots, -s}_{q+1} \end{array} \right. \right) \qquad (9.65)$$

$$+ \sum_{k=1}^{L} \operatorname{Res}\left(\frac{\Gamma(w)^{q-N+1}}{\Gamma(w+1)^{q+1}} \chi(w) \, z^{s-w}, w = s_k \right).$$

When $s = \frac{r}{2}$, we have

$$
\sum_{k=1}^{\infty} \frac{\alpha_k}{\lambda_k^{\frac{r}{2}}} \, G_{q+1,q+1}^{q+1,0}\left(z\lambda_k \,\middle|\, \overset{\overbrace{q+1}}{\frac{r}{2}+1,\ldots,\frac{r}{2}+1}\atop \underset{q+1}{\underbrace{\frac{r}{2},\ldots,\frac{r}{2}}} \right)
$$

$$
= \sum_{k=1}^{\infty} \frac{\beta_k}{\mu_k^{\frac{r}{2}}} \, G_{q-N+1,q+N+1}^{N,q-N+1}\left(\frac{\mu_k}{z} \,\middle|\, \overset{\overbrace{q-N+1}}{1-\frac{r}{2},\ldots,1-\frac{r}{2}}\atop \underset{N}{\underbrace{\frac{r}{2},\ldots,\frac{r}{2}}},\underset{q+1}{\underbrace{-\frac{r}{2},\ldots,-\frac{r}{2}}} \right) \tag{9.66}
$$

$$
+ \sum_{k=1}^{L} \mathrm{Res}\left(\frac{\Gamma(w)^{q-N+1}}{\Gamma(w+1)^{q+1}} \, \chi(w) \, z^{\frac{r}{2}-w}, w = s_k \right).
$$

Using (9.46) with (9.52) and (9.50), we see that

$$
G_{q-N+1,q+N+1}^{N,q-N+1}\left(z \,\middle|\, \overset{\overbrace{q-N+1}}{1-\frac{r}{2},\ldots,1-\frac{r}{2}}\atop \underset{N}{\underbrace{\frac{r}{2},\ldots,\frac{r}{2}}},\underset{q+1}{\underbrace{-\frac{r}{2},\ldots,-\frac{r}{2}}} \right)
$$

$$
= \int_0^{\infty} G_{0,2N}^{N,0}\left(\frac{t}{2^{2N}} \,\middle|\, \overset{-}{\underset{N}{\underbrace{\frac{r}{2},\ldots,\frac{r}{2}}},\underset{N}{\underbrace{-\frac{r}{2},\ldots,-\frac{r}{2}}}} \right)
$$

$$
\times\, G_{q-N+1,q-N+1}^{0,q-N+1}\left(\frac{2^{2N}z}{t} \,\middle|\, \overset{\overbrace{q-N+1}}{1-\frac{r}{2},\ldots,1-\frac{r}{2}}\atop \underset{q-N+1}{\underbrace{-\frac{r}{2},\ldots,-\frac{r}{2}}} \right) \frac{dt}{t}
$$

$$
= 2^q \int_0^{\infty} 2^{-N+1} G_{0,2N}^{N,0}\left(\left(\frac{u}{2^N}\right)^2 \,\middle|\, \overset{-}{\underset{N}{\underbrace{\frac{r}{2},\ldots,\frac{r}{2}}},\underset{N}{\underbrace{-\frac{r}{2},\ldots,-\frac{r}{2}}}} \right)
$$

$$
\times \frac{1}{2^{q-N}} G_{q-N+1,q-N+1}^{0,q-N+1}\left(\left(\frac{2^N\sqrt{z}}{u}\right)^2 \,\middle|\, \overset{\overbrace{q-N+1}}{1-\frac{r}{2},\ldots,1-\frac{r}{2}}\atop \underset{q-N+1}{\underbrace{-\frac{r}{2},\ldots,-\frac{r}{2}}} \right) \frac{du}{u}
$$

$$
= 2^q \int_0^{2^N\sqrt{z}} K_r\left(u;r;N\right) \frac{1}{(q-N)!} \left(\frac{2^N\sqrt{z}}{u}\right)^{-r} \log^{q-N}\!\left(\frac{2^N\sqrt{z}}{u}\right) \frac{du}{u}.
$$

Hence using also (9.51), we have

$$\frac{1}{q!} \sum_{\lambda_k < \frac{1}{z}} \alpha_k \log^q \left(\frac{1}{z\lambda_k} \right)$$

$$= 2^{q-Nr} \sum_{k=1}^{\infty} \frac{\beta_k}{\mu_k^r} \int_0^{2^N \sqrt{\frac{\mu_k}{z}}} u^{r-1} K_r\left(u; r; N\right) \frac{1}{(q-N)!} \log^{q-N} \left(\frac{2^N \sqrt{\frac{\mu_k}{z}}}{u} \right) du$$

$$+ \sum_{k=1}^{L} \operatorname{Res} \left(\frac{\Gamma(w)^{q-N+1}}{\Gamma(w+1)^{q+1}} \chi(w) \, z^{-w}, w = s_k \right),$$

and the substitution $\frac{1}{z}$ for z, yields Berndt's Theorem 3 ([BeII]) on the **logarithmic Riesz sum**:

$$\frac{1}{q!} \sum_{\lambda_k < z} \alpha_k \log^q \left(\frac{z}{\lambda_k} \right)$$

$$= 2^{q-Nr} \sum_{k=1}^{\infty} \frac{\beta_k}{\mu_k^r} \int_0^{2^N \sqrt{z\mu_k}} u^{r-1} K_r\left(u; r; N\right) \frac{1}{(q-N)!} \log^{q-N} \left(\frac{2^N \sqrt{z\mu_k}}{u} \right) du$$

$$+ \sum_{k=1}^{L} \operatorname{Res} \left(\frac{\Gamma(w)^{q-N+1}}{\Gamma(w+1)^{q+1}} \chi(w) \, z^{w}, w = s_k \right), \tag{9.67}$$

complementing earlier results e.g. [Opp] on $d(n)$.

Fig. 9.1 Carl Friedlich Gauss

Chapter 10

Miscellany

We assemble here miscellaneous remarks on the modular relation which are of interest in their own right. Some of them are still in progress and we will be rather speculative.

10.1 Future projects

10.1.1 Rankin-Selberg convolution

Let $\varphi(z)$ denote a holomorphic cusp form of weight κ with respect to the full modular group, and let $c(n)$ the n-th Fourier coefficients. Further suppose that $\varphi(z)$ is a normalized Hecke eigenform, i.e. $c(1) = 1$, $T(n)\varphi = c(n)\varphi$ for all $n \in \mathbb{N}$, where $T(n)$ indicates the Hecke operator. Let c_n be the convolution defined by

$$c_n = n^{1-\kappa} \sum_{d^2 \mid n} d^{2(\kappa-1)} \left| c\left(\frac{n}{d^2}\right) \right|^2. \tag{10.1}$$

Then the generating function defined by

$$L(s, f \otimes f) = \sum_{k=1}^{\infty} \frac{c_k}{k^s} \tag{10.2}$$

is absolutely convergent for $\sigma > 1$ by Deligne's estimate and satisfies the functional equation

$$\Gamma(s + \kappa - 1)\Gamma(s)Z(s) = (2\pi)^s \Gamma(\kappa - s)\Gamma(1 - s)Z(1 - s). \tag{10.3}$$

$Z(s) = L(s, f \otimes \bar{f})$ is the Rankin-Selberg square, which is a special case of the Rankin-Selberg convolution or L-function. This class of L-functions will surely turn out to be a rich source of functional equations. Cf. e.g. [IcM], [Iv], [WLi] [Ra], [Shm2], [Zag] etc.. Cf. §3.3.4.

We digress a little and refer to the recent MS [RZZ]. Let

$$F(s, \chi) = \frac{L(s, \chi)}{\zeta^2(s)} = \sum_{n=1}^{\infty} \frac{b(n)}{n^s}, \tag{10.4}$$

so that $b = \chi * \mu * \mu$. From $F(s, \bar{\chi})$, we similarly obtain $\tilde{b} = \bar{\chi} * \mu * \mu$. They satisfy the functional equation

$$\Gamma\left(\frac{1-s}{2}\right) F(s, \chi) = \sqrt{\pi} \epsilon_\chi (\pi q)^{-s} \Gamma\left(\frac{s}{2}\right) F(1-s, \bar{\chi}) \tag{10.5}$$

with the constant $|\epsilon_\chi| = 1$. Thus this falls under the category of the zeta-function satisfying the functional equations with A < 0 ([ChN3, p. 112]), where only omega results are obtained. The authors [RZZ] went further to obtain a modular relation with the infinite sum of residues. As can be expected, the main result is the modular relations between two Lambert series.

We briefly mention a result of [LiM] which is a generalization of [HS] and [RZZ]. With the data

$$\lambda_n = \mu_n = 4\pi n, \tag{10.6}$$

$$\alpha_n = c^2(n), \beta_n = \frac{1}{\sqrt{2\pi}} \sum_{d_1^2 d_2^2 d_3} d_1^{2k-2} d_2^{2k-1} \mu(d_2) c^2(d_3)$$

let

$$\varphi(s) = \sum_{n=1}^{\infty} \frac{\alpha_n}{\lambda_n^s}, \psi(s) = \sum_{n=1}^{\infty} \frac{\beta_n}{\mu_n^s}. \tag{10.7}$$

Then the functional equation takes the form

$$\Gamma(s)\varphi(s) = \frac{\Gamma(2k-1-s)\Gamma(k-s)}{\Gamma\left(\frac{2k-1}{2}-s\right)} \frac{\zeta(2(k-s))}{\zeta(2k-1-2s)} \psi(2k-1-s), \tag{10.8}$$

which is a variant of Rankin's result [RR].

Since the second factor on the right of (10.8) can be expanded into a Dirichlet series absolutely convergent for $\sigma < 1 - k$, we may appeal to the modular relation with

$$G_{1,2}^{2,0}\left(z \left| \begin{matrix} k - \frac{1}{2} \\ 2k-1, k \end{matrix} \right.\right) = e^{-z} z^{2k-1} U\left(-\frac{1}{2}, k, z\right), \tag{10.9}$$

provided that the infinite sum of the residues at non-trivial zeros of the Riemann zeta-function is made meaningful. Here naturally we need to extend the modular relation to the case where there are infinitely many poles in the critical strip and is postponed in the coming volumes.

10.1.2 *Maass forms*

For more details of the theory of Maass forms we refer to [Bak], [JYL], [MaTIFR], [Mot], [Meu], etc. For weak harmonic Maass forms cf. [BOn]. We show that [MaTIFR] can be interpreted as a modular relation as follows.

Theorem 10.1 *Assume the functional equation*

$$\chi(s) \tag{10.10}$$

$$= \begin{cases} G_{2,2}^{2,2}\left(\frac{1}{2}\,\bigg|\,\begin{matrix}1,a+b\\s,b\end{matrix}\right)\displaystyle\sum_{k=1}^{\infty}\frac{\alpha_k}{\lambda_k^s} + G_{2,2}^{2,2}\left(\frac{1}{2}\,\bigg|\,\begin{matrix}1,a+b\\s,a\end{matrix}\right)\displaystyle\sum_{k=1}^{\infty}\frac{\alpha_k'}{\lambda_k'^{\,s}}, \\[4mm] G_{2,2}^{2,2}\left(\frac{1}{2}\,\bigg|\,\begin{matrix}1,a+b\\r-s,b\end{matrix}\right)\displaystyle\sum_{k=1}^{\infty}\frac{\beta_k}{\mu_k^{r-s}} + G_{2,2}^{2,2}\left(\frac{1}{2}\,\bigg|\,\begin{matrix}1,a+b\\r-s,a\end{matrix}\right)\displaystyle\sum_{k=1}^{\infty}\frac{\beta_k'}{\mu_k'^{\,r-s}}. \end{cases}$$

Let **X** *denote the key-function*

$$X(z) = \frac{1}{2\pi i}\int_L \chi(w)\,z^{-w}\mathrm{d}w. \tag{10.11}$$

Then we have the modular relation

$$X(z) = \begin{cases} \displaystyle\sum_{k=1}^{\infty}\alpha_k G_{1,2}^{2,1}\left(2z\lambda_k\,\bigg|\,\begin{matrix}1-b\\0,1-a-b\end{matrix}\right)e^{-z\lambda_k} \\[2mm] \quad + \displaystyle\sum_{k=1}^{\infty}\alpha_k' G_{1,2}^{2,1}\left(2z\lambda_k'\,\bigg|\,\begin{matrix}1-a\\0,1-a-b\end{matrix}\right)e^{-z\lambda_k'}, \\[5mm] z^{-r}\displaystyle\sum_{k=1}^{\infty}\beta_k G_{1,2}^{2,1}\left(\frac{2\mu_k}{z}\,\bigg|\,\begin{matrix}1-b\\0,1-a-b\end{matrix}\right)e^{-\frac{\mu_k}{z}} \\[2mm] \quad + z^{-r}\displaystyle\sum_{k=1}^{\infty}\beta_k' G_{1,2}^{2,1}\left(\frac{2\mu_k'}{z}\,\bigg|\,\begin{matrix}1-a\\0,1-a-b\end{matrix}\right)e^{-\frac{\mu_k'}{z}} + \text{Res}. \end{cases} \tag{10.12}$$

In Maass's notation

$$\Gamma(s;a,b) = \frac{2^{\frac{1}{2}(a+b)}}{\Gamma(b)\Gamma(1-a)}\,G_{2,2}^{2,2}\left(\frac{1}{2}\,\bigg|\,\begin{matrix}1,a+b\\s,b\end{matrix}\right) \tag{10.13}$$

$$W(z; a, b) = \frac{2^{\frac{1}{2}(a+b)}}{\Gamma(b)\Gamma(1-a)} G_{1,2}^{2,1}\left(2z \left| \begin{matrix} 1-b \\ 0, 1-a-b \end{matrix} \right.\right) e^{-z}. \tag{10.14}$$

With this notation, Theorem 10.1 asserts that

$$\chi(s) = \begin{cases} \Gamma(s; a, b) \displaystyle\sum_{k=1}^{\infty} \frac{\alpha_k}{\lambda_k^s} + \Gamma(s, b, a) \sum_{k=1}^{\infty} \frac{\alpha_k'}{\lambda_k'^{\,s}}, \\[3mm] \Gamma(r-s; a, b) \displaystyle\sum_{k=1}^{\infty} \frac{\beta_k}{\mu_k^{r-s}} + \Gamma(r-s; b, a) \sum_{k=1}^{\infty} \frac{\beta_k'}{\mu_k'^{\,r-s}}. \end{cases} \tag{10.15}$$

$$X(z) \tag{10.16}$$
$$= \begin{cases} \displaystyle\sum_{k=1}^{\infty} \alpha_k W(z\lambda_k; a, b) + \sum_{k=1}^{\infty} \alpha_k' W(z\lambda_k'; b, a), \\[3mm] \displaystyle z^{-r} \sum_{k=1}^{\infty} \beta_k W\left(\frac{\mu_k}{z}; a, b\right) + z^{-r} \sum_{k=1}^{\infty} \beta_k' W\left(\frac{\mu_k'}{z}; b, a\right) + \text{Res}. \end{cases}$$

10.1.3 *G-functions of two variables*

To deduce Berndt [BeIII, (9.1)] ([BK, Theorem 8.7, p. 124]) in a natural way from the functional equation we need G-function of two variables whose theory will be developed subsequently in coming volumes. Cf. [NgY].

Definition 10.1 We define G-functions of two variables by

$$G_{0,1;1,0;0,1}^{1,0;0,1;1,0}\left(\begin{matrix} a^{-1}z \\ bz \end{matrix} \left| \begin{matrix} - \\ \nu \end{matrix} \right| \begin{matrix} 1 \\ - \end{matrix} \left| \begin{matrix} - \\ 0 \end{matrix} \right.\right)$$
$$= \frac{1}{2\pi i} \int_{L_2} \left(\frac{1}{2\pi i} \int_{L_1} \Gamma(\nu+s+t)\, \Gamma(-s)\, \Gamma(t) \left(\frac{z}{a}\right)^{-s} (bz)^{-t}\, ds \right) dt \tag{10.17}$$
$$= \frac{1}{2\pi i} \int_{L_2} \left(\frac{1}{2\pi i} \int_{L_1} \Gamma(\nu+s+t)\, \Gamma(-s)\, \Gamma(t)\, z^{-s-t} a^s b^{-t} ds \right) dt$$

or by

$$G_{0,1;1,0;0,1}^{1,0;0,1;1,0}\left(\begin{matrix} z \\ w \end{matrix} \left| \begin{matrix} - \\ \nu \end{matrix} \right| \begin{matrix} 1 \\ - \end{matrix} \left| \begin{matrix} - \\ 0 \end{matrix} \right.\right)$$
$$= \int_0^{\infty} G_{0,1}^{1,0}\left(\frac{1}{x} \left| \begin{matrix} - \\ \nu \end{matrix} \right.\right) G_{1,0}^{0,1}\left(zx \left| \begin{matrix} 1 \\ - \end{matrix} \right.\right) G_{0,1}^{1,0}\left(wx \left| \begin{matrix} - \\ 0 \end{matrix} \right.\right) \frac{dx}{x}. \tag{10.18}$$

Exercise 10.1

$$G_{0,1;0,1;0,1}^{1,0;1,0;1,0}\left(\begin{array}{c}z\\w\end{array}\left|\begin{array}{c}-\\\nu\end{array}\right|\begin{array}{c}-\\0\end{array}\left|\begin{array}{c}-\\0\end{array}\right.\right)$$

$$= \int_0^\infty G_{0,1}^{1,0}\left(\frac{1}{x}\left|\begin{array}{c}-\\\nu\end{array}\right.\right) G_{0,1}^{1,0}\left(zx\left|\begin{array}{c}-\\0\end{array}\right.\right) G_{0,1}^{1,0}\left(wx\left|\begin{array}{c}-\\0\end{array}\right.\right)\frac{dx}{x}$$

$$= \int_0^\infty x^{-\nu}\exp\left(-\frac{1}{x}\right)\exp(-zx)\exp(-wx)\frac{dx}{x} \qquad (10.19)$$

$$= \int_0^\infty x^{\nu}\exp(-x)\exp\left(-\frac{z}{x}\right)\exp\left(-\frac{w}{x}\right)\frac{dx}{x}$$

$$= 2\left(\frac{1}{z+w}\right)^{-\frac{\nu}{2}}K_\nu\left(2\sqrt{z+w}\right)$$

$$G_{0,1;1,0;0,1}^{1,0;0,1;1,0}\left(\begin{array}{c}z\\w\end{array}\left|\begin{array}{c}-\\\nu\end{array}\right|\begin{array}{c}1\\-\end{array}\left|\begin{array}{c}-\\0\end{array}\right.\right)$$

$$= \int_0^\infty G_{0,1}^{1,0}\left(\frac{1}{x}\left|\begin{array}{c}-\\\nu\end{array}\right.\right) G_{1,0}^{0,1}\left(zx\left|\begin{array}{c}1\\-\end{array}\right.\right) G_{0,1}^{1,0}\left(wx\left|\begin{array}{c}-\\0\end{array}\right.\right)\frac{dx}{x}$$

$$= \int_0^\infty G_{0,1}^{1,0}\left(\frac{1}{x}\left|\begin{array}{c}-\\\nu\end{array}\right.\right) G_{0,1}^{1,0}\left(\frac{1}{zx}\left|\begin{array}{c}-\\0\end{array}\right.\right) G_{0,1}^{1,0}\left(wx\left|\begin{array}{c}-\\0\end{array}\right.\right)\frac{dx}{x}$$

$$= \int_0^\infty x^{-\nu}\exp\left(-\frac{1}{x}\right)\exp\left(-\frac{1}{zx}\right)\exp(-wx)\frac{dx}{x} \qquad (10.20)$$

$$= \int_0^\infty x^{\nu}\exp(-x)\exp\left(-\frac{x}{z}\right)\exp\left(-\frac{w}{x}\right)\frac{dx}{x}$$

$$= 2\left(\frac{1+z^{-1}}{w}\right)^{-\frac{\nu}{2}}K_\nu\left(2\sqrt{(1+z^{-1})\,w}\right).$$

Theorem 10.2   *Assuming a functional equation of Hecke type*

$$\Gamma(s)\,\varphi(s) = \Gamma(r-s)\,\psi(r-s) \qquad (10.21)$$

for Dirichlet series $\varphi(s),\psi(s)$ *we have the following*

$$\sum_{k=1}^\infty \frac{\alpha_k}{\lambda_k^\nu}\, G_{0,1;1,0;0,1}^{1,0;0,1;1,0}\left(\begin{array}{c}a^{-1}\lambda_k\\b\lambda_k\end{array}\left|\begin{array}{c}-\\\nu\end{array}\right|\begin{array}{c}1\\-\end{array}\left|\begin{array}{c}-\\0\end{array}\right.\right)$$

$$= \sum_{k=1}^\infty \frac{\beta_k}{\mu_k^{r-\nu}}\, G_{0,1;1,0;0,1}^{1,0;0,1;1,0}\left(\begin{array}{c}b^{-1}\mu_k\\a\mu_k\end{array}\left|\begin{array}{c}-\\r-\nu\end{array}\right|\begin{array}{c}1\\-\end{array}\left|\begin{array}{c}-\\0\end{array}\right.\right) + \text{Res} \qquad (10.22)$$

or

$$2 \sum_{k=1}^{\infty} \alpha_k \left(\frac{\lambda_k + a}{b} \right)^{-\frac{\nu}{2}} K_\nu \left(2\sqrt{(\lambda_k + a)\, b} \right)$$

$$= 2 \sum_{k=1}^{\infty} \beta_k \left(\frac{\mu_k + b}{a} \right)^{-\frac{r-\nu}{2}} K_{r-\nu} \left(2\sqrt{(\mu_k + b)\, a} \right) + \text{Res}. \tag{10.23}$$

Proof.

$$\sum_{k=1}^{\infty} \frac{\alpha_k}{\lambda_k^\nu} G_{0,1;1,0;0,1}^{1,0;0,1;1,0} \left(\begin{matrix} a^{-1}\lambda_k \\ b\lambda_k \end{matrix} \left| \begin{matrix} - \\ \nu \end{matrix} \right| \begin{matrix} 1 \\ - \end{matrix} \left| \begin{matrix} - \\ 0 \end{matrix} \right. \right) \tag{10.24}$$

$$= \sum_{k=1}^{\infty} \frac{\alpha_k}{\lambda_k^\nu} \frac{1}{2\pi i} \int_{L_2} \left(\frac{1}{2\pi i} \int_{L_1} \Gamma(\nu + s + t)\, \Gamma(-s)\, \Gamma(t)\, \lambda_k^{-s-t} a^s b^{-t} ds \right) dt$$

$$= \frac{1}{2\pi i} \int_{L_2} \left(\frac{1}{2\pi i} \int_{L_1} \Gamma(\nu + s + t)\, \Gamma(-s)\, \Gamma(t) \sum_{k=1}^{\infty} \frac{\alpha_k}{\lambda_k^{\nu+s+t}}\, a^s b^{-t} ds \right) dt$$

$$= \frac{1}{2\pi i} \int_{L_2} \left(\frac{1}{2\pi i} \int_{L_1} \Gamma(\nu + s + t)\, \Gamma(-s)\, \Gamma(t)\, \varphi(\nu + s + t)\, a^s b^{-t} ds \right) dt.$$

Use the functional equation here.

$$= \frac{1}{2\pi i} \int_{L_2'} \left(\frac{1}{2\pi i} \int_{L_1'} \Gamma(r - \nu - s - t)\, \Gamma(-s)\, \Gamma(t)\, \psi(r - \nu - s - t)\, a^s b^{-t} ds \right) dt$$

$$+ \text{Res}$$

$$= \frac{1}{2\pi i} \int_{L_2'} \left(\frac{1}{2\pi i} \int_{L_1'} \Gamma(r - \nu - s - t)\, \Gamma(-s)\, \Gamma(t) \sum_{k=1}^{\infty} \frac{\beta_k}{\mu_k^{r-\nu-s-t}}\, a^s b^{-t} ds \right) dt$$

$$+ \text{Res}$$

$$= \sum_{k=1}^{\infty} \frac{\beta_k}{\mu_k^{r-\nu}} \frac{1}{2\pi i} \int_{L_2'} \left(\frac{1}{2\pi i} \int_{L_1'} \Gamma(r - \nu - s - t)\, \Gamma(-s)\, \Gamma(t)\, \mu_k^{s+t} a^s b^{-t} ds \right) dt$$

$$+ \text{Res}$$

$$= \sum_{k=1}^{\infty} \frac{\beta_k}{\mu_k^{r-\nu}} G_{1,0;1,0;0,1}^{0,1;0,1;1,0} \left(\begin{matrix} a^{-1}/\mu_k \\ b/\mu_k \end{matrix} \left| \begin{matrix} 1 - r + \nu \\ - \end{matrix} \right| \begin{matrix} 1 \\ - \end{matrix} \left| \begin{matrix} - \\ 0 \end{matrix} \right. \right) + \text{Res}. \tag{10.25}$$

Now use formulas for *G*-function of 2 variables

$$G_{1,0;1,0;0,1}^{0,1;0,1;1,0}\left(\begin{array}{c} a^{-1}/z \\ b/z \end{array} \middle| \begin{array}{c} 1-r+\nu \\ - \end{array} \middle| \begin{array}{c} 1 \\ - \end{array} \middle| \begin{array}{c} - \\ 0 \end{array} \right)$$

$$= G_{0,1;0,1;1,0}^{1,0;1,0;0,1}\left(\begin{array}{c} az \\ b^{-1}z \end{array} \middle| \begin{array}{c} - \\ r-\nu \end{array} \middle| \begin{array}{c} - \\ 0 \end{array} \middle| \begin{array}{c} 1 \\ - \end{array} \right) \tag{10.26}$$

$$= G_{0,1;1,0;0,1}^{1,0;0,1;1,0}\left(\begin{array}{c} b^{-1}z \\ az \end{array} \middle| \begin{array}{c} - \\ r-\nu \end{array} \middle| \begin{array}{c} 1 \\ - \end{array} \middle| \begin{array}{c} - \\ 0 \end{array} \right)$$

we have

$$= \sum_{k=1}^{\infty} \frac{\beta_k}{\mu_k^{r-\nu}} G_{0,1;1,0;0,1}^{1,0;0,1;1,0}\left(\begin{array}{c} b^{-1}\mu_k \\ a\mu_k \end{array} \middle| \begin{array}{c} - \\ r-\nu \end{array} \middle| \begin{array}{c} 1 \\ - \end{array} \middle| \begin{array}{c} - \\ 0 \end{array} \right) + \text{Res.} \tag{10.27}$$

Therefore, using

$$G_{0,1;1,0;0,1}^{1,0;0,1;1,0}\left(\begin{array}{c} a^{-1}z \\ bz \end{array} \middle| \begin{array}{c} - \\ \nu \end{array} \middle| \begin{array}{c} 1 \\ - \end{array} \middle| \begin{array}{c} - \\ 0 \end{array} \right)$$

$$= 2 \left(\frac{1+a/z}{bz} \right)^{-\frac{\nu}{2}} K_\nu\left(2\sqrt{(1+a/z)\,bz} \right) \tag{10.28}$$

$$= 2\,z^\nu \left(\frac{z+a}{b} \right)^{-\frac{\nu}{2}} K_\nu\left(2\sqrt{(z+a)b} \right)$$

we deduce (10.23). □

10.1.4 *Plausible general form*

A general form

Assume a functional equation:

$$\chi(s) = \begin{cases} \Gamma(s\,|\Delta)\,\varphi(s) \\ \Gamma(r-s\,|\Xi)\,\psi(r-s) \end{cases} \tag{10.29}$$

for Dirichlet series

$$\varphi(s) = \sum_{k=1}^{\infty} \frac{\alpha_k}{\lambda_k^s}, \quad \psi(s) = \sum_{k=1}^{\infty} \frac{\beta_k}{\mu_k^s}. \tag{10.30}$$

Define

$$X\left(s, \begin{matrix} a \\ b \end{matrix} \middle| \Delta_1 \middle| \Delta_2 \right) \tag{10.31}$$

$$= \frac{1}{2\pi i} \int_{L_2} \left(\frac{1}{2\pi i} \int_{L_1} \Gamma(w_1 | \Delta_1) \, \Gamma(w_2 | \Delta_2) \, \chi(s + w_1 + w_2) \, a^{-w_1} b^{-w_2} \, dw_1 \right) dw_2$$

then, we have

$$X\left(s, \begin{matrix} a \\ b \end{matrix} \middle| \Delta_1 \middle| \Delta_2 \right) \tag{10.32}$$

$$= \begin{cases} \displaystyle\sum_{k=1}^{\infty} \frac{\alpha_k}{\lambda_k^s} H\left(\begin{matrix} a\lambda_k \\ b\lambda_k \end{matrix} \middle| \Delta + s \middle| \Delta_1 \middle| \Delta_2 \right) \\[4mm] \displaystyle\sum_{k=1}^{\infty} \frac{\beta_k}{\mu_k^{r-s}} H\left(\begin{matrix} a^{-1}\mu_k \\ b^{-1}\mu_k \end{matrix} \middle| \Xi + r - s \middle| \Delta_1^* \middle| \Delta_2^* \right) + \text{Res.} \end{cases}$$

In particular

$$X\left(s, \begin{matrix} a^{-1} \\ b \end{matrix} \middle| \Delta_2^* \middle| \Delta_2 \right) \tag{10.33}$$

$$= \begin{cases} \displaystyle\sum_{k=1}^{\infty} \frac{\alpha_k}{\lambda_k^s} H\left(\begin{matrix} a^{-1}\lambda_k \\ b\lambda_k \end{matrix} \middle| \Delta + s \middle| \Delta_2^* \middle| \Delta_2 \right) \\[4mm] \displaystyle\sum_{k=1}^{\infty} \frac{\beta_k}{\mu_k^{r-s}} H\left(\begin{matrix} a\mu_k \\ b^{-1}\mu_k \end{matrix} \middle| \Xi + r - s \middle| \Delta_2 \middle| \Delta_2^* \right) + \text{Res} \end{cases}$$

which is converted to a more symmetric form

$$X\left(s, \begin{matrix} a^{-1} \\ b \end{matrix} \middle| \Delta_2^* \middle| \Delta_2 \right) \tag{10.34}$$

$$= \begin{cases} \displaystyle\sum_{k=1}^{\infty} \frac{\alpha_k}{\lambda_k^s} H\left(\begin{matrix} a^{-1}\lambda_k \\ b\lambda_k \end{matrix} \middle| \Delta + s \middle| \Delta_2^* \middle| \Delta_2 \right) \\[4mm] \displaystyle\sum_{k=1}^{\infty} \frac{\beta_k}{\mu_k^{r-s}} H\left(\begin{matrix} b^{-1}\mu_k \\ a\mu_k \end{matrix} \middle| \Xi + r - s \middle| \Delta_2^* \middle| \Delta_2 \right) + \text{Res.} \end{cases}$$

10.2 Quellenangaben

10.2.1 *Berndt-Knopp book and Berndt's series of papers*

This subsection partly depends on the corresponding subsection in [KTY7]. Meanwhile the Berndt and M. Knopp book [BK] has been published whose two chapters contain equivalent relation to the functional equation, outside of the Hecke correspondence, i.e. Chapter 7 is about Bochner's generalization of Hecke theory and Chapter 8 assembles some results from Berndt's series of papers and some related ones.

Since the treatment in [BK, Chapter 8] is rather sketchy, we shall refer to the above-mentioned Berndt's series of 7 papers. They are one of the main researches after the work of Bochner-Chandrasekharan-Narasimhan along with M. I. Knopp, Lavrik, and Terras, unifying many of the existing results under the functional equation and summability.

We shall now briefly review the contents which are relevant to our present view point but not touched upon above.

Most of the results in [BeI] and [BeII] are incorporated in §9.2.

In [BK], they mention the logarithmic Riesz sum [BK, p.84] of the $r(n)$, which is referred to as the Müller-Carlitz-Ayoub-Chowla-Redmond-Berndt theorem [IK99]. Similar results are obtained [Ber75-2, p.434] from periodic analogues of summation formulas, which in turn is a development of [BeV]. We note a passing remark [Ber75-2, p.435] that [BeII, p.372] Stirling's formula is shown to be a consequence of the functional equation of the Riemann zeta-function (as a Hecke type: $\varphi(s) = \psi(s) = \pi^{-s}\zeta(2s)$, $\lambda_n = \mu_n = \pi n^2$). This has been further developed by Katsuarada in his series of papers and then elucidated in, and culminated to, our view that the theory of the gamma function may be based on that of the Riemann zeta-function [Vista, Chapter 5] (a là Lerch's formula).

[BeIII] deals with the special cases where the gamma factors are the product of two gamma functions of given form and obtains identities for both Riesz sums and logarithmic Riesz sums as in Theorem 9.2. The following are relevant to the present book.

An identity involving an exponential integral (§7) as a prototype of the Ewald expansion, thirdly a general modular relation involving Laguerre polynomials (§5.1), where Laguerre polynomials are defined by (5.16) and coincide with the confluent hypergeometric functions (5.21).

Berndt's contribution to perturbed Dirichlet series has been expounded in §4.7.

[BeV] is concerned with the Voronoǐ summation formula which contains, as a special case, the modular relation (cf. Example 1). General results are obtained for functions smooth or of bounded variation.

[BeVI] develops the theory of Epstein zeta-functions by using his theory of perturbed Dirichlet series which in turn depends on the beta transform or the modular relation. This has been incorporated in [Vista] and [Vi2] and also briefly referred to in Chapters 4 and 5 of the present book.

[BeVII] gives an alternate proof, using only classical results in the theory of Bessel functions, of a theorem of Chandrasekharan and Narasimhan on the Bessel series expression for the Riesz sum of the arithmetic functions generated by the zeta-function which satisfies the Hecke type functional equation.

Subsequent works have been done by Hafner [Haf], Dixit [ADi] (more papers on the Wev), Benrdt-Kim-Zaharescu [BKZI], [BKZII] et al.

"And therefore as a stranger give it welcome. There are more things in Heaven and Earth, Horatio, than are dreamed of in our philosophy."

— Hamlet, Act One, Scene 5.

It was to the west, however, that the conspicuous grandeur lay, for there the mountains rose in such splendor that when men saw them they gasped. Row upon row the marvelous peaks marched north and south, so many and so varied that the eye could never tire of them. ⋯ The good part was that close up, these splendid ranges were just as impressive as they had been from a distance. They dominated the plains and served as a backdrop to extraordinary beauty.

— J. Michener, 'Centennial', pp. 131-132.

Fascinated by the beauty of the scenery, we have come up here. It is time to turn the horse back. The last sentence of T. Takagi, 'Class field theory'.

10.2.2   Corrections to "Number Theory and its Applications"

Since the book [NTA] has a lot of relevance to the present book, it would not be too inadequate to include the corrections to it.

The following is the list of corrections and modifications.

p.9, ll.1-3, for "*G*" read "*H*".

p.18, l.15, for "*m* = 0 or 1" read "*m* = 0".

p.27, l.15, for "$\theta(\varphi(X))$" read "$\theta(= \varphi(X))$".

p.29, ll.18-19 for "which is the field consisting of all algebraic elements of (an algebraically closed) extension of *K*" read "which is the field consisting of all algebraic elements of an algebraically closed extension of *K*" or more precisely, "which is the union of all algebraic extensions of an algebraically closed extension of *K*" or "which is an algebraic extension of *K* which is algebraically"

p.30, l.2, for "α, β coincide" read "α, β to coincide".

p.31, l.3, for "as" read "is".

p.43, l.10 from below, for "$q := p - \nu$" read "$q := p^{\nu}$".

p.45, in (1.57), for "$(-1)^{\frac{p-1}{2}}(-1)^{\frac{q-1}{2}}$" read "$(-1)^{\frac{p-1}{2}\frac{q-1}{2}}$".

p.46, l.5, for "$q^* = (-1)^{\frac{p-1}{2}}p.$ $\frac{p-1}{2}$ being an integer," read "$q^* = (-1)^{\frac{p-1}{2}}q$". Since $\frac{p-1}{2}$ is an integer,"

p.56, l.12, for "$a_i \in \iota_k$ " read "$a_i \in \mathfrak{a}_k$".

p.99, l.1 from below, for "for the following general convolution" read "for a general convolution ".

p.171, l.6, New line from "Now".

p.171, l.8, for "below" read "on p. 170".

p.171, Figure should appear on p. 164.

p.171, ll.9-10, for "*K*" read "*C*".

p.173, l.13, delete "of".

p.173, l.3 from below, for "$I_t^{\lambda} D_t^{-\lambda}$ " read "$I_t^{\lambda} = D_t^{-\lambda}$".

Fig. 10.1   Bernhard Riemann

Bibliography

[Ahm] M. Ahmed, On the expansion of a residual function as series of certain special forms, *Ann. of Math.* (2) **63** (1956), 549–564.

[Apo1] T. M. Apostol, Generalized Dedekind sums and the transformation formula of certain Lambert series, *Duke Math. J.* **17** (1950), 147–157.

[AIK] T. Arakawa, T. Ibukiyama, and M. Kaneko, *Bernoulli numbers and the zeta functions*, Springer Verl., Berlin etc. 2014.

[Asa] M. Asano, Report on multiple zeta-functions and Dedekind sums, Masters' thesis, Nagoya University 2003 (in Japanese).

[Atk1] F.V. Atkinson, The mean value of the zeta-function on the critical line, *Proc. London Math. Soc.* (2) **47** (1942), 174–200.

[Atk2] F.V. Atkinson, The mean value of the Riemann zeta-function, *Acta Math.* **81** (1949), 353–376.

[Atk3] F.V. Atkinson, The Riemann zeta-function, *Duke Math. J.* **17** (1950), 63–38.

[Ay] R. Ayoub, Euler and the zeta function, *Amer. Math. Monthly* **81** (1974), 1067–1086.

[Bak] R. Baker, *Kloostermann sums and Maass forms I,* Kendrick Press, Heber 2003.

[Bel1] R. Bellman, An analog of an identity due to Wilton, *Duke Math. J.* **16** (1949), 539–545.

[Bel2] R. Bellman, Wigert's approximate functional equation and the Riemann zeta-function, *Duke Math. J.* **16** (1949), 547–552.

[Bel3] R. Bellman, Generalized Eisenstein series and non-analytic automorphic functions, *Proc. Nat. Acad. Sci., USA* **36** (1950), 356–359.

[Bel4] R. Bellman, On the functional equations of the Dirichlet series derived from Siegel modular forms, *Proc. Nat. Acad. Sci., USA* **37** (1951), 84–87.

[Bel5] R. Bellman, On a class of functional equations of modular type, *Proc. Nat. Acad. Sci., USA* **42** (1956), 84–87.

[Be1] B. C. Berndt, Generalized Dirichlet series and Hecke's functional equation, *Proc. Edinburgh Math. Soc.* **15** (1967), 309–313.

[Be2] B. C. Berndt, Arithmetical identities and Hecke's functional equation, *Proc. Edinburgh Math. Soc.* **16** (1968/69), 309–313.

[BeI] B. C. Berndt, Identities involving the coefficients of a class of Dirichlet series I, *Trans. Amer. Math. Soc.* **137** (1969), 345–359.

[BeII] B. C. Berndt, Identities involving the coefficients of a class of Dirichlet series II, *Trans. Amer. Math. Soc.* **137** (1969), 361–374.

[BeIII] B. C. Berndt, Identities involving the coefficients of a class of Dirichlet series III, *Trans. Amer. Math. Soc.* **146** (1969), 323–348.

[BeIV] B. C. Berndt, Identities involving the coefficients of a class of Dirichlet series IV, *Trans. Amer. Math. Soc.* **149** (1970), 179–185.

[BeV] B. C. Berndt, Identities involving the coefficients of a class of Dirichlet series V, *Trans. Amer. Math. Soc.* **160** (1971), 139–156.

[BeVI] B. C. Berndt, Identities involving the coefficients of a class of Dirichlet series VI, *Trans. Amer. Math. Soc.* **160** (1971), 157–167.

[BeVII] B. C. Berndt, Identities involving the coefficients of a class of Dirichlet series VI, *Trans. Amer. Math. Soc.* **201** (1975), 247–261.

[Ber3] B. C. Berndt, Generalized Dedekind eta-function and generalized Dedekind sums, *Trans. Amer. Math. Soc.* **178** (1973), 495–508.

[Be11] B. C. Berndt, Modular transformations and generalizations of several formulae of Ramanujan, *Rocky Mountain J. Math.* **7** (1977), 147–189.

[BKZI] B. C. Berndt, S. Kim, and A. Zaharescu, Weighted divisor sums and Bessel function series, *Math. Ann.* **335** (2006), 249-283.

[BKZII] B. C. Berndt, S. Kim, and A. Zaharescu Weighted divisor sums and Bessel function series, II , *Adv. Math.* **229** (2012), 2055–2097.

[BK] B. C. Berndt and M. I. Knopp, Hecke's theory of modular forms and Dirichlet series, World Sci., Singapore etc., 2008.

[Bo1] S. Bochner, Some properties of modular relations, *Ann. of Math.* (2) **53** (1951), 332–363=*Collected Papers of Salomon Bochner*, Part II, Amer. Math. Soc., Providence, RI 1991, 665–696.

[Bo2] S. Bochner, Gamma factors in functional equations, *Proc. Nat. Acad. Sci.* **42** (1956), 85-89=*Collected Papers of Salomon Bochner*, Part II, Amer. Math. Soc., Providence, RI 1991, 711–714.

[Bo3] S. Bochner, On Riemann's functional equation with multiple gamma factors, *Ann. of Math.* (2) **67** (1958), 29–41=*Collected Papers of Salomon Bochner*, Part II, Amer. Math. Soc., Providence, RI 1991, 763–775.

[BCh1] S. Bochner and K. Chandrasekharan, On Riemann's functional equation, *Ann. of Math.* (2) **63** (1956), 336–360=*Collected Papers of Salomon Bochner*, Part II, Amer. Math. Soc., Providence, RI 1991, 715–739.

[BB1] J. M. Borwein and P. B. Borwein, *Pi and the AGM: A study in analytic number theory and computational complexity*, Wiley, 1987.

[BB2] D. Borwein and J. M. Borwein, On some trigonometric and exponential lattice sums, *J. Math. Analy. Appl.* **188** (1994), 209–218.

[BOn] J. H. Bruinier and K. Ono, Heegner divisors, L-functions and harmonic weak Maass forms, *Ann. of Math.* (2) **172** (2010), 2135–2181.

[Bu1] D. Bump, Automorphic forms of GL(3, $\mathbb{R}$), Lect. Notes Math. 1083, Springer Verl., Berlin etc. 1984.

[Bu2] D. Bump, Barnes' second lemma and its application to Rankin-Selberg convolutions, *Amer. J. Math.* **110** (1988), 179–185.

[Bu3] D. Bump, The Rankin-Selberg method: an introduction and survey, **110** (1988), 179–185.

[CP1] A. N. Chaba and R. K. Pathria, Evaluation of a class of lattice sums in arbitrary dimensions, *J. Math. Phys.* **16** (1975), 1457–1460.

[CP2] A. N. Chaba and R. K. Pathria, Evaluation of a class of lattice sums using Poisson's summation formula. II, *J. Phys. A: Math. Gen.* **9** (1976), 1411–1423.

[CP3] A. N. Chaba and R. K. Pathria, Evaluation of a class of lattice sums using Poisson's summation formula. III, ibid. **9** (1976), 1801–1810.

[CP4] A. N. Chaba and R. K. Pathria, Evaluation of a class of lattice sums using Poisson's summation formula. IV, ibid. **10** (1977), 1823–1831.

[CKK] K. Chakraborty, S. Kanemitsu and T. Kuzumaki, Finite expressions for higher derivatives of the Dirichlet *L*-function and the Deninger *R*-function, *Hardy-Ramanujan J.* **32** (2009), 38–53.

[Vi2] K. Chakraborty, S. Kanemitsu and H. Tsukada, *Vistas of Special Functions II*, World Scientific, London-Singapore-New Jersey, 2009.

[CKT1] K. Chakraborty, S. Kanemitsu and H. Tsukada, On arithmetical Fourier series and the modular relation, *Kyushu J. Math.* **66** (2012), 411–427.

[ChR] F. A. C.C. Chalub and J. F. Rodrigues, *The mathematics of Darwin's legacy*, Birkhäuser, Basel 2011.

[HH] H. H. Chan, On the equivalence of Ramanujan's partition identities and a connection with the Rogers-Ramanujan continued fraction, *J. Math. Anal. Appli.* **198** (1996), 111–120.

[Ch] K. Chandrasekharan, *Arithmetical Functions*, Springer Verl., Berlin-New York etc. 1969.

[CJ] K. Chandrasekharan and H. Joris, Dirichlet series with functional equations and related arithmetical functions, *Acta Arith.* **24** (1973), 165–191.

[ChM] K. Chandrasekharan and S. Mandelbrojt, On Riemann's functional equation, *Ann. of Math.* (2) **66** (1957), 285–296.

[ChTyp] K. Chandrasekharan and S. Minakshisundaram, *Typical means*, Oxford UP, Oxford 1952.

[ChN1] K. Chandrasekharan and Raghavan Narasimhan, Hecke's functional equation and the average order of arithmetical functions, *Acta Arith.* **6** (1961), 487–503.

[ChN2] K. Chandrasekharan and Raghavan Narasimhan, Hecke's functional equation and arithmetical identities, *Ann. of Math.* (2) **74** (1961), no. 2, 1–23.

[ChN3] K. Chandrasekharan and Raghavan Narasimhan, Functional equations with multiple gamma factors and the average order of arithmetical functions. *Ann. of Math.* (2) **76** (1962), 93–136.

[CZ] M. A. Chaudhry and S. M. Zubair, *On a Class of Incomplete Gamma Functions with Applications*, Chapman & Hall/CRC, New York 2001.

[Cho] S. Chowla, Remarks on class invariants and related topics, Calcutta Math. Soc. Golden Jubilee Comm. Vol. (1958/59), Part II, 361–372; Lect. Notes Math., Springer Verl. Berlin 1966 (Seminar on complex multiplication, VI-1 to VI-15): *The Collected Papers of Sarvadaman Chowla II*, CRM, 1999, 1008–1019.

[Coh] H. Cohen, *q*-identities for Maass waveforms, *Invent. Math.* **91**(1998), 409–422.

[Dav] H. Davenport, Multiplicative number theory, 1st ed. Markham, Chicago 1967, 2nd ed. Springer, New York etc. 1980.

[DeM] B. Debica and J. Mehta, Linearized product of zeta-functions, Proc. Japan Acad. Ser. A Math., to appear.

[De] R. Dedekind, Erläuterungen zu zwei Fragmenten von Riemann, Math. Werke Bd. 1, 1930, Braunschweich, 159–173 (in: Bernhard Riemanns Ges. Math. Werke und Wiss. Nachlass, 2, Aufl., 1892, 466–478).

[ADi] A. Dixit, Transformation formulas associated with integrals involving the Riemann *Xi*-function, *Monatsh. Math.* **164** (2011), 133–156.

[DM] K. Doi and T. Miyake, *The theory of modular forms*, Springer

[Ep] P. Epstein, Zur Theorie allgemeiner Zetafunktionen, I, II *Math. Ann.* **56** (1903), 615–644; **62** (1906), 205–246.

[ErdH] A. Erdélyi, W. Magnus, F. Oberhettinger and F. G. Tricomi, *Higher Transcendental Functions*, Vols I–III, Based, in part, on notes left by Harry Bateman, McGraw-Hill, New York 1953.

[ErdT] A. Erdélyi, W. Magnus, F. Oberhettinger and F. G. Tricomi, *Tables of Integral Transforms. Vols. I–II* Based, in part, on notes left by Harry Bateman, McGraw-Hill Book Company, Inc., New York-Toronto-London, 1954.

[Eu] L. Euler, Remarques sur un beau rapport entre les séries des puissances tant directes que réciproques, 1749.

[Ew] P. P. Ewald, Die Berechnung optischer und elektrostatischer Gitterpotentiale, *Ann. d. Phys.* **64** (1921), 253–287.

[F] C. Fox, The *G* and *H* functions as symmetrical Fourier Kernels, *Trans. Amer. Math. Soc.* **98** (1961), 395–429.

[Gel] A. O. Gel'fond, Some functional equations implied by equations of Riemann type, *Izv. Akad. Nauk SSSR Ser. Mat.* **24** (1960), 469–474.

[GJ] R. Godement and H. Jacquet, *Zeta Functions of Simple Algebras*, Lect. Notes Math. Vol. 260, Springer Verl., Berlin etc. 1972.

[Gs] D. Goss, A simple approach to the analytic continuation and values at negative integers for Riemann's zeta-function, *Proc. Amer. Math. Soc.* **81** (1981), 513–517.

[GR] I. S. Gradshteyn and I. M. Ryzhik, *Table of Integrals, Series, and Products.* Translated from the fourth Russian edition. Fifth edition. Translation edited and with a preface by Alan Jeffrey. Academic Press, Inc., Boston, MA, 1994.

[Gr1] E. Grosswald, Die Werte der Riemannschen Zetafunktion an ungeraden Argumentstellen, *Nachr. Akad. Wiss. Göttingen Math.-Phys.* Kl. II (1970), 9–13.

[Gr2] E. Grosswald, Comments on some formulae of Ramanujan. *Acta Arith.* **21** (1972), 25–34.

[Gr3] E. Grosswald, Relations between the values at integral arguments of Dirichlet series that satisfy functional equations. *Proc. Sympos. Pure Math.*, Vol. 24, Amer. Math. Soc., Providence, 1973, 111–122.

[Gui1] A. P. Guinand, Functional equations and self-reciprocal functions connected with Lambert series. *Quart. J. Math., Oxford Ser.* **15** (1944), 11–23.

[Gui2] A. P. Guinand, Some rapidly convergent series for the Riemann ξ-function. *Quart. J. Math., Oxford Ser.* (2) **6** (1955), 156–160.

[Haf] J. L. Hafner, On the representation of the summatory functions of a class of arithmetical functions, Lecture Notes Math. **899**, Springer Verl. Berlin-Heidelberg, 1981, 148–165.

[HS] J. Hafner and J. Stopple, A heat kernel associated to Ramanujan's tau function, *Ramanujan J.* **4** (2000), 123–128.

[Hal] G. L. Hall, Asymptotic properties of generalized Chaba and Pathria lattice sums, *J. Math. Phys.* **17** (1976), 259–260.

[HarMB] G. H. Hardy, Some multiple integrals, *Quart. J. Math. (Oxford)* (2) **5** (1908), 357–375; *Collected Papers. Vol. V* (1972), 434–452, Comments 453.

[HarCP] G. H. Hardy, On the expression of a number as the sum of two squares, *Quart. J. Math. (Oxford)* **46** (1915), 263–283; *Collected Papers. Vol. II*, Oxford UP, Oxford 1967, 243–263.

[HarDP] G. H. Hardy, On Dirichlet's divisor problem, *Proc. London Math. Soc.* (2) **15** (1916), 1–25; *Collected Papers. Vol. II*, Oxford UP, Oxford 1967, 268–292, Comments 293.

[HarAO] G. H. Hardy, The average order of the arithmetical functions P(x) and $\Delta(x)$, *Proc. London Math. Soc.* (2) **15** (1916), 1–25; *Collected Papers. Vol. II*, Oxford UP, Oxford 1967, 294–315.

[HarRam] G. H. Hardy, *Ramanujan*, CUP. Cambridge 1940.

[HarRie] G. H. Hardy and M. Riesz, *The general theory of Dirichlet's series*, CUP. Cambridge 1915; reprint, Hafner, New York 1972.

[HaWr] G. H. Hardy and E. M. Wright, *An introduction to the theory of numbers*, Oxford UP. Oxford 1915; reprint, Hafner, New York 1972.

[Has] H. Hasse, Ein Summierungsverfahren für die Riemannsche ζ-Reihe, *Math. Z.* **32** (1930), 458–464.

[Hau] A. Hautot, A new method for the evaluation of slowly convergent series, *J. Math. Phys.* **15** (1974), 1722–1727.

[Heck1] E. Hecke, Eine neue Art von Zetafunktionen und ihre Beziehungen zur Verteilung der Primzahlen, I, *Math. Z.* **1** (1918), 357–376 =Mathematische Werke, 215–234, Vandenhoeck u. Ruprecht, Göttingen 1959.

[Heck2] E. Hecke, Eine neue Art von Zetafunktionen und ihre Beziehungen zur Verteilung der Primzahlen, II, *Math. Z.* **6** (1920), 11–51 =Mathematische Werke, 249–289, Vandenhoeck u. Ruprecht, Göttingen 1959.

[HeckW] E. Hecke, Mathematische Werke, Vandenhoeck u. Ruprecht, Göttingen 1959.

[Heck3] E. Hecke, *Lectures on Dirichlet Series, Modular Functions and Quadratic Forms*, ed. by B. Schoenberg in coll. with W. Maak, Vandenhoeck u. Ruprecht, Göttingen, 1983 (first edition: *Dirichlet Series*, Planographed Lecture Notes, Princeton IAS, Edwards Brothers, Ann Arbor, 1938).

[Hei] H. Heilbronn, On Dirichlet series which satisfy a certain functional equation, *Quart. J. Math. (Oxford)* (2) **9** (1938), 194–195=The Coll. Papers of Hans Arnold Heilbronn, ed. by E. Kani and R. A. Smith, John Wiley & Sons, New York-Chichester-Brisbane-Toronto-Singapore, 1988, 343–344 (commentary p. 599).

[Hey] K. Hey, Analytische Zahlentheorie in Systemen hyperkomplexer Zahlen, Hamburg, Diss., 1929, 49 pages.

[HiTa] T. Hiramatsu and I. Takada, Dedekind sums and automorphic forms, *RIMS Kokyuroku* **572** (1985), 151–175.

[KH] K. Hirokane, Evening comets, *Big Comic Original* **866** (2003), 79.

[IcM] Y. Ichihara and K. Matsumoto (2008). On the Siegel-Tatuzawa theorem for a class of L-functions, Kyushu J. Math. **62** (2008), 201–215.

[Ing] A. E. Ingham, *The distribution of prime numbers.* Cambridge Tracts Math. Math. Phys., No. 30 Stechert-Hafner, Inc., New York 1964.

[TIk] T. Ikeda, On the gamma factor of the triple L-function I, Duke Math. J. **24** (1957), 653–662.

[SI] Shô Iseki, The transformation formula for the Dedekind modular function and related functional equation, Duke Math. J. **97** (1999), 301–318.

[IO] T. Ishii and T. Oda, A short history on investigation of the special values of zeta and L-functions of totally real number fields, in *Proc. of the conf. in memory of Tsuneo Aarakawa, Automorphic Forms and Zeta Functions*, World Sci., Singapore etc., 2006.

[Iv] A. Ivić, On some mean square estimates in the Rankin-Selberg problem, *Appl. Anal. Discr. Math.* **1** (2007), 111–121.

[IwKo] H. Iwaniec and E. Kowalski, *Analytic number theory*, AMS, Providence, R. I. 2004.

[Joh] B. R. Johnson, Generalized Lerch zeta function,*Pacific J. Math.* **53** (1974), 189–193.

[Joy] D. Joyner, *Distribution theorems of L-functions*, Longman, Essex 1986.

[Jut] M. Jutila, *Lectures on a method in the theory of exponential sums*, Lect. Math. Phys. **80** (1987), TIFR, Bombay 1987.

[KP7] J. Kaczorowski and A. Perelli, On the structure of the Selberg class, VII, *Ann. of Math.* (2) **173** (2011), 1397–1441.

[Ka] S. Kanemitsu A note on the general divisor problem, *Mem. Fac. Sci. Kyushu Univ. Ser. A* **32** (1978), 211–221.

[KaR] S. Kanemitsu, On the Riesz sums of some arithmetical functions, *in p-adic L-functions and algebraic number thery Surikaiseki Kenkyusho Kokyuroku* **411** (1981), 109–120.

[KKY1] S. Kanemitsu, M. Katsurada and M. Yoshimoto, On the Hurwitz-Lerch zeta function, *Aeq. Math.* **59** (2000), 1–19.

[KKY2] S. Kanemitsu, H. Kumagai and M. Yoshimoto, Sums involving the Hurwitz zeta function, *The Ramanujan J.* **5** (2001), 5–19.

[KKY3] S. Kanemitsu, H. Kumagai and M. Yoshimoto, On rapidly convergent series expressions for zeta- and L-values, and log sine integrals. *The Ramanujan J.* **5** (2001), 91–104.

[KK] S. Kanemitsu and T. Kuzumaki, Transformation formulas for Lambert series, *Siaulai Math. Sem.* **4** (2009), 105–123.

[KLW] S. Kanemitsu, H.-L. Li and N.-L. Wang, Weighted short-interval character sums, *Proc. Amer. Math. Soc.* **139** (2011), 1521–1532.

[KTT1] S. Kanemitsu, Y. Tanigawa and H. Tsukada, A generalization of Bochner's formula, *Hardy-Ramanujan J.*, **27** (2004), 28–46.

[KTT2] S. Kanemitsu, Y. Tanigawa and H. Tsukada, Some number theoretic applications of a general modular relation, *International J. Number Theory* **2** (2006), 599–615.

[KTT3] S. Kanemitsu, Y. Tanigawa and H. Tsukada, On Kronecker's limit formula and the hypergeometirc function, *Hardy-Ramanujan J.* **31** (2008), 28–40.

[CZS2] S. Kanemitsu, Y. Tanigawa, H. Tsukada, and M. Yoshimoto, *On Bessel series expressions for some lattice sums II*, J. Phys. A: Math. Gen **37** (2004), 719–734.

[CZS3] S. Kanemitsu, Y. Tanigawa, H. Tsukada and M. Yoshimoto, Crystal symmetry viewed as zeta symmetry, *Proceedings of Kinki University Symposium "Zeta Functions, Topology and Quantum Physics 2003"* (Developments in Mathematics, Vol.14) 91–129, Springer Verl., Berlin etc. 2005.

[KTY1] S. Kanemitsu, Y. Tanigawa and M. Yoshimoto, On rapidly convergent series for the Riemann-zeta values via the modular relation, *Abh. Math. Sem. Univ. Hamburg* **72** (2002), 187–206.

[KTY2] S. Kanemitsu, Y. Tanigawa and M. Yoshimoto, On rapidly convergent series for Dirichlet *L*-function values via the modular relation, *Proc. of the Intern. Conf. on Number Theory and Discrete Mathematics in honor of Srinivasa Ramanujan*, Oct. 2–6, 2000, Hindustan Book Agency 2002, 113–133.

[KTY3] S. Kanemitsu, Y. Tanigawa, and M. Yoshimoto, On zeta- and *l*-function values at special rational arguments via the modular relation, *Proc. Int. Conf. SSFA*, Vol. I (2001), 21–42.

[KTY4] S. Kanemitsu, Y. Tanigawa and M. Yoshimoto, On multiple Hurwitz zeta-function values at rational arguments, *Acta Arith.* **107** (2003), 45–67.

[KTY5] S. Kanemitsu, Y. Tanigawa and M. Yoshimoto, On the values of the Riemann zeta-function at rational arguments, *The Hardy-Ramanujan J.*, **24** (2001), 10–18.

[KTY6] S. Kanemitsu, Y. Tanigawa and M. Yoshimoto, On Dirichlet L-function values at rational arguments, Proc. of the Intern. Conf. on Analytic Number Theory held at IMSc, Jan. 2–3, 2002, *Lect. Notes Ser. No.* **1** (2004), 31–37.

[KTY7] S. Kanemitsu, Y. Tanigawa and M. Yoshimoto, Ramanujan's formula and modular forms, *Number-theoretic Methods – Future Trends, Proceedings of a conference held in Iizuka* (S. Kanemitsu and C. Jia, eds.), Kluwer, Dordrecht, 2002, pp. 159–212.

[CZS1] S. Kanemitsu, Y. Tanigawa and W.-P. Zhang, On Bessel series expressions for some lattice sums, *Chebyshëvskii Mat. Sb.* **5** (2004), 128–137.

[Vista] S. Kanemitsu and H. Tsukada, *Vistas of Special Functions*, World Scientific, Singapore etc. 2007.

[CZS4] S. Kanemitsu and H. Tsukada, Crystal symmetry viewed as zeta symmetry II, in "The legacy of Alladi Ramakrishnan in the mathematical sciences" ed. by K. Alladi, J. Klauder, C. R. Rao, Springer Verl., Berlin etc. 2010, 275–292.

[KUW] S. Kanemitsu, J. Urbanowicz and N.-L. Wang, On some new congruences for generalized Bernoulli numbers, *Acta Arith.*, **155** (2012), 247–258.

[Kat1] M. Katsurada, On Mellin-Barnes type of integrals and sums associated with the Riemann zeta-function, *Publ. Inst. Math.* **62** (1997), 13–25.

[Kat2] M. Katsurada, Power series and asymptotic series associated with the Lerch zeta-function, *Proc. Japan Acad. Ser. A* **74** (1998), 167–170.

[Kat3] M. Katsurada, Rapidly convergent series representations for $\zeta(2n+1)$ and their χ-analogue, *Acta Arith.* **40** (1999), 79–89.

[Kat4] M. Katsurada, On an asymptotic formula of Ramanujan for a certain theta-type series, *Acta Arith.* **97** (2001), 157–172.

[Kat5] M. Katsurada, Complete asymptotic expansions associated with Epstein zeta-functions, *Ramanujan J.* **14** (2007), 249–275.

[Kat6] M. Katsurada, Asymptotic expansions for double Shintani zeta-functions of several variables, Amer. Inst. Phys., New York 2011, 58–72.

[KatM1] M. Katsurada and K. Matsumoto, Explicit formulas and asymptotic expansions for certain mean sqaure of Hurwitz zeta functions I, *Math. Scand.* **78** (1996), 161–177.

[KatM2] M. Katsurada and K. Matsumoto, Explicit formulas and asymptotic expansions for certain mean sqaure of Hurwitz zeta functions II, Analytic and probabilistic methods in number theory, New Trends in Probab. and Stats., Vol. 4, A. Laurinčikas (eds), VSP/TEV, 1997, 119–134.

[KatM3] M. Katsurada and K. Matsumoto, Explicit formulas and asymptotic expansions for certain mean sqaure of Hurwitz zeta functions III, *Compos. Math.* **131** (2002), 239–266.

[KatNo] M. Katsurada and T. Noda, On generalized Lipschitz-Lerch formulae and application II, Amer. Inst. Phys., New York 2011, 73–86.

[Kaw] Y. Kawada, *Reports on Algebra Sympos.: Number Theory on Discontinuous Groups: Zeta-functions*, Math. Soc. Japan, Tokyo 1966, 80pp.

[Kl1] D. Klusch, On the Taylto expansion for the Lerch zeta function, *J. Math. Anal. Appl.* **170** (1992), 513–523.

[Kl] D. Klush, On Entry 8 of Chapter 15 of Ramanujan's Notebook II, *Acta Arith.* **58** (1991), 59–64.

[Kn1] K. Knopp, Über das Eulersche Summierungsverfahren II, *Math. Z.* **18** (1923), 125–156.

[MKn0] M. I. Knopp, *Modular Functions in Analytic Number Theory*, Markham, Chicago 1970.

[MKn1] M. I. Knopp, On Dirichlet series satisfying Riemann's functional equation, *Invent. Math.* **117** (1994), 361–372.

[MKn2] M. Knopp, Hamburger's theorem on $\zeta(s)$ and the abundance principle for Dirichlet series with functional equations, *Number Theory* (ed. by R. P. Bambah et al.), 201–216, Hindustan Book Agency, New Delhi, 2000.

[Kob1] H. Kober, Transformationen gewisser Besselscher Reihen, Beziehungen zu Zeta-Fuktionen, *Math. Z.* **39** (1935), 609–624.

[Koc] M. Koecher, Über Dirichlet-Reihen mit Funktionalgleichung, *J. Reine Angew. Math.* **192** (1953), 1–23.

[KMaT] Y. Komori, K. Matsumoto and H. Tsumura, Barnes multiple zeta-functions, Ramanujan's formula and relevant series involving hypergeometric functions, *J. ramanujan Math. Soc.* **28** (2013), 49–69.

[Kosum] N. S. Koshlyakov, Application of the theory of sum-formulae to the investigation of a class of one-valued analytical functions in the theory of numbers, *Mess. Math.* **58** (1928/29), 1–23.

[Kosvor] N. S. Koshlyakov, On Voronoï's sum formula, *Mess. Math.* **58** (1928/29), 30–32.

[KoshI] N. S. Koshlyakov, Investigation of some questions of analytic theory of the rational and quadratic fields, I (Russian), *Izv. Akad. Nauk SSSR, Ser. Mat.* **18** (1954), 113–144, Errata: ibid. **19** (1955), 271 (in Russian).

[KoshII] N. S. Koshlyakov, Investigation of some questions of analytic theory of the rational and quadratic fields, II (Russian), *Izv. Akad. Nauk SSSR, Ser. Mat.* **18** (1954), 213–260, Errata: ibid. **19** (1955), 271 (in Russian).

[KoshIII] N. S. Koshlyakov, Investigation of some questions of analytic theory of the rational and quadratic fields, III (Russian), *Izv. Akad. Nauk SSSR, Ser. Mat.* **18** (1954), 307–326, Errata: ibid. **19** (1955), 271 (in Russian).

[KoshE] N. S. Koshlyakov, Letter to the editor, *Izv. Akad. Nauk SSSR, Ser. Mat.* **19** (1955), 271 (in Russian).

[Kuz'1] R. Kuz'min, Contribution to the theory of a class of Dirichlet series (Russian), *Izv. Akad. Nauk SSSR, Ser. Math. Nat. Sci.*, **7** (1930), 115–124.

[Kuz'2] R. Kuz'min, On the roots of Dirichlet series, *Izv. Akad. Nauk SSSR, Ser. Math. Nat. Sci.*, **7** (1934), 1471–1491.

[Kuz] T. Kuzumaki, Asymptotic expansions for a class of zeta-functions, *Ramanujan J.* **24** (2011), 331–343.

[LanE] E. Landau, Euler und Fuktionalgleichung der Riemannschen Zetafunktion, *Bibliotheca Math.* (3) **7** (1906), 69–79=Collected Works, Vol. 1, Thales Verl., Essen 1985, 335–345.

[LanMe] E. Landau, Zur analytischen Zahlentheorie der definiten quadratischen Formen (Über die Gitterpunkte in einem mehrdimensionalen Ellipsoid), *Sitzungsber. Preuß. Akad. Wiss.* **31** (1915), 458–476=Collected Works, Vol. 6, Thales Verl., Essen 1985, 200–218.

[LanII] E. Landau, Über die Anzahl der Gitterpunkte in gewissen Bereichen (Zweite Mit.), *Nachr. Ges. Wiss. Göttingen, Math.-Phys. Kl.* (1915), 209–243=Collected Works Vol. 6, Thales Verl., Essen 1985, 308–342.

[LanW] E. Landau, Über die Wigertsche asymptotische Funktionalgleichung für die Lambertische Reihe, *Arch. Math. Physics* (3) **29** (1918), 144–146=Collected Works, Vol. 6, Thales Verl., Essen 1985, 135–137.

[LW1] E. Landau, Bemerkungen zu der Arbeit des Herrn Walfisz: Über das Piltzsche Teilerproblem in algebraishcen Zahlkörpern, *Math. Z.* **22** (1925), 189–205=Collected Works Vol. 8, Thales Verl., Essen 1985, 211–228.

[LanA] E. Landau (herausg. by A. Z. Walfisz), *Ausgewählte Abhandlungen zur Gitterpinktlehre*, VEB Deutsch. Verl. Wiss., Berlin 1962.

[LanZT] E. Landau, Vorlesungen über Zahlentheorie, Teubner, Leipzig 1921 and reprint Chelsea, New York 1967.

[LG] A. Laurinchikas and R. Garunkstis, *The Lerch Zeta-Function*, Kluwer Academic Publ., Dordrecht-Boston-London 2002.

[Lav1] A. F. Lavrik, Approximate functional equations of Dirichlet functions (Russian). *Izv. Akad. Nauk SSSR*, Ser. Mat. **32** (1968), 134–185; English translation in *Math. USSR-Izv.* **2** (1968), 129–179.

[Lav2] A. F. Lavrik, An approximate functional equation for the Dirichlet *L*-function, *Trudy Moskov Mat. Obšč.* **18** (1968), 91–104 = *Trans. Moscow Math. Soc.* **18** (1968), 101–115.

[Lav3] A. F. Lavrik, The principle of the theory of nonstandard functional equation for Dirichlet functions, consequences and applications of it, *Trudy Mat. Inst. Steklov* **132** (1973), 70–76= *Proc. Steklov Int. Math.* **132** (1973), 77–85.

[Lav4] A. F. Lavrik, Closed triplicity of functional equations of zeta functions (Russian). *Dokl. Akad. Nauk SSSR* **308** (1989), 1044–1046; English translation in *Soviet Math. Dokl.* **40** (1990), 403–405.

[Lav5] A. F. Lavrik, Functional equations with a parameter of zeta-functions (Russian). *Izv. Akad. Nauk SSSR Ser. Mat.* **54** (1990), 501–521; English translation in *Math. USSR-Izv.* **36** (1991), 519–540.

[Lep] H. Leptin, Die Funktionalgleichung der Zetafunktion einer enfachen Algebra, *Abh. Math. Sem. Univ. Hamburg* **19** (1955), 198–220.

[LWKman] F.-H. Li, N.-L. Wang and S. Kanemitsu, Manifestations of the general modular relation, *Šiaulai Math. Sem.* **7** (2012), 59–77.

[NTA] F.-H. Li, N.-L. Wang and S. Kanemitsu, *Number Theory and its Applications*, World Scientific, Singapore etc. 2013.

[Li] H.-L. Li, *Number Theory and Special Functions*, Science Press, Beijing 2011.

[LKT] H.-L. Li, S. Kanemitsu and H. Tsukada, Modular relation interpretation of the series involving the Riemann zeta values, *Proc. Japan Acad. Ser. A*, **84** (2008), 154–158.

[LKW] H.-L. Li, S. Kanemitsu and X.-H. Wang, Plana's summation formula as a modular relation and applications, *Int. Transf. Special Functions* **23** (2012), 183–190.

[LiM] H.-L. Li and J. Mehta, Modular relations associated to the Rankin-Selberg *L*-function, to appear.

[WLi] W.-C. W. Li, *L*-series of Rankin type and their functional equations, *Math. Ann.* **244** (1979), 135–166.

[Lin1] Ju. V. Linnik, An asymptotic formula in an additive problem of Hardy and Littlewood, *Izv. Akad. Nauk SSSR Ser. Mat.* **24** (1960), 629–706; English transl., *Amer. Math. Soc. Transl.* (2) **46** (1965), 65–148.

[Lin] Ju. V. Linnik, All large numbers are sum of a prime and two squares (A problem of Hardy and Littlewood) II, *Mat. Sb.* **53** (1961), 3–38; English transl., *Amer. Math. Soc. Transl.* (2) **37** (1964), 197–240.

[Lip1] R. Lipschitz, Untersuchung einer aus vier Elementen gebildeten Reihe, *J. Reine. Angew. Math.* **54** (1857), 313–328.

[Lip2] R. Lipschitz, Untersuchung der Eigenschaften einer Gattung von unendlichen Reihen, *J. Reine. Angew. Math.* **105** (1889), 127–156.

[Lit] J. E. Littlewood, *A Mathematician's Miscellany*, Cambridge UP.

[JYL] J.-Y. Liu, *A quick introduction to Maass forms*, in preparation.

[Mawave] H. Maass, Über eine neue Art von nichtanalytischen automorphen Funktionen und die Bestimmung Dirichletscher Reihen durch Funktionalgelichungen, *Math. Ann.* **121** (1949), 141–183.

[Maa1] H. Maass, Spherical functions and quadratic forms, *J. Indian Math. Soc.*

(N.S.) **20** (1956), 117–162.

[Maa2] H. Maass, Zetafunktionen mit Grösencharakteren und Kugelfunktionen, *Math. Ann.* **134** (1957), 1–32.

[Maa3] H. Maass, Über die räumliche Verteilung der Punkte in Gittern mit indefiniter Metrik, *Math. Ann.* **138** (1959), 287–135.

[MaTIFR] H. Maass, *Lectures on modular functions of one complex variable*, Tata Inst., Bombay 1964.

[Maa4] H. Maass, *Siegel's Modular Forms and Dirichlet Series*, Lect. Notes Math. **216** Springer Verl., Berlin etc., 1971.

[Mapa] F. Mainardi and G. Pagnini, Salvatore Pincherle: the pioneer of the Mellin-Barnes integrals, *J. Comp. Appl. Math.* **153** (2003), 331–342.

[Masa] A. M. Mathai and R. K. Saxena, *The H-Function with Applications in Statistics and Other Disciplines*, Wiley, New Delhi, 1978.

[Mat] K. Matsumoto, Recent developments in the mean square theory of the Riemann zeta and other zeta-functions, *Number Theory ed. by R. P. Bambah et al*, Hindustan Books Agency, 2000, 241–286.

[M1] C. S. Meijer, Über Whittakersche bezw. Besselsche Funktionen und deren Produkte, *Nieuw Archief voor Wiskunde* (2) **18** (4tes Heft) (1936), 10–39.

[M2] C. S. Meijer, On the G-function I–VIII, *Proc. Nederl. Akad. Wetensch.* **49** (1946), 227–237, 344–356, 457–469, 632–641, 765–772, 936–943, 1063–1072, 1165–1175.

[Mel] Hj. Mellin, Über eine Verallgemeinerung der Riemannshcen Function $\zeta(s)$, *Acta Soc. Sci. Fenn.* **24** (1898), 1–50.

[Mel1] Hj. Mellin, Die Dirchletscher Reihen, die zahlentheoretische Funktionen und die unendlichen Produkte von endlichen Geschlecht, *Acta Soc. Sci. Fenn.* **31** (1902), 1–48.

[Mel2] Hj. Mellin, Die Dirchletscher Reihen, die zahlentheoretische Funktionen und die unendlichen Produkte von endlichen Geschlecht, *Acta Math.* **28** (1903), 37–64.

[Meu] T. Meurman, On exponential sums involving the Fourier coefficients of Maass forms, *J. Reine Angew. Math.* **384** (1988), 192–207.

[Mik] M. Mikolás, Mellinsche Transformation und Orthogonalität bei $\zeta(s, u)$; Verallgemeinerung der Riemannschen Funktionalgleichung von $\zeta(s)$, *Acta Sci. Math. (Szeged)* **17** (1956), 143–164.

[MitK] D. S. Mitrinović and J. D. Kečkić, *The Cauchy method of residues Theory and Applications*, Reidel, Dordrecht-Boston-Lancaster 1984.

[Mor] Y. Morita, On the Hurwitz-Lerch L-function, *J. Fac. Sci. Univ. Tokyo Sect. IA Math.* **24** (1977), 29–43.

[Mot] Y. Motohashi, *Spectral theory of the Riemann zeta-function*, Cambridge UP, Cambridge 1997.

[MS] M. Ram Murty and Kaneenika Sinha, Multiple Hurwitz zeta functions, *Proc. Sympos. Pure Math.* **75** (2006), 135–156.

[Nai] H. Naito, The p-adic Hurwitz L-function, *Tohoku Math. J.* **34** (1982), 553–558.

[Nak1] M. Nakajima, Shifted divisor problem and random divisor problem, *Proc. Japan Acad. Ser. A Math.* **69** (1993), 19–22.

[Nak2] M. Nakajima, A new expression for the product of two Dirichlet series, *Proc. Japan Acad. Ser. A Math.* **79** (2003), 49–52.

[NgY] Nguen Thanh Hai and S. B. Yakubovich, *The double Mwllin-Barnes type integrals and their applications to covolution theory*, World Sci., Singapore etc. 1992.

[Ober] F. Oberhettinger, Note on the Lerch zeta function, *Pacific J. Math.* **6** (1956), 117–128.

[OS] F. Oberhettinger and K. L. Soni, On some relations which are equivalent to the functional equations involving the Riemann zeta function, *Math. Z.* **127** (1972), 17–34.

[OggM] A. P. Ogg, *Modular Forms and Dirichlet Series*, Benjamin, New York 1969.

[Ogg] A. P. Ogg, On product expansions of theta-functions, *Proc. Amer. Math. Soc.* **21** (1969), 365–368.

[OTW] Y. Ohno, T. Taniguchi and S. Wakatsuki, A conjecture on coincidence among the zeta functions associated with binary cubic forms, *Amer. J. Math.* **119** (1997), 1083–1094.

[Oh] Y. Ohno, Relations among Dirichlet series whose coefficients are class numbers of binary cubic forms, *Amer. J. Math.* **131** (2009), 1525–1541.

[Opp] A. Oppenheim, Some identities in the theory of numbers, *Proc. London Math. Soc.* (2)**26** (1927), 295–350.

[PK] R. B. Paris and D. Kaminski, *Asymptotics and Mellin-Barnes Integrals*, Cambridge UP, Cambridge 2001.

[PP] P. C. Pasles and W. d. A Pribitkin, A generalization of the Lipschitz summation formula and some applications, *Proc. Amer. Math. Soc.* **129** (2001), 3177–3184.

[Pra] K. Prachar, *Primzahlverteilng*, Springer Verl., Berlin etc. 2001.

[PBM] A. P. Prudnikov, Yu. A. Bychkov and O. I. Marichev, *Integrals and Series, Supplementary Chapters*, Izd. Nauka, Moscow 1986.

[RR] R. A. Rankin, Contributions to the theory of Ramanujan's function $\tau(n)$ and similar arithmetical functions, I, II, Proc. Cambridge Philos. Soc. **36** (1939), 351–372.

[Ran] R. A. Rankin, *Modular Forms*, Ellis Horwood Ltd, Chichester, 1984.

[Ric] H.-E. Richert, Über Dirichlet Reihen mit Fuktionalgleichung, *Acad. Serbe Sci.Publ. Inst. Math.* **11** (1957), 73–124.

[Rie0] B. Riemann, *Collected Works of Bernhard Riemann*, ed. by H. Weber, 2nd ed. Dover, New York 1953.

[Rie2] B. Riemann, Über die Anzahl der Primzahlen unter einer gegebenen Grösse, *Monatsber. Berlin. Akad.* (1859), 671–680.

[RZZ] A. Roy, A. Zaharescu and M. Zaki, Some identities involving convolutions of Dirichlet characters and the Möbius function, preprint.

[Sa] F. Sato, Searching for the origin of prehomogeneous vector spaces, at annual meeting of the Math. Soc. Japan 1992 (in Japanese).

[MS] M. Sato and T. Shintani, On zeta-functions associated with prehomogeneous vector spaces, *Ann. Math.* (2) **100** (1974), 131–170.

[Sch] O. Schlömilch, Uebungsaufgaben für Schuler, Lehrsatz von dem Herrn Prof.

Dr. Schlömilch, *Arch. Math. u. Phys. (Grunert's Archiv)* **12** (1849) 415.

[Ser] J. -P. Serre, *A course in arithmetic*, Springer Verl., New York-Heidelberg-Berlin 1973.

[Shm1] G. Shimura, *Introduction to the theory of automorphic functions*, Princeton University Press, Princeton, N. J. (1971).

[Shm2] G. Shimura, On the holomorphy of certain Dirichlet series, *Proc. London Math. Soc., Ser.* (3) **31** (1975), 79–98.

[Sie1] C. L. Siegel, Bemerkungen zu einem Satz von Hamburger über die Funktionalgleichung der Riemannschen Zetafunktion, *Math. Ann.* **85** (1922), 276–279=*Ges. Abh.*, Bd. I, Springer Verl. Berlin-Heidelberg, 1966, 154–156.

[Sie2] C. L. Siegel, Über Riemanns Nachlaßzur analytischen Zahlentheorie, *Quellen u. Studien zur Geschichte der Math., Astr. Phys.,* **2** (1932), 45–80 =*Ges. Abh.*, Bd. I, 275–310, Springer Verl., 1966.

[Sie3] C. L. Siegel, Über die Zetafunktionen indefiniter quadratische Formen, I, II *Math. Z.* **43** (1938), 682–708; **44** (1939), 398–426 =*Ges. Abh.*, Bd. II, Springer Verl. Berlin-Heidelberg, 1966, 41–67; 68–96.

[Sie4] C. L. Siegel, Indefinite quadratische Formen und Funktionentheorie, I, *Math. Ann.* **124** (1951), 17–54=*Ges. Abh.*, Bd. III, Springer Verl. Berlin-Heidelberg, 1966, 105–142.

[Sie5] C. L. Siegel, A simple proof of $\eta(-1/\tau) = \eta(\tau)\sqrt{\tau/i}$, *Mathematika,* **1** (1954), p.4=*Ges. Abh.*, Bd. III, Springer Verl. Berlin-Heidelberg-New York, 1966, 188.

[SieTa] C. L. Siegel, *Lectures on Advanced Analytic Number Theory*, Tata Inst., Bombay 1961, 2nd ed. *ditto* 1980.

[Sl] L. J. Slater, *Generalized Hypergeometric Functions* Cambridge UP, Cambridge 1966.

[So] J. Sondow, Analytic continuation of Riemann's zeta function and values at negative integers via Euler's transformation of series, Proc. Amer. Math. Soc. **120** (1994), 421–424.

[Son] K. Soni, Some relations associated with an extension of Koshkiakov's formula, Proc. Amer. Math. Soc. **17** (1966), 543–551.

[SpH] Robert Spira, *History of Zeta-Functions.* Vols. 1–3, Quartz Press, Ahsland, OR 1999.

[Spl] W. Splettstösser, Some aspects of the reconstruction of sampled signal functions, 126–142.

[SCh] H. M. Srivastava and J. Choi, *Series Associated with the Zeta and Related Functions*, Kluwer Academic Publishers, Dordrecht-Boston-London, 2001.

[Sta] H. M. Stark, Dirichlet's class number formula revisited, Contemp Math. **143** (1993), 571–577.

[St] S. W. P. Steen, Divisor functions: their differential equations and recurrence formulas, *Proc. London Math. Soc.* (2) **31** (1930), 47–80.

[Sue] Z. Suetuna, *Analytic number theory*, Iwanami-Shoten, Tokyo.

[SzLP] G. Szegö, Beiträge zur Theorie der Laguerreschen Polynome. Zahlentheoretische Anwendungen, *Math. Z.* **25** (1926), 388–404.

[SzW] G. Szegö and A. Z. Walfisz, Über das Piltzsche Teilerproblem in algebraishcen Zahlkörpern, *Math. Z.* **26** (1927), 138–156, 467–486.

[SUZ] J. Szmidt, J. Urbanowicz and D. Zagier, Congruences among generalized Bernoulli numbers, *Acta Arith.* **3** (1995), 273–278.

[Tem] N.M. Temme, *Special Functions. An Introduction to Classical Functions in Mathematical Physics,* Wiley-Interscience, New York etc. 1996.

[Te1] A. Terras, Bessel series expansion of the Epstein zeta function and the functional equation, *Trans. Amer. Math. Soc.* **183** (1973), 477–486.

[Te2] A. Terras, Some formulas for the Riemann zeta function at odd integer argument resulting from Fourier expansions of the Epstein zeta function, *Acta Arith.* **29** (1976), 181–189.

[Te3] A. Terras, The Fourier expansion of Epstein's zeta function over an algebraic number field and its consequences for algebraic number theory. *Acta Arith.* **32** (1977), 37–53.

[Te4] A. Terras, The minima of quadratic forms and the behavior of Epstein and Dedekind zeta functions, *J. Number Theory* **12** (1980), 258–272.

[Te5] A. Terras, *Harmonic Analysis on Symmetric Spaces and Applications I, II,* Springer Verl., New York etc. 1985.

[Tip] F. Tipler, *The Physics of Immortality,* Pan Books, London 1994.

[Titr] E. C. Titchmarsh, *The Theory of the Riemann Zeta-Function,* (second edition revised by D. R. Heath-Brown), OUP 1986.

[Ts] H. Tsukada, A general modular relation in analytic number theory. *Number Theory: Sailing on the sea of number theory, Proc. 4th China-Japan Seminar on number theory 2006* 214–236, World Sci., Singapore etc. 2007.

[UN] K. Ueno and M. Nishizawa, Quantum groups and zeta-functions, in *Quantum Groups: Formalism and Applications, Proc. of the Thirtieth Karpacz Winter School (Karpacz, 1994)* (J. Lubkierski et al., Editors), 115–126, Polish Sci. Publ. PWN, Warsaw 1995.

[Ur1] J. Urbanowicz, On some further generalizations of congruences obtained by Z.-H. Sun, to appear.

[Ve1] A. B. Venkov, Spectral theory of automorphic functions, *Trudy Mat. Inst. Steklov* **153** (1981), 3–171=Proc. Inst. Math. Steklov (1982), issue **4**, 1–163.

[Ve2] A. B. Venkov, *Spectral Theory of Automorphic Functions and its Applications,* Kluwer Acad. Publ., Dordrecht etc. 1990.

[Vor1] G. F. Voronoï, Sur une fonction transcendente et ses applications à la sommation de qulques séries, Ann. École Norm. Sup. (3) **21** (1904), 459–533=Sob. Soč. II, Izd. Akad. Nauk Ukr. SSR, Kiev 1952, 51–165.

[Vor1] G. F. Voronoï, Sur le développement, à l'aide des fonctions cylindriques, des sommes doubles $\sum f(pm^2 + 2qmn + rn^2)$, où $pm^2 + 2qmn + rn^2$ est une forme positive à coefficients entiers, Verh. dritten Internat. Math.-Kongr., Teubner, Leipzig 1905, 241–245=Sob. Soč. II, Izd. Akad. Nauk Ukr. SSR, Kiev 1952, 166–170.

[AAW1] A. A. Walfisz, On sums of coefficients of some Dirichlet series, *Soobšč. Akad. Nauk Grundz. SSR* **26** (1961), 9–16.

[AAW2] A. A. Walfisz, On the theory of a class of Dirichlet series, *ibid.* **27** (1961), 9–16.

[AAW3] A. A. Walfisz, Über die summatorischen Funktionen einiger Dirichletscher Reihen II, *Acta Arith.* **10** (1964/65), 73–120.

[AAW4] A. A. Walfisz, The Fourier-Poisson formula a class of Dirichlet series, *Trudy Tbilis. Mat. Inst. im. A. M. Razmadze* **29** (1963), 1–13.

[AZW1] A. Z. Walfisz, Über die summatorischen Funktionen einiger Dirichletscher Reihen, Inaugural Diss. Göttingnen 1923.

[AZW2] A. Z. Walfisz, Über das Piltzsche Teilerproblem in algebraishcen Zahlkörpern, *Math. Z.* **22** (1925), 153–188 (cf. [LW1]).

[AZW3] A. Z. Walfisz, Über das Piltzsche Teilerproblem in algebraishcen Zahlkörpern II, *Math. Z.* **23** (1927), 487–494.

[XHW] X.-H. Wang, Analytic continuation of the Riemann zeta-function, to appear.

[WW] X.-H. Wang and N.-L. Wang, Modular-relation-theoretic interpretation of M. Katsurada's results, to appear.

[Wa] G. N. Watson, Some self-reciprocal functions, *Quart. J. Math. Oxford Ser.* **2** (1931), 298–309.

[Wat] G. N. Watson, *A Treatise on the Theory of Bessel functions*, 2nd ed., Cambridge UP, Cambridge 1944.

[We1] A. Weil, Sur une formule classique, *J. Math. Soc. Japan* **20** (1968), 400–402 = *Coll. Papers, III*, Springer Verl., 1980, Berlin etc., 198–200.

[We3] A. Weil, Remarks on Hecke's lemma and its use, *Algebraic Number Theory*, Intern. Symposium Kyoto 1976, S. Iyanaga (ed.), Jap. Soc. for the Promotion of Science 1977, pp. 267–274= *Coll. Papers, III*, 405–412, Springer Verl., Berlin etc. 1979.

[We4] A. Weil, On Eisenstein's copy of the Disquisitiones, *Advanced Studies in Pure Mathematics* **17**, 1989 Algebraic Number Theory – in honor of K. Iwasawa, 463–469.

[WeIII] A. Weil, *Coll. Papers, III*, Springer Verl., Berlin etc. 1979.

[WW] E. T. Whittaker and G. N. Watson, *A Course in Modern Analysis*, 4th ed., Cambridge UP, Cambridge 1927.

[Wil0] J. R. Wilton, A proof of Burnside's formula for $\log \Gamma(x + 1)$ and certain allied properties of Riemann's ζ-function, *Mess. Math.* **55** (1922/1923), 90–93.

[Wil1] J. R. Wilton, A series of Bessel functions connected with the theory of lattice points, *Proc. London Math. Soc.* (2) **29** (1928), 166–188.

[Wil2] J. R. Wilton, An approximate functional equation for the product of two ζ-functions, *Proc. London Math. Soc.* (2) **31** (1930), 11–17.

[Wil3] J. R. Wilton, On Voronoï's summation formula, *Quart. J. Math. (Oxford Ser.)* (2) **3** (1932), 26–32.

[Wil4] J. R. Wilton, An extended form of Dirichlet's divisor problem, *Proc. London Math. Soc.* (2) **36** (1933), 391–426.

[Win3] A. Wintner, On Riemann's fragment concerning elliptic modular functions, *Amer. J. Math.* **63** (1941), 628–634.

[Wis] J. Wishart, The generalized product moment distributions in samples from a normal multiplicative population, *Biometrika* **20** (1928), 32–43.

[Ya] Y. Yamamoto, Dirichlet series with periodic coefficients, *Proc. Intern. Sympos. "Algebraic Number Theory"*, Kyoto 1976, 275–289. JSPS, Tokyo 1977.

[Zag] D. Zagier, The Rankin-Selberg method for automorphic functions which are not of rapid decay, *J. Fac. Sci. Univ. Tokyo IA Math.* **28** (1981), 415–437.

[Zu] W. Zudilin, Arithmetic of linear forms involving odd zeta values, *J. Théorie Nombres Bordeaux* **16** (2004), 251–291.

Papers on the Chowla-Selberg formula

[BG] P. Bateman and E. Grosswald, On the Epstein zeta-function, *Acta Arith.* **9** (1964), 365–373.

[Berg] A. Berger, Sur une sommation de quelques séries *Nova Acta Reg. Soc. Sci. Upsal.* (3) **12** (1883), 29–31.

[CS] S. Chowla and A. Selberg, On Epstein's zeta-function (I), *Proc. Nat. Acad. Sci. USA* **35** (1949), 371–374; *Collected Papers of Atle Selberg I*, Springer Verlag, 1989, 367–370. *The Collected Papers of Sarvadaman Chowla II*, CRM, 1999, 719–722.

[Den] C. Deninger, On the analogue of the formula of Chowla and Selberg for real quadratic fields, *J. Reine Angew, Math.* **351** (1984), 171–191.

[Gut] M. Gut, Die Zetafunktion, die Klassenzahl und die Kronecker'shce Grenzformel eines beliebiges Kreikörpers, *Comment. Math. Helv.* **1** (1929), 160–226.

[Kur] N. Kurokawa, One hundred years of the zeta regularized products, The 39th Algebra Sympos. 1994, 153–166.

[LanCS] E. Landau, Über die zu einem algebraischen Zahlkörper gehörige Zetafunktion und die Ausdehnung der Tschebyscheffschen Primzahltheorie auf das Problem der Verteilung der Primideale, *J. Reine. Angew. Math.* **125** (1903), 176–188=Collected Works, Vol. I, Thales Verl., Essen 1985, 201–325.

[Ler1] M. Lerch, Dalši studie v oboru Malmsténovských řad, *Rozpravy České Akad.* **3** (1894), no. 28, 1–61 (Czech).

[Ler3] M. Lerch, Sur quelques formules relatvies au nombre des classes, *Bull. Sci. Math.* **21** (1897), 290–304.

[Ma] C. J. Malmstén, De integralibus quibusdam definitis, seriebusque infinitis, *J. Reine Angew. Math.* **38** (1849), 1–39.

[SC] A. Selberg and S. Chowla, On Epstein's zeta-function, *J. Reine Angew, Math.* **227** (1967), 86–110; *Collected Papers of Atle Selberg I*, Springer Verlag, 1989, 521–545; *The Collected Papers of Sarvadaman Chowla II*, CRM, 1999, 1101–1125.

[WeE] A. Weil, Elliptic functions according to Eisenstein and Kronecker, Springer Verl., 1976, Berlin-Heidelberg-New York.

Papers on Lambert series

[AdR] J. Arias-de-Reyna, Riemann's fragments on limit values of elliptic modular functions, *Ramanujan J.* **8** (2004), 57–123.

[Berndt7] B. C. Berndt, *Ramanujan's Notebooks Part I,* Springer Verl., Berlin etc. 1985.

[Berndt8] B. C. Berndt, *Ramanujan's Notebooks Part II,* Springer Verl., Berlin etc. 1989.

[Berndt9] B. C. Berndt, *Ramanujan's Notebooks Part IV*, Springer Verl., Berlin etc. 1994.

[Br] D. M. Bradley, Series acceleration formulas for Dirichlet series with periodic coefficients, *Ramanujan J.* **6** (2002), 331–346.

[Ded] R. Dedekind, Erläuterungen zu zwei Fragmenten von Riemann, *Math. Werke Bd. 1*, 159–173, Braunschweig, (1930) (in: Bernhard Riemanns gesammelte mathematische Werke und wissenschaftlichen Nachlass, 2, Aufl., 466–478, (1892)).

[Fl] T. M. Flett, On the function $\sum \frac{1}{n} \sin \frac{x}{n}$, *J. London Math. Soc.* **25** (1950), 5–19.

[HarLam] G. H. Hardy, Note on the limiting values of the elliptic modular-functions, *Quart. J. Math.* (2) **34** (1903), 76–86; Collected Papers of G. H. Hardy IV, Oxford UP, Oxford 1966, 351–361.

[HarLLam] G. H. Hardy and J. E. Littlewood, Note on the theory of series (XX): On Lambert series, *Proc. London Math. Soc* (2) **41** (1936), 257–270; Collected Papers of G. H. Hardy IV, Oxford UP, Oxford 1969.

[Ja] C. G. J. Jacobi, *Fundamenta Nova*, Königsberg, 1829, in Latin. Reprinted with corrections in: C. Jacobi. *Ges. Werke*. 8 volumes, Berlin. 1881–1891. Vol. 1. 49–239. Reprinted New York (Chelsea, 1969).

[KKLam] S. Kanemitsu and T. Kuzumaki, Transformation formulas for Lambert series, Šiaulai Math. Sem. **4** (2009), 105–123.

[KMT2] S. Kanemitsu, J. Ma and Y. Tanigawa, Corrigenda and addenda to "Arithmetical identities and zeta-functions, *Math. Nachr.* **284** (2011), 287–297", to appear.

[KnL] K. Knopp, Über Lambertsche Reihen, *J. Reine Angew. Math.* **142** (1913), 283–315.

[Kn2] K. Knopp, *Theory and applications of infinite series*, Blackie and Sons, London 1951 (first German edition published in 1921).

[Lam] J. H. Lambert, *Anlage zur Archtektonik oder Theorie des Einfachen und Ersten in der philosophischen und mathematischen Erkenntnis*, 2 Bände, Riga, Vol. 2, 1771, §875 (p.507).

[Mik2] M. Mikolás, Über gewisse Lambertsche Reihen, I: Verallgemeinerung der Modulfunktionen $\eta(\tau)$ und ihrer Dedekindschen Transformationformel, *Math. Z.* **68** (1957), 100–110.

[Ra1] H. Rademacher, Zur Theorie der Modulfunktionen, *J. Reine Angew. Math.* **167** (1931), 312–336; Collected Papers of H. Rademacher I, 1974, 652–677.

[Ra] H. Rademacher, Collected Papers of Hans Rademacher I, II, MIT Press, *Cambridge Mass.* (1974).

[RG] H. Rademacher and E. Grosswald, Dedekind sums, MAA, New York 1972.

[Rie1] B. Riemann, Über die Darstellbarkeit einer Function durch eine trigonometrische Reihe, in [Rie0], 227–265.

[Rie2] B. Riemann, Fragmente über Grenzfälle der ellipitischen Modulfunctionen, in [Rie0], 455–465.

[Seg] S. L. Segal, On $\sum 1/n \sin(x/n)$ and $\sum 1/n \cos(x/n)$, *J. London Math. Soc.* (2) **4** (1972), 385–393.

[Tit] E. C. Titchmarsh, On a series of Lambert's type, *J. London Math. Soc.* **12**

(1938), 248–253.

[NLW] N.-L. Wang, On Riemann's posthumous fragment II on the limit values of the elliptic modular functions, *Ramanujan J.*, **24** (2011), 129–145.

[Wig1] S. Wigert, Sur la série de Lambert et son application à la théorie des nombres, Acta Math. **41** (1918), 197–218.

[Win1] A. Wintner, On a trigonometrical series of Riemann, *Amer. J. Math.* (3) **59** (1937), 629–634.

[Win2] A. Wintner, On Riemann's fragment concerning elliptic modular functions, *Amer. J. Math.* **63** (1941), 628–634.

Papers on arithmetical Fourier series

[ApoMA] T. M. Apostol, *Mathematical analysis*, Addison-Wesley, Reading, 1957.

[CW] S. Chowla and A. Walfisz, Über eine Riemannsche Identität, *Acta Arith.* **1** (1935), 87–112=Collected Papers of Sarvadaman Chowla, Vol. I, 267–279.

[Ch] S. Chowla, On some infinite series involving arithmetic functions, Proc. Indian Acad. Sci. A **5** (1937), 511–513=Collected Papers of Sarvadaman Chowla, Vol. II, 489–491.

[D1] H. Davenport, On some infinite series involving arithmetic functions, *Quart. J. Math.* **8** (1937), 8–13=Collected papers of H. Davenport IV, Academic Press London etc. 1977, 1781–1786.

[D2] H. Davenport, On some infinite series involving arithmetic functions II, *Quart. J. Math.* **8** (1937), 313–320=Collected papers of H. Davenport, IV, Academic Press London etc. 1977, 1787–1794.

[BT] R. de la Bretèche and G. Tennenbaum, Séries trigonométriques à coefficients arithmétiques, *J. Anal, Math.* **92** (2004), 1–79.

[Ger1] J. Gerver, The differentiability of the Riemann function at certain rational multiple of π, *Amer. J. Math.* **92** (1970), 33–55.

[Ger2] J. Gerver, More on the differentiability of the Riemann function, *Amer. J. Math.* **92** (1971), 33–41.

[H] G. H. Hardy, Weierstrass's non-differentiable function, *Trans. Amer. Math. Soc.* (3) **17** (1916), 301–325=Collected Papers of G. H. Hardy Vol. I, Oxford University Press, Oxford 1966, 477–501.

[HL1] G. H. Hardy and J. E. Littlewood, Some problems of Diophantine approximations, I. The fractional part of $n^k\theta$, *Acta Math.* **37** (1914), 155–191=Collected Papers of G. H. Hardy Vol. I, Oxford University Press, Oxford, 28–63 (the original page 191 containing the contents of the paper is omitted in the Collected Papers).

[HL2] G. H. Hardy and J. E. Littlewood, Some problems of Diophantine approximation, The trigonometrical series associated with the elliptic θ-function, *Acta Math.* **37** (1914), 193–239=Collected Papers of G. H. Hardy Vol. I, Oxford University Press, Oxford 1966, 67–112 (page 239 in the original paper which contains contents of the paper is omitted and as a result, the page numbers are stated in the Collected papers incorrectly as 193–239).

[HL3] G. H. Hardy and J. E. Littlewood, Some problems of Diophantine approximation, *Trans. Cambridge Philos. Soc.* **27** (1923), 519–534=Collected

Papers of G. H. Hardy Vol. I, Oxford UP, Oxford 1966, 212–226 (page 534 in the original paper which contains contents of the paper is omitted and as a result, the page numbers are stated in the Collected papers incorrectly as 519–533).

[HW] P. Hartman and A. Wintner, On certain Fourier series involving sums of divisors, *Trudy Tbillis. Mat. Obsc. im. G. A. Razmadze* **3** (1938), 113–119.

[HT] M. Holschneider and Ph. Tchamichain, Pointwise analysis of Riemann's "nondifferentiable" function, *Invent. Math.* **105** (1991), 157–175.

[I] S. Itatsu, Differentiability of Riemann's function, *Proc. Japan Acad. Ser. A Math. Sci.* (10) **57** (1981), 492–495.

[KMT1] S. Kanemitsu, J. Ma and Y. Tanigawa, Arithmetical identities and zeta-functions, Math. Nachr. **284** (2011), 287–297.

[KTZ] S. Kanemitsu, Y. Tanigawa and J. -H. Zhang, Evaluation of Spannen-integrals of the product of zeta-functions, Integral Transform and Special Functions (2007) (to appear).

[KY] S. Kanemitsu and M. Yoshimoto, Euler products, Farey series, and the Riemann hypothesis, *Publ. Math. Debrecen* **56** (2000), 431–449.

[Kano] T. Kano, On the size of $\sum d(n)e(n\theta)$, in *Analytic Number Theory and Diophantine Problems* (Stillwater, OK, 1984), Progr. Math. **70**, Birkhäuser Boston, Boston 1987, 283–290.

[Kan] T. Kano, On the Bessel series expression for $\sum 1/n \sin(x/n)$, *Math. J. Okayama Univ.* **16** (1974), 129–136.

[Ko] J. F. Koksma, Diophantische Approximationen. (German), Reprint. Springer Verl., Berlin-New York, 1974.

[LMZ] H.-L. Li, J. Ma and W.-P. Zhang, On some Diophantine Fourier series, *Acta Math. Sinica*, English Ser. **26** (2010), 1125–1132.

[L] W. Luther, The differentiability of Fourier gap series and "Riemann's example" of a continuous, nondifferentiable function, *J. Approx. Theory* (3) **48** (1986), 303–321.

[Mey] Y. Meyer, Le traitement du signal et l'analyse mathématique, *Ann. Inst. Fourier* **50** (2000), 593–632.

[Neu] E. Neuenschwander, Riemann's example of a continuous, 'nondifferentiable' function, *Math. Intelligencer* (1) **1** (1978/79), 40–44.

[On] L. A. Onufrieva, The Chebyshev interpolation method in the case of a large number of data, (Russian) *Istor.-Mat. Issled.* **27** (1983), 259–274.

[Que1] M. Queffelec, Dérivbilité de cetains sommes de séries de Fourier lacunnaires *Note au C.R. Séan. Acad. Sci., Sér. A* **273** (1983), 291–293.

[Que2] M. Queffelec, Dérivbilité de cetains sommes de séries de Fourier lacunnaires *Centre de Orssay, Univ. Paris-Sud Thèse* , 1971.

[Ri2] B. Riemann, Über die Darstellbarkeit einer Function durch eine trigonometrische Reihe, in *Riemann's Collected Works*, 1854, 227–265 2nd ed. and the supplement, *Dover, New York.*, 1953, 227–265.

[Ro2] N. P. Romanoff, Hilbert spaces and the theory of numbers. II. (Russian). *Izv. Akad. Nauk SSSR, Ser. Mat.* **15** (1951), 131–152.

[S1] S. L. Segal, On an identity between infinite series of arithmetic functions, *Acta Arith.* (4) **28** (1975/76), 345–348.

[S2] S. L. Segal, Riemann's example of a continuous 'nondifferentiable' function. II, *Math. Intelligencer* (2) **1** (1978/79), 81–82.

[Sm] H. J. S. Smith, On some discontinuous series considered by Riemann: Messenger of Math. (2) **9** (1881), 1–11=Smith's Collected Works. Vol. I, Chelsea, New York, 1965, 312–320.

[AS] A. Smith, The differentiability of Riemann's function, *Proc. Amer. Math. Soc.* **34** (1972), 463–468.

[Wal] H. Walum, Multiplication formulae for periodic functions. *Pacific J. Math.* **149**, no. 2 (1991), 383–396.

[Wi0] A. Wintner, A note on the non-differentiable function of Wererstrass, *Amer. J. Math.* (3) **59** (1933), 603–605.

[Wi3] A. Wintner, Diophantine approximations and Hilbert's space, *Amer. J. Math.* (3) **66** (1944), 564–578.

Papers on zeta-functions with periodic coefficients

[AC] R. Ayoub and S. Chowla, On a theorem of Müller and Carlitz, *J. Number Theory* **2** (1970), 342–344.

[Ber75-1] B.C. Berndt, Periodic Bernoulli numbers, summation formulas and applications, *Theory and application of special functions*, Academic press, New York, 1975.

[Ber75-2] B.C. Berndt, Character analogues of the Poisson and Euler-Maclaurin summation formula with applications, *J. Number Theory* **7** (1975), 413–445.

[Car1] L. Carlitz, A formula connected with lattice points in a circle, *Abh. Math. Sem. Univ. Hamburg* **21** (1957), 87–89.

[IK99] S. Kanemitsu and M. Ishibashi, Dirichlet series with periodic coefficients, *Res. Math.* **35** (1999), 70–88.

[Mue] C. Müller, Eine Formel in der analytischen Zahlentheorie, *Abh. Math. Sem. Univ. Hamburg* **19** (1954), 62–65.

[Now] W.-G. Nowak, Letter to the first author, Dec. 6 1984.

[Red1] D. Redmond, A generalization of a theorem of Ayoub and Chowla, *Proc. Amer. Math. Soc.* **85** (1982), 574–580.

[Red2] D. Redmond, Corrections and additions to "A generalization of a theorem of Ayoub and Chowla", *Proc. Amer. Math. Soc.* **90** (1984), 345–346.

[Scn1930] W. Schnee, Die Funktionalgelichung der Zetafunktion und der Dirichletschen Reihen mit periodischen Koeffizienten, *Math. Z.* **31** (1930), 378–390.

[BSm] R. A. Smith, Dirichlet series with periodic coefficients, *J. Reine Angew. Math.* **35** (1980), 70–88.

Papers on the approximate functional equation. Esp. on the Wigert-Bellman divisor problem

[Bel1] R. Bellman, Wigert's approximate functional equation and the Riemann zeta-function, Duke Math. J. **16** (1949), 547–552.

[Lan1] E. Landau, Über die Wigertsche asymptotische Funktionalgelichung für die Lambertsche Reihe, Arch. Math. Phys. (3) **27** (1918), 144–146=Coll. Papers,

[Spi] R. Spira, Approximate functional approximations and the Riemann hypothesis, Proc. Amer. Math. Soc. **17** (1966), 314–317.

[Wig1] S. Wigert, Sur la série de Lambert et son application à la théorie des nombres, Acta Math. **41** (1918), 197–218.

[Wil2] J. R. Wilton, An approximate functional equation for the product of two ζ-functions, Proc. London Math. Soc. (2) **31** (1930), 11–17.

[Wil5] J. R. Wilton, An approximate functional equation with applications to a problem of Diophantine approximation, J. Reine Angew. Math. **169** (1933), 219–237.

Papers on Hamburger's theorem

[BochCh] S. Bochner and K. Chandrasekharan, On Riemann's functional equation, *Ann. of Math.* (2) **63** (1957), 336–360.

[Boch1957] S. Bochner, On Riemann's functional equation with multiple gamma factors, *Ann. of Math.* (2) **67** (1958), 29–41.

[ChM1] K. Chandrasekharan and S. Mandelbrojt, On Riemann's functional equation, *Ann. of Math.* (2) **65** (1957), 285–296.

[ChM2] K. Chandrasekharan and S. Mandelbrojt, On solutions of Riemann's functional equation, *Bull. Amer. Math. Soc.* **65** (1959), 358–362.

[EK] L. Ehrenpreis and T. Kawai, Poisson's summation formula and Hamburger's theorem, *Publ. RIMS, Kyoto Univ.* **18** (1982), 413–426.

[GV] D. Goldfeld and C. Viola, Mean values of *L*-functions associated to elliptic, Fermat and other curves at the center of the critical strip, *J. Number Theory* **11** (1979), 305–320.

[GoRa] L. J. Goldstein and M. Razar, The theory of Hecke integrals, *Nagoya Math. J.* **63** (1976), 93–121.

[Ham1] H. Hamburger, Über die Riemannschen Funktionalgleichung der ζ-Funktion, *Math. Z.* **10** (1921), 240–254. (Erste Mitteilung)

[Ham2] H. Hamburger, Über einige Beziehungen, die mit der Funktionalgleichung der Riemannschen ζ-Funktion äquivalent sind, *Math. Ann.* **85** (1922), 129–140.

[Hec1] E. Hecke, Über die Lösungen der Riemannschen Funktionalgleichung, *Math. Z.* **16** (1923), 301–307=Mathematische Werke, 374–380, Vandenhoeck u. Ruprecht, Göttingen 1959.

[HecB] E. Hecke, Über die Bestimmung Dirichletscher Reihen durch ihre Funktionalgleichung, *Math. Ann.* **112** (1936), 664–669=Mathematische Werke, 591–626, Vandenhoeck u. Ruprecht, Göttingen 1959.

[HecE] E. Hecke, Herleitung des Euler-Produktes der Zetafunktion und einiger *L*-Reihen aus ihrer Funktionalgleichung, *Math. Ann.* **119** (1944), 266–287=Mathematische Werke, 919–940, Vandenhoeck u. Ruprecht, Göttingen 1959.

[KnSh] M. I. Knopp and M. Sheingorn, On Dirichlet series and Hecke triangle groups of infinite volume, *Acta Arith.* **76** (1996), 227–244.

[Kob2] H. Kober, Eine der Riemannschen verwandte Funktionalgleichung, *Math. Z.* **39** (1935), 630–633.

[Lev] N. Levinson, *Gap and density theorems*, AMS Colloq. Publ. Vol. 26, AMS, R.I. 1940.

[Rya] C. Ryavec, The analytic continuation of Euler products with applications to asymptotic formulae, *Illinois J. Math.* **17** (1973), 608–618.

[Sar] P. Sarnak, Fourth moments of Grossencharakteren zeta functions, *Comm. Pure Appl. Math.* **38** (1985), 167–178.

[FSato] F. Sato, The Hamburger theorem for the Epstein zeta functions, Algebraic Analysis, Vol. II, Academic Press, 1988, 789–807.

[Siegl] C. L. Siegel, Bemerkungen zu einem Satz von Hamburger über die Funktionalgleichung der Riemannschen Zetafunktion, *Math. Ann.* **85** (1922), 276–279=Ges. Abh., Bd. I, Springer Verl. Berlin-Heidelberg, 1966, 154–156.

[Vig] M. -F. Vignéras, Factuer gamma et Équations fonctionnelles, Intern. Summer School on Modular Functions, Bonn, 1976, Lect. Notes Math. **627**, *Modular functions of one variable VI*, Springer Verl., Berlin-Heidelberg-New York 1977, 79–103 (J. -P. Serre, Appendice, Relations entre facteurs gamm, 99–103).

[We2] A. Weil, Über die Bestimmung Dirichletscher Reihen durch Funktionalgleichungen, *Math. Ann.*, **168** (1967), 149–156 = *Coll. Papers, III*, 165–172, Springer Verl., Berlin etc. 1979.

[Weds] A. Weil, *Dirichlet series and automorphic forms*, Lect. Notes Math. 189, Springer Verl., Berlin-Heidelberg-New York 1971.

[AY] A. Yoshimoto, On a generalization of Hamburger's theorem, *Nagoya Math. J.* **98** (1985), 67–76.

Index

Printed in the United States
By Bookmasters